T5-AFS-466

Mosby's
Biomedical Science Series

Understanding Neuroscience

Mosby's

Biomedical Science Series

Understanding Neuroscience

William R. Klemm, DVM, PhD
Professor of Neuroscience
Professor of Veterinary Anatomy & Public Health
Texas A & M University
College Station, Texas

with 88 illustrations

Mosby

St. Louis Baltimore Boston Carlsbad Chicago Naples New York Philadelphia Portland
London Madrid Mexico City Singapore Sydney Tokyo Toronto Wiesbaden

Mosby
Dedicated to Publishing Excellence

A Times Mirror Company

Publisher: Don Ladig
Executive Editor: Paul W. Pratt, VMD
Developmental Editor: Jolynn Gower
Project Manager: Carol Sullivan Weis
Design Manager: Sheilah Barrett
Cover Designer: Jeanne Wolfgeher
Manufacturing Supervisor: Karen Lewis

Copyright © 1996 by Mosby-Year Book, Inc.

All rights reserved. No part of this publication may be reproduced, stored in a retrieval system, or transmitted, in any form or by any means, electronic, mechanical, photocopying, recording, or otherwise, without written permission of the publisher.

Permission to photocopy or reproduce solely for internal or personal use is permitted for libraries or other users registered with the Copyright Clearance Center, provided that the base fee of $4.00 per chapter plus $.10 per page is paid directly to the Copyright Clearance Center, 27 Congress Street, Salem, MA 01970. This consent does not extend to other kinds of copying, such as copying for general distribution, for advertising or promotional purposes, for creating new collected works, or for resale.

Printed in the United States of America
Composition by Nucomp
Printing/binding by Phoenix

Mosby-Year Book, Inc.
11830 Westline Industrial Drive
St. Louis, Missouri 63146

International Standard Book Number 0-8151-5223-X

96 97 98 99 00 / 9 8 7 6 5 4 3 2 1

JAN 22 1996 $25.00

Series Preface

Science textbooks are commonly entitled "Principles of ..." Yet almost all of these books go far beyond principles to specify so much detail that they often obscure the principles. Too often the emphasis is on currently accepted facts, with principles used only to illustrate the facts, rather than using the minimum amount of fact to illustrate principles.

This situation argues for a new kind of textbook, one that focuses on the first principles of the discipline. In addition to encyclopedic tomes for each discipline, we need small texts that give the big picture and explicitly describe the basic foundational principles of the discipline—and no more!

As a start to what I hope will be a new movement in science education we are initiating a series of biomedical science teaching books that really are about principles. These books aim to be quick, yet elegant, overviews of the essence of the respective disciplines. The newcomer to the subject should find the preparation needed to understand the more comprehensive and detailed traditional textbooks and research journals. Maybe even highly specialized experts can find some useful perspectives from this approach. Senior people tend to get wrapped up in the details of their specialty and sometimes take the principles for granted.

We must have some kind of working definition of "principle." While many ways can express the idea, this Series of texts uses the following working definition:

> A principle must go beyond a collection of observations. Principles integrate multiple observations and help to explain these observations, providing understanding and insight. A principle embodies the underlying rules or mechanisms of structure, organization, or operation that give rise to the observations. We distinguish principles from concepts only in the sense that concepts often embody more than one principle.

In identifying these principles, we must recognize that they are basic tenets; i.e., commonly held *beliefs* about what is true and fundamental. Not everyone will agree that each of these "principles" deserves such lofty status. However, there is a fine line between principle and theory. Theories serve to inspire better theories that can establish principles more firmly.

Many of the statements of principle are incomplete. They may also lack sufficient qualification. Statements of principles serve to inspire more complete or precise exposition. The effort to identify principles comes at a cost: arbitrariness, uncertainty, controversy—but the cost is worth the price. *The search for principles is the Holy Grail of science.*

The practical value of such texts may lie in their pedagogical approach, which is the opposite from the tack taken in many textbooks

University of Arkansas
LIBRARY
for Medical Sciences

on biological and medical subjects. Students are fed up with encyclopedic subjects. Students and professors alike are tired of an educational process devoted to pouring information into one ear while it spills out the other. The exponential expansion of new knowledge is causing cognitive overload in students and professors, short-circuiting their ability to sustain perspective about the whole of biomedical science and to think coherently about the details of how the body works in both health and disease. We are learning more and more about less and less, and that causes a progressive loss of capacity for synthesis and ability to think about the larger meanings of biomedical science. The texts in this series require the student to be actively involved in completing missing detail and providing the qualifications of special relevance.

Why These Books are Needed

- Traditional texts are too big, too detailed, too indigestible
- Biomedical information is accumulating faster than students can handle
- Students are learning more and more about less and less
- Students, and even teachers, have trouble discerning the "must know" from the "nice to know"
- The new emphasis in teaching of biomedical sciences will be on how to manage, integrate, and apply information. That requires identification and understanding of the basic principles of the discipline.

The advantages of books in this Series, as I see it, are as follows:

Advantages to Students:

- Students more likely to "see the forest instead of the trees"
  - What is important is made explicit
  - Cognitive overload can't obscure perspective and insight
- "Less is More"
  - Less material is easier to comprehend and remember
  - It is easier to remember unifying principles and concepts
  - Understanding principles empowers students to get more from new information
- Promotes active learning, critical thinking, insight, understanding
- Very useful in self-paced or collaborative learning paradigms
  - Helps assure that students have the required background
  - Students are better prepared to understand what they read in journals and reference books—with minimal help from professors
- Books are smaller, cost less, are more portable, and are easier to peruse

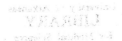

LIBRARY

- High portability and condensed nature encourage frequent review
- Longer half-life
  - Traditional texts are out of date as soon as printed
  - Principles are "eternal"

Benefits to Professors

- Allows specialized instruction with less fear that basics are being missed
- After mastering principles, students are equally prepared for subsequent instruction
- Enlivens the lecture period. Meaningful discussion and debate facilitated

Benefits to Graduates

- Quick way to review latest concepts, especially in fields in which they have not kept up
- "User-friendly" access to specialties or to especially complex disciplines

As professors increasingly recognize, the proliferation of research literature has made it difficult to decide what is fundamental about their discipline that must be taught to students. The books in this Series aim to help professors identify that core understanding. Even where a professor may disagree with certain statements of principle, the books provide a focus and a stimulus to the professor to refine those statements of principle with which they disagree.

Where professors believe that more factual detail is needed, they can provide it or direct students to it with some reassurance that students still understand the central core of the discipline. Thus, these books can substitute for the standard, fact-filled textbook and give professors the flexibility to use other instructional media, such as computer-assisted instruction, journal articles, and even standard textbooks as reference sources to complement the Mosby's Biomedical Science Series.

Books in this Series would seem especially important for curricula that stress problem- or case-based learning (PBL). In recognition of the cognitive overload problem some medical schools (McMasters, Harvard, Southern Illinois, Bowman Gray, University of New Mexico) have pioneered in converting the traditional lecture-based curriculum to a tutorial, group-based learning format where critical thinking and information management skills are emphasized rather than rote memory. Some veterinary colleges are also making similar curricular changes. This trend will surely grow, because it is aimed at teaching students to *manage* and *integrate* an ever-expanding biomedical data base.

However, many institutions have thoughtfully considered but rejected PBL-type curricula out of fear that students will have huge gaps in their understanding of the core biomedical science disciplines. Books in this Series not only help PBL students to know what the fundamental principles are, but present them in a quick and easily digestible form. Understanding these principles increases the likelihood that PBL students will truly understand what they read in journals and reference books as they try to apply it to the clinical cases or academic problems.

Organization of These Books

The first—and most difficult—part of writing this kind of text is the identification and terse statement of the first principles of a discipline. Then principles are consolidated, if necessary, and grouped into categories to make it easier to organize and remember them.

Each category has an Introduction that states the principles in that category. There is also a "concept map," which is a diagram that shows the interrelationship of the principles in that category. The authors treat each principle as a module that states the principle and identifies its category. Then the principle is explained, including the use of one or more examples, accompanied by one or two diagrams or pictures. Another section defines key terms, and yet another lists other principles that are most directly related. Finally, a reference section lists two or three key references, along with a list of "Citation Classics,"* where possible. At the end of the modules in a given category, a review section presents some open-ended questions for review and debate.

W. R. (Bill) Klemm
Texas A & M University
College Station, Texas

*A "Classic" is a highly cited publication (on the order of 500 or more citations) as identified by the *Science Citation Index,* published by ISI Press, Philadelphia. For some of these, there is an associated publication that appeared in the ISI publication, *Current Contents* that contains a history that led up to the research that enabled the publication to have such a major impact. In some cases, the author has taken the liberty to list a publication that in his opinion should have been accorded formal "Classic" status.

Contents

List of Neuroscience Principles

Introduction

"Science is not a collection of facts or of unquestionable generalizations, but a logically connected network of hypotheses that represent our current opinion about what the real world is like."

—P.B. Medawar

What the brain produces is a kind of mental model of the world, a system for handling the information that flows from sense organs to the generation of appropriate responses. The integration of the sensory data is central to monitoring the world "out there" and to creating a model of it "in here." The "in here" becomes the real world as far as animals and people experience it. To explain what we know about how all this occurs in the brain is no trivial task. Too many books, in my opinion, fall short of clear explanations because they are so heavily laden with technical detail that the essence of understanding is often obscured. Students are easily confused over what they *must* know as opposed to what is *nice to know*. What a student-oriented text should be is one that focuses on principles and concepts.

Many of the textbooks that claim to be about "principles" are really about principles only in the sense that a forest is about ecology or equations are about mathematics. The standard science textbooks have gotten progressively larger, serving a better role as a reference book than as a textbook. This trend is especially pervasive in such active research areas as neuroscience, where students and teachers alike are swamped with new information.

How bad have things gotten? We can best illustrate the problem by the annual meeting for the Society of Neuroscience. For years now, the attendance has been running about 18,000 and the number of papers presented hovers around 10,000 each year. The professors can't keep up with all that. How can students do so?

This learning problem is especially acute for professionals who are not neuroscientists but whose work must be informed by neuroscience. These professionals include physicians, osteopaths, veterinarians, dentists, clinical and experimental psychologists, computer scientists, bioengineers, animal behaviorists, biologists, nurses, and allied health workers.

An early, well-known neuroscience principles book is *Elements of Neurophysiology* (by Ochs, 1965, Wiley & Sons). This book took 621 pages to specify the "Elements" (principles). Now, the most popular book, Principles of Neural Science (3rd ed.), is by Kandel et al., 1991, Elsevier. This book is **1,135 pages long.** Clearly, such a book tells most readers more than they want or need to know.

The situation argues for a new kind of textbook, one that is focused on the first principles of the discipline. In addition to encyclopedic tomes for each discipline, we need small **supplementary** texts that give the big picture and explicitly describe the basic foundational principles of the discipline—and no more!

Most of what everybody needs to know about the nervous system can be summarized in a list of about 80 principles. This book of principles is aimed at people who are not neuroscientists but who need to understand the basics. The book hopefully gives an overview of what is most important for our understanding of how the nervous system works. Readers in other scientific disciplines or newcomers to neuroscience need a concise presentation of all the relevant principles of the nervous system.

For the student or the newcomer to a given area in neuroscience, the learning tactic should be to study this book *first*. Then the reader is better prepared to understand and view with better perspective what is read in other textbooks, reviews, and primary literature. Even professional neuroscientist need the opportunity to step back from their myopic subspecialty perspective and view the nervous system less reductionistically and more comprehensively—at least that is what I discovered as I developed this book of neuroscience principles.

This book is written to help the learner, not to impress fellow neuroscientists. As such, many professional neuroscientists may fault the book for not providing enough factual material and experimental evidence—in short, it is not the encyclopedia that we have come to expect in textbooks. My justification is that books should have a focus, and you can't simultaneously focus on the "forest" and on the "trees." Excellent neuroscience books already exist that focus on trees.

A more serious criticism is that some of the neuroscience principles I have identified are not universally accepted as such. In other words, some critics will complain that the book is too speculative and theoretical. But science is driven by theory and controversy over theory. What can be wrong about encouraging students to participate in such controversy? Indeed, I think that a major value of the book is that it encourages teacher and student alike to think critically and creatively. One practical way that teachers can exploit this feature of the book is to let the stated principles serve as a focus for classroom analysis and debate or serve as a point of reference for evaluating current journal articles.

Organization of the Book

Principles are grouped into categories to make it easier to organize and remember them. The categories are:

Overview

Cell Biology

Senses

Information Processing

States of Consciousness

Emotions

Learning and Memory

Motor Output and Control

Development/Trophism

Each category section begins with a short introduction that presents an overview of the principles in that category. There is a "concept map" that helps to display graphically the relationships among the various principles. A succession of modules follows, each of which covers one of the principles in that category. Each module has a principle name, followed by a terse and explicit statement of the principle. Then the principle is explained and one or more examples are given, usually accompanied by one or two diagrams or pictures. Another section contains a definition of key terms, and there is also a listing of other principles that are most directly related. Finally, a reference section lists two or three key references, along with a list of "Citation Classics."*

W. R. (Bill) Klemm
Texas A & M University
College Station, Texas

*A "Classic" is a highly cited publication, as identified by the Science Citation Index, published by ISI Press, Philadelphia. For some of the classics, there is an associated publication that appeared in the ISI publication, Current Contents, that contains a history of the research that enabled the publication to have such a major impact. In some cases, the author has taken the liberty of including some references in the "Classic" section that have not been officially "anointed" by ISI's data. Sometimes it is because these were landmark papers that were published before ISI began keeping citation records. In a few other cases, the choice is strictly the author's opinion.

CHAPTER 1

Overview

Absolute truth is like a mirage; it tends to disappear when you approach it ...
Passionately though I may seek certain answers, some will remain, like the mirage, forever beyond my reach

— Richard Leakey and Roger Lewin

Perhaps nowhere in science is this quote more appropriate than in neuroscience. We live in the "Decade of the Brain," so-called because we believe that a critical mass of information and understanding now exists that tempts us to believe we can understand the great mysteries of brain and mind. Yet our search for full and absolute truth may well prove to be forever beyond our reach.

A beginning point in this search is to ask, "What is a nervous system for?" Plants don't have one, and plant species generally seem to survive just fine. Clearly, a nervous system is not necessary for evolutionary success—at least for organisms that do not move about in their environment. But creatures that move about in their environment have the opportunity to change their environment, unlike plants, which have no choice. That ability allows such organisms to have more options for survival. In short, there are ecological niches for organisms that are flexible enough to change their environment. And that is why those niches have been filled with organisms with nervous systems (Figure 1-1).

To be able to move about and change the environment, organisms have certain special needs not required by plants. Obviously, they must have a mechanism to move their protoplasm around in the environment, and it certainly helps to have a control system for coordinating movements. Additionally, such organisms need an array of sensors that inform them of environmental conditions, as well as a processing network that decides whether the environment is optimal or whether "better" conditions should be "sought."

Even the most primitive one-celled animals, such as Paramecia, have some of these capabilities. The evolution of higher life forms could be regarded as Nature's way of evolving more efficient and powerful ner-

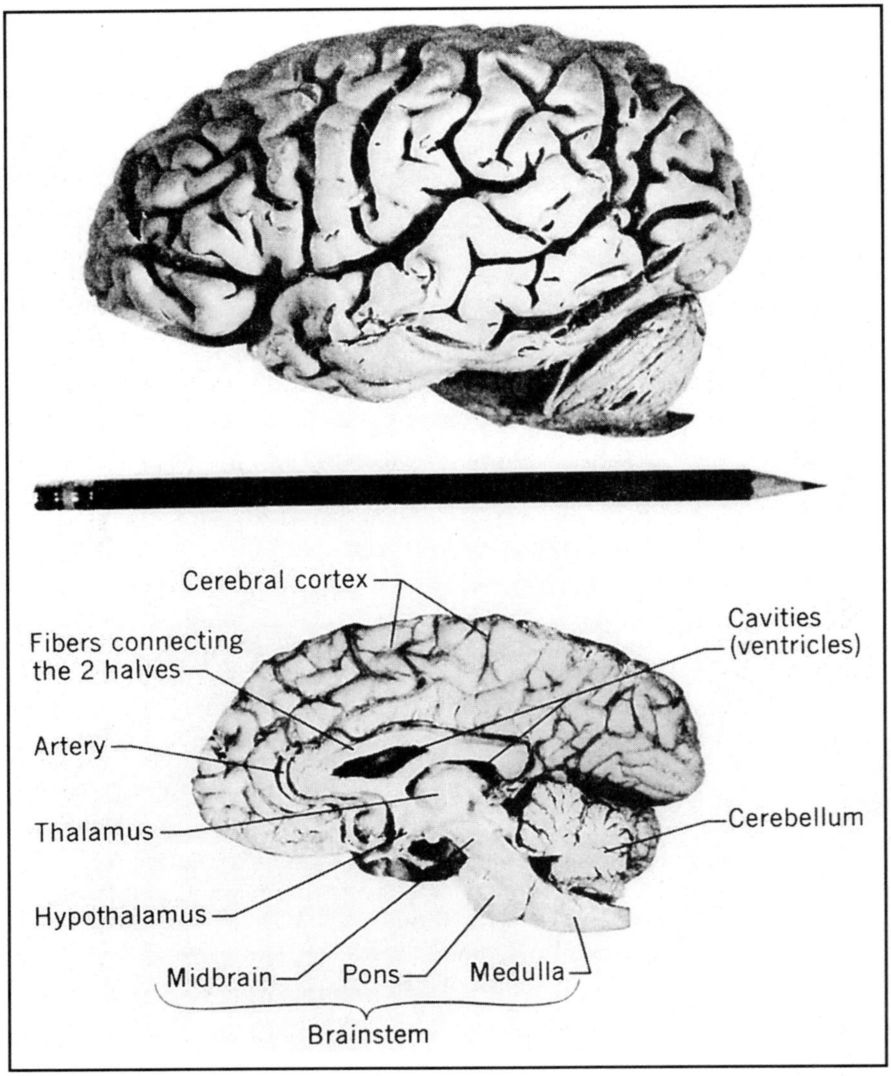

Fig. 1-1 The human brain.

vous systems. Indeed a progressive refinement of nervous systems is a major theme of animal evolution, culminating in the extraordinary mental powers of humans.

In this overview category of twelve nervous system principles, we begin by identifying the basic operational unit of the system: the **Neuron.** Next, we introduce the principle of **Neuron Numbers and Types,** the idea of differing kinds of neurons and the importance of their occurrence in large numbers in the higher animals and humans. Then, we deal with the perplexing issue of **Brain Size,** which seems to be related, but only incompletely so, to the computational power of nervous systems.

Operations of neurons paradoxically exhibit both **Stochastic and Deterministic** (random and non-random) properties that give animals a basis for self-generating and initiating function, as well as a sensitive ability to change function in response to changing conditions.

The neurons are organized into various complex **Circuit Designs,** which route information flow. Organization of these circuits tends to have **Symmetry and Hemispheric Lateralization,** yet another paradox in which seemingly incompatible functions co-exist. We describe how many of these circuits are organized to act with **Modularity** and thus constitute a system of interacting modules.

Then, we discuss **Topographical Mapping,** the idea that much of the world, including an organism's own body, is mapped in the brain. By such mapping, the brain has a way to represent the world within its circuits and a way to issue output commands that are appropriate to that world. Next we identify a principle of **Hierarchical Control** in the nervous system, one that provides efficiency of operation typical of hierarchies in general. But the hierarchical organization of complex nervous systems is adjustable to biological demands, and thus the nervous system can be flexible. **Homeostasis** is a process of servo-regulating control that keeps the various nervous operations in balance. A key to the ability to exert homeostatic control is **Neuro-hormonal Control.**

The collective influence of these various principles leads to our last principle in the overview category, **Behavior.** We can regard behavior from a simple mechanistic perspective of patterned activity of glands and muscles.

Any overview of nervous system principles rightfully begins with the basic operational unit (neuron) and the ultimate product (behavior) (Figure 1-2). Once these principles are understood, the reader should have the perspective to appreciate and understand the other categories of nervous system principles that provide a more detailed perspective on how nervous systems actually enable their host organisms to move about and change their environments (see box on pp. 4 and 5).

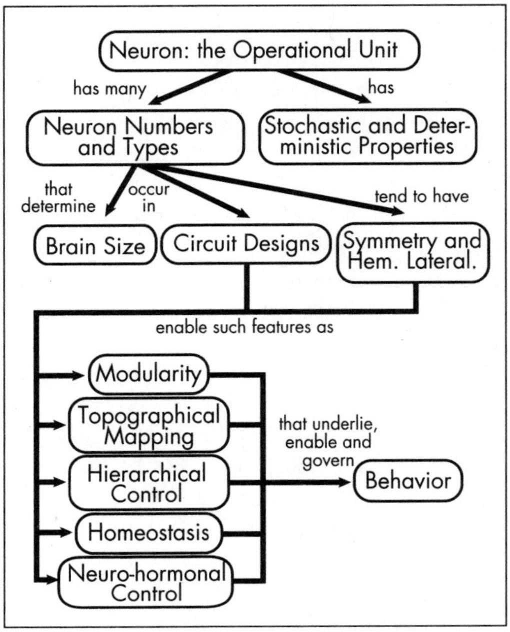

Fig. 1-2 Concept map for the principles that provide an overview of the nervous system.

List of Principles

Neuron: The Operational Unit The basic cellular unit that mediates the information processing actions of the nervous system is the single cell type called a neuron. The essence of neurons can be captured in three words: they are **specialized, numerous,** and **hyperdense** in their interconnections.

Neuron Numbers and Types Brains have enormous numbers of "computing" elements (neurons) that accomplish sophisticated computation because of their large numbers, extensive interconnections, and their high degree of specialization into different types of neurons.

Brain Size Brain size and neuron number are related to mental and behavioral capabilities, but not always in any clear, simple, or linear way.

Stochastic and Deterministic Properties The brain is a highly complex system that has both stochastic and nonlinear, deterministic properties. These are big words, loaded with meaning. But they provide a crucial perspective from which to comprehend how the brain operates.

Circuit Design Neural circuits are organized in certain basic ways: converging, diverging, parallel, and feedback. This provides an anatomical basis of distributed, parallel processing.

Symmetry and Hemispheric Lateralization The brain is basically bilaterally symmetrical, which is a fundamental biological principle of vertebrate structure. Many functions in the brain are not bilaterally symmetrical, but rather are controlled by neuronal groups in one or the other hemisphere. These lateralizations seem to involve higher nervous system functions only and seem to involve only cortical regions of the brain.

Modularity The nervous system is organized as interacting subsystem assemblies of neuronal ensembles.

Topographical Mapping Major sensory and motor systems are topographically mapped. That is, the body, both inside and out, is mapped by the nervous system. Major sensory systems map the external world within their own circuitry. Likewise, the nervous system contains a mapped control over the muscles of the body. Mapped regions may have different inputs or outputs or may share the same ones. Maps are interconnected so that projections from one map to another trigger a back projection to the first map. Mapping can persist at all levels in a given pathway.

Neurohormonal Control A major function of the nervous system is to release certain chemicals into the bloodstream that act as hormones to regulate various hormone-producing glands.

Hierarchical Control The nervous system functions as a hierarchy of semiautonomous subsystems whose rank order is variable. There is no permanent "supervisor" neuron or population of neurons. Any subsystem may take part in many types of interrelationships. Whichever subsystem happens to dominate a situation, each subsystem is independent only to a certain extent, being subordinate to the unit above it and modulated by the inputs from its own subordinate subsystems and by other subsystems whose position in the hierarchy is ill-determined. This design feature of the mammalian nervous system provides maximum flexibility and is probably the basis for the brain's marvelous effectiveness.

Homeostasis The brain regulates the bodily internal milieu through coordinated control over hormones and the nerves that supply viscera.

Behavior Behavior is what emerges from the nervous system's output to glands and muscles, particularly muscles.

NEURON: The Operational Unit

The basic cellular unit that mediates the information processing actions of the nervous system is the single cell type called a neuron. The essence of neurons can be captured in three words: they are specialized, numerous, and hyperdense in their inter-connections. Neurons are also polarized, both in the sense of their input/output relations and in terms of being electronegative, inside relative to the outside of the cell.

Explanation

The neuron is a single cell that is the basic functional building block of the nervous system. Most neurons have many cytoplasmic processes that give them the appearance of a bush or a tree. These branches are called "dendrites" and "axons." Dendrites bring electrochemical stimuli into the neuron, and axons carry electrochemical output away from the neuron toward other target cells. The target cells of a neuron are muscles, glands, or—within the nervous system itself—other neurons (Figure 1-3).

Axons terminate in branches that interact with branches of other neurons. These *synaptic* contacts permit impulse transmission between neurons, typically from one neuron's axonal branches to the short dendrites of an adjacent neuron body. The active region of a synaptic membrane is presumed to be the isolated dark patches near the membrane that are seen in electron micrographs.

Neuron organelles function similarly to those of any cell: chromosomes carry "ancestral wisdom," mitochondria regulate energy supply, microsomal particles control biochemical synthesis, and cell membranes surround the cytoplasm and regulate transport of solutes. Mature neurons are unusual in that they normally do not divide. We don't know why mature neurons do not divide, but it may relate to one or more of the other unusual features of neurons: (1) they express a higher fraction of the genome than other cells, (2) the RNA is clustered as "Nissl substance" on endoplasmic reticulum, (3) they are closely invested by supporting cells (glia), (4) they continuously exhibit pulsatile membrane voltage changes and associated ionic fluxes, and (5) they have an extensive cytoskeleton of tubules and filaments for transporting chemicals throughout an extensive proliferation of protoplasmic processes. Finally, a special pigment, lipofuscin, accumulates in the cytoplasm as the neuron ages, or if its mitochondria or lysosomes are damaged.

The neuron is "polarized," in more than one sense of the word. In one sense, the neuron is polarized so that it can more or less simultaneously respond to input and deliver an output. In an electrical sense, the membrane of a neuron is polarized, having a voltage difference (about 70 mV) between the inside of the neuron and the extracellular fluids that surround it. The inside is electronegative. Responsiveness to input is achieved by

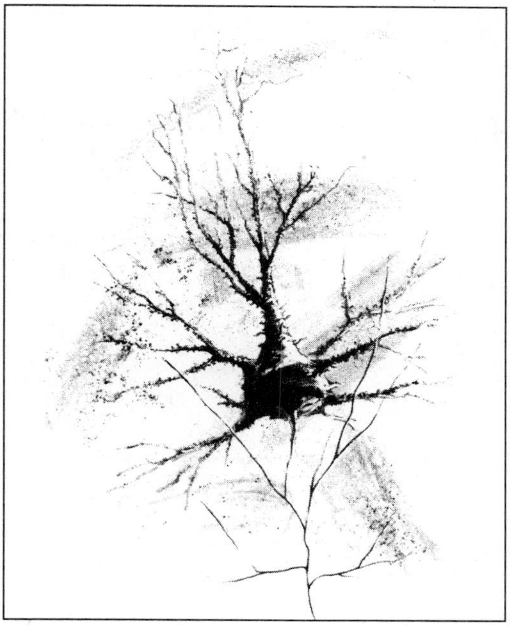

Fig. 1-3 Diagrammatic, three-dimensional representation of a neuron. Dendrites have fuzzy extensions (spines) on their surface. Emerging from the bottom of the cell body is a single axon, which divides into many branches. (From Cotman and McGaugh, 1980, with permission.)

changes in the voltage difference between the inside and outside of the cell. If an external stimulus produces depolarizing changes (inside less negative), the neuron becomes unstable, and at some point a threshold voltage is reached at which a pulsed voltage discharge occurs, serving as an output. If hyperpolarization occurs, a neuron is less likely to discharge and it takes more than the usual amount of excitatory input to elicit discharge.

Examples

Figure 1-3 illustrates the basic structure of neurons. In the diagrammatic three-dimensional representation, we see a single axon emerging from the lower side of the neuronal cell body, and that axon is seen to have a few lateral branches. Note that these branches lack the fuzzy appearing nodules ("spines") that are seen on all the other branches, which are dendrites. These spines are actually points of synaptic contact with axons from other cells.

TERMS

Axons Cytoplasmic extensions of a neuron that provide an output, in the form of electrical pulses and/or chemical release, to target structures (other neurons, glands, or muscles).

Dendrites Cytoplasmic extensions of neurons that receive input, in the form of chemical or electrical stimuli, from other neurons. Some neurons that are specialized to act as sensory receptors for environmental stimuli have functional equivalents of dendrites.

Dendritic Spines Small outgrowths on the membrane surfaces of dendrites. These are sites of synaptic contact with other axons and even dendrites.

Glia The other basic cell type found in the nervous system. These are actually more numerous than neurons, but they are not known to participate directly in the information processing reactions of the neurons. Glial cells do have metabolic supporting functions and, by surrounding neuronal processes with high-resistance membranes, provide some electrical insulation and isolation.

Synapse Points of functional contact between the membranes of two or more neurons. Electrochemical activity in one neuron affects the activity of a target neuron via the synapse.

Related Principles

Neuron Numbers and Types
... see also Cell Biology and Development principles

References

Bray, D. and Gilbert, D. 1981. Cytoskeletal elements in neurons. Ann. Rev. Neurosci. 4:505-523.
Craig, A.M. and Banker, G. 1994. Neuronal polarity. Ann. Rev. Neuroscience. 17:267-310.
Harris, K. and Kater, S.B. 1994. Dendritic spines: cellular specializations imparting both stability and flexibility to synaptic function. Ann. Rev. Neuroscience. 17:341-372.
Joseph, R. 1993. The Naked Neuron. The Evolution of the Languages of the Body and Brain. Plenum Press, New York.
Landis, D.M.D. 1994. The early reactions of non-neuronal cells to brain injury. Ann. Rev. Neurosci. 17:133-152.
Steindler, D.A. 1993. Glial boundaries in the developing nervous system. Ann. Rev. Neurosci. 16:445-470.
Yu, A.c.H., Hertz, L., Norenberg, M.D., Sydová, and Waxman, S.G. 1992. Neuronal-astrocytic Interactions. Implications for Normal and Pathological CNS Function. Progress in Brain Research. Vol. 94. Elsevier, Amsterdam.

Citation Classics

Bodian, D. 1936. A new method for staining nerve fibers and nerve endings in mounted paraffin sections. Anat. Rec. 65:89-97.
Cajal, R.S. 1937. Recollections of My Life. (E. Horne Craigie, translator). Mem. Am. Phil. Soc. 8.
Cowan, W.M., Gottlieb, D.I., Hendrickson, A.E., Price, J.L., and Woolsey, T.A. 1972. The autoradiographic demonstration of axonal connections in the central nervous system. Brain Res. 37:21-51.
Gray, E.G. 1959. Axo-somatic and axo-dendritic synapses of the cerebral cortex; an electron microscope study. J. Anat. 93:420-433.

Jones, E.G. and Leavitt, R.Y. 1974. Retrograde axonal transport and the demonstration of non-specific projections to the cerebral cortex and striatum from thalamic intralaminar nuclei in the rat, cat, and monkey. J. Comp. Neurol. 154:349-377.

Kluver, H. and Barrera, E. *1953*. A method for the combined staining of cells and fibers in the nervous system. J. Neuropathol. 12:400-410.

LaVall, J.H. and LaVall, M.M. *1972*. Retrograde axonal transport in the central nervous system. Science 176:1416-1417.

Nauta, W.J.H. and Gygax, P.A. 1954. Silver impregnation of degenerating axons in the central nervous system: a modified technique. Stain Technol. 29:91-93.

Nicholls, J.G. and Kuffler, S.W. 1964. Extracellular space as a pathway for exchange between blood and neurons in the central nervous system of the leech: ionic composition of glial cells and neurons. J. Neurophysiology. 27:645-671.
(Reviewed in *Current Contents,* Sept. 18, 1989: Nicholls, J.G. Exploring extracellular space.)

Peters, A., Palay, S.L., and Webster, H. deF. 1970. The Fine Structure of the Nervous System. The Neurons and Supporting Cells. Harper and Row, New York, and The Fine Structure of the Nervous System. The Neurons and Supporting Cells, Saunders, Philadelphia.
(Reviewed in *Current Contents,* June 3, 1991: Peters, A. A guide to the cellular structure of the nervous system.)
(Reviewed in *Current Contents,* July 11, 1983: LaVail, J.H. Untitled)
(Reviewed in *Current Contents,* July 12, 1982: Cowan, W.M. Untitled)

Raff, M.C. 1989. Glial cell diversification in the rat optic nerve. Science. 243:1450-1455.

Neuron Numbers and Types

Brains have enormous numbers of "computing" elements (neurons) that accomplish sophisticated computation because of their large numbers, extensive inter-connections, and their high degree of specialization into different types of neurons.

Explanation

Neurons differ in size, shape, and the length of their cytoplasmic processes. They also differ in terms of the kinds of chemicals that they secrete. Because neurons are so specialized, anatomically, physiologically, and biochemically, they can interact with each other in various ways. This variety quantitatively magnifies the number and kinds of interactions and creates flexibility in information processing that cannot be matched, by digital computers, for example.

Examples

The human brain has at least 10^{11} neurons. Each of these makes functional contact with about 1,000 other neurons. At the unit level, speed at which the nervous system operates is rather slow, compared to artificial electronic systems. A given neuron can, for example, change its output state at a maximum rate of about 100 times per second. However, because there are so many interconnections, the system-level speed is impressive even by modern electronic standards. There are about 10^{16} functional interactions/sec. Despite that overall functional capacity, the living "computer" of humans weighs only 3 lbs.

Unlike most electronic neural networks where the basic computational units are exactly the same, neurons of the brain and spinal cord vary dramatically in their structure. The input regions of the cells, called dendrites, can be extensively branched, as seen in A, B, and C of Figure 1-4. The output of the neuron usually occurs via a single process, called an axon. The axon, however, often has many terminal branches (not shown) that make contact on their targets. Axonal targets are either other neurons or "output devices" such as muscle or glands.

Neurons are also specialized in terms of the neurochemicals that they secrete, which serve as chemical signals to other neurons. As many as 100 or more chemicals are considered to act as information carriers in neuronal synapses. This is discussed in more detail in connection with other principles in the cell biology category.

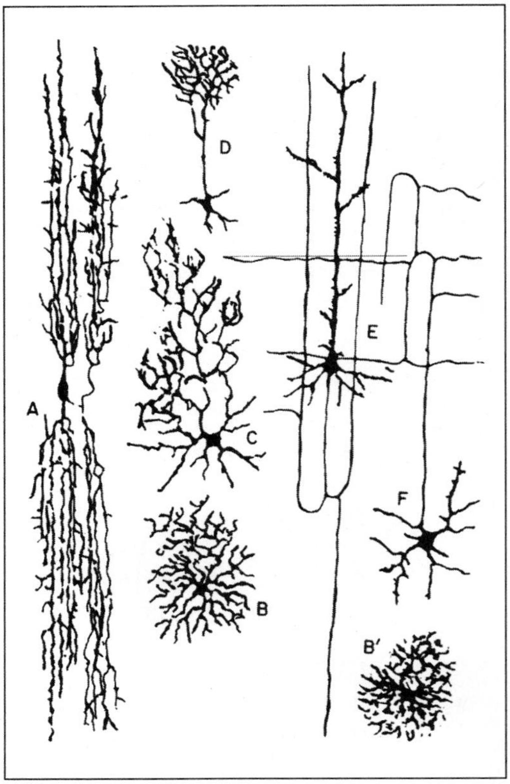

Fig. 1-4 Collage of drawings of representative structural types of neurons that are found in the nervous system. (From Ramon y Cajal.)

TERMS

Neurotransmitters	Chemical secretions that are released from neurons in the immediate vicinity of other neurons. These chemicals are called transmitters because they transmit information from one neuron to another, in terms of exciting, inhibiting, or modulating the activity of target neurons.

Related Principles
Brain Size
Neuron: the Operational Unit

References

Citation Classics
Cajal, Ramon y. 1909. Histologie du systeme nerveaux de l'homme et des vertebrates. Vol. 1. Maloine, Paris. Republished 1952. Instituto Ramon y Cajal, Madrid.
Cajal, Ramon y. 1911. Histologie du systeme nerveaux de l'homme et des vertebrates. Vol. 2. Maloine, Paris. Republished 1952. Instituto Ramon y Cajal, Madrid.
Lorente de No, R. 1934. Studies on the structure of the cerebral cortex. II. Continuation of the study of the ammonic system. J. Psychologie and Neurologie. 46: 113-177.

Brain Size

Brain size and neuron number are related to mental and behavioral capabilities, but not always in any clear, simple, or linear way.

Explanation

One might expect that bigger brains have more neurons, and that is true. One might also expect, therefore, that larger brains are more intelligent and more capable. That is not necessarily true, although typically it is. What seems to be most relevant is the *ratio* of brain size to body size, the idea being that large bodies require more neurons to process sensory input over a larger spatial domain and to control the movements of more muscle fibers (Figure 1-5).

Actually, the microstructure, as opposed to gross size and weight, is more crucially related to mental and behavioral capacity.

Also, the ecological niche and natural selection pressures that have led to the evolution of a given species dictate neuronal number, brain size, and functional capabilities.

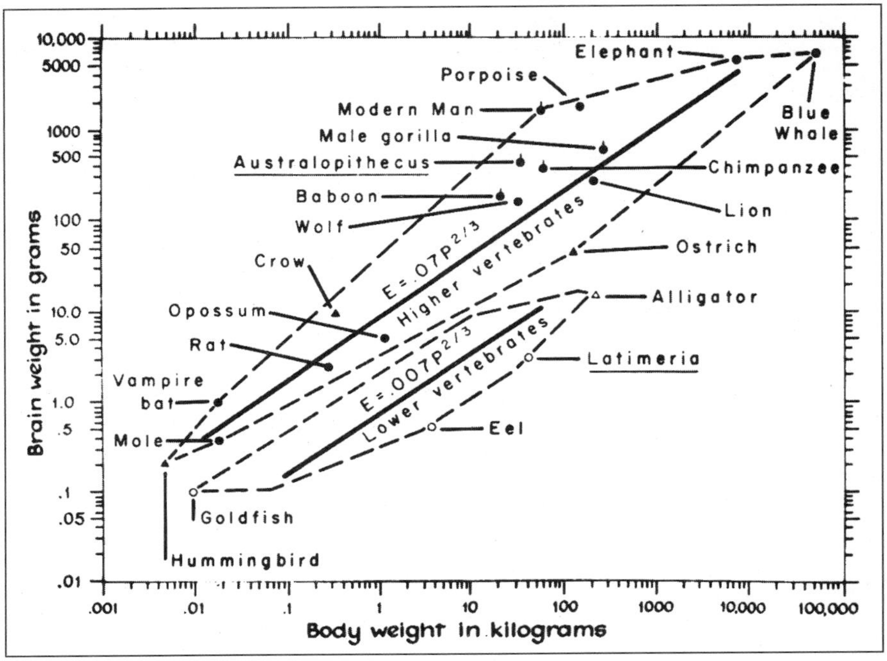

Fig. 1-5 Graphical map of relation of brain size to body weight, showing a general increase in the ratio of brain-to-body weight in higher animals. The visually fitted slopes of two-thirds is the same for both vertebrates and invertebrates. Note that whales and elephants have significantly larger brains than humans, but they also have a much larger body for their brain to control. Interestingly, porpoises of the same approximate weight as humans have brains just as large. (From Jerison, 1973.)

Examples

Large animals, such as elephants and whales, have proportionately larger brains than small animals such as moles. Mammals, as a Class, have larger brains, even when expressed in terms of body weight, than do lower vertebrates. Note that humans deviate the most from the regression line, indicating that they have a disproportionate relationship of brain weight to body weight. The porpoise also deviates conspicuously from the norm.

With regard to microstructure, neurons are less densely packed in larger brains than in smaller brains. As a consequence, there is more space for proliferation of neuronal processes, particularly dendritic trees. Because neurons communicate with each other via their processes, it is evident that functional capacity should be enhanced in the larger brains that have more axonal and dendritic contacts.

The length of dendritic processes is also proportional to brain weight. Compare, for example, rabbits to mice. Packing density, i.e., number of neurons/mm^3, is inversely proportional to brain weight. Compare, for example, the neuron density of mice to the much lower density of rabbits (Figure 1-6).

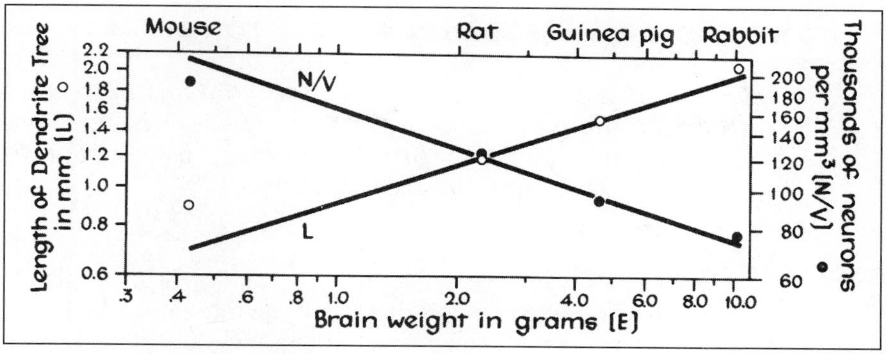

Fig. 1-6 Correlation's between brain size (x axis), length of dendritic trees (left Y axis), and packing density of neurons (right Y axis). The species comparisons show that mice have the smallest brains, the shortest dendritic trees, and the greatest packing density. Rabbits are at the other end of this spectrum, with the largest brains, longest dendritic trees, and smallest packing density. (Graph from Jerison, 1973, based on data of Bok, 1959.)

TERMS

Dendritic Trees The branching patterns of some neurons are so extensive that they resemble the arborization of trees, hence the terms dendritic trees.

Microstructure The histological structure of the cells and their cytoplasmic processes (dendrites and axons).

Related Principles

Neuron: the Operational Unit
Neuronal Induction and Trophism (Development)
Neuronal Numbers and Types
Behavior

References

Bok, S.T. 1959. Histonomy of the Cerebral Cortex. Van Nostrand-Reinhold. Princeton, New Jersey.

Citation Classic

Jerison, H.J. 1973. Evolution of the brain and intelligence. Academic Press, New York.

Stochastic and Deterministic Properties

The brain is a highly complex system that has both stochastic and nonlinear, deterministic properties. These are big words, loaded with meaning. But they provide a crucial perspective from which to comprehend how the brain operates.

Explanation

Stochastic properties are those that operate on the laws of probability. In phenomena that change over time, such as the incidence of action potentials when a neuron discharges, exact predictions are not possible if the process is stochastic. Rather, future occurrences have a probability distribution that is conditioned by a knowledge of past values. In systems that are deterministic, future events can be predicted on the basis of the sequence of preceding events.

Linear systems are predictable. In linear systems, a steady increase in input leads a proportional increase in output. The fundamental units of the nervous system, the neurons, do not operate this way. As explained in other modules in this Overview section and in the Cell Biology section, neurons exhibit discontinuous output. That is, progressive excitatory input, for example, at first leads to no output. Only after a certain threshold of input excitation is reached is there any output, and then it occurs as a voltage pulse that is typically the same size, irrespective of how much the input excitation increases. Although a given neuron always produces the same size voltage pulses, increasing the stimulus input may cause some neurons (but not all) to increase the frequency or pattern of firing. Even here, the change in discharge frequency is often not linearly related to stimulus intensity.

The nonlinear deterministic nature of brain function arises most fundamentally from anatomy. Neurons do not project output to all other neurons, nor do they receive input from all other neurons. Each neuron has its place in a circuit, albeit a complex circuit. The consequence of activity within a circuit is therefore nonlinear and deterministic. Formally, deterministic mechanisms are those that theoretically can be characterized by non-linear differential equations. In the brain, of course, no one knows what those equations are.

Examples

The intervals between action potentials often have an incidence distribution that is characteristic of a given neuron. For example, the intervals of some neurons have a Poisson distribution, while others have a Gaussian or bimodal distribution (Figure 1-7).

One indicator of brain function that has been at the root of much of the debate about the brain's nature is the electroencephalogram, or EEG. The EEG is a time-varying electrical signal, just the kind that is conve-

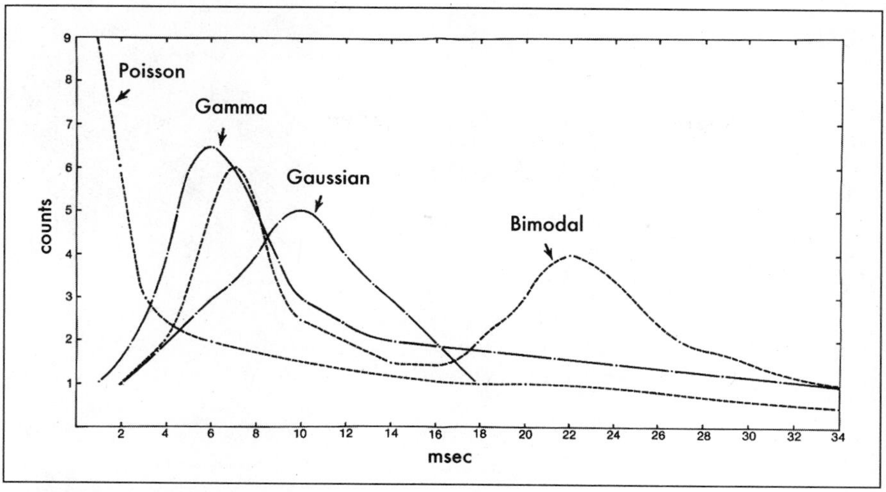

Fig. 1-7 Interval histogram comparison of the various kinds of interval distributions among neurons. Some neuronal trains of action potentials have intervals that approximate a Poisson distribution, whereas others are more like gamma or Gaussian distributions. Some others have various forms of bimodal distributions. Y axis = number of intervals x 1000; X axis = interval duration, in milliseconds.

nient for statisticians and mathematicians to study from both stochastic and deterministic perspectives.

In recent years, the EEG has been perceived in the light of "chaos theory," the leading theory for evaluating complex, nonlinear dynamical systems. In recent years, theoreticians have been applying chaos theory toward understanding how the brain works. Chaos theory mathematically describes systems that change continuously over time and that are extremely sensitive to initial conditions. A small change in initial conditions can theoretically give rise to exponentially larger changes in outcome, thus giving an impression of chaos or randomness. This impression is misleading because chaotic systems are deterministic in that the behavior can be completely specified at all times. Over sufficient time, chaotic systems will settle down into predictable patterns of behavior (Figure 1-8).

Chaotic systems SEEM to be stochastic because their output SEEMS to vary unpredictably, resembling random noise. Human alpha EEG activity may be an example of chaos in the sense that it waxes and wanes in amplitude and its on-set and offset are not always predictable. It is too rhythmic and frequency bounded to be noise. Epileptic discharges are commonly believed to meet the criteria for chaotic behavior.

Related Principles

Action Potentials (Cell Biology)
Cortical Columns (Information Processing)

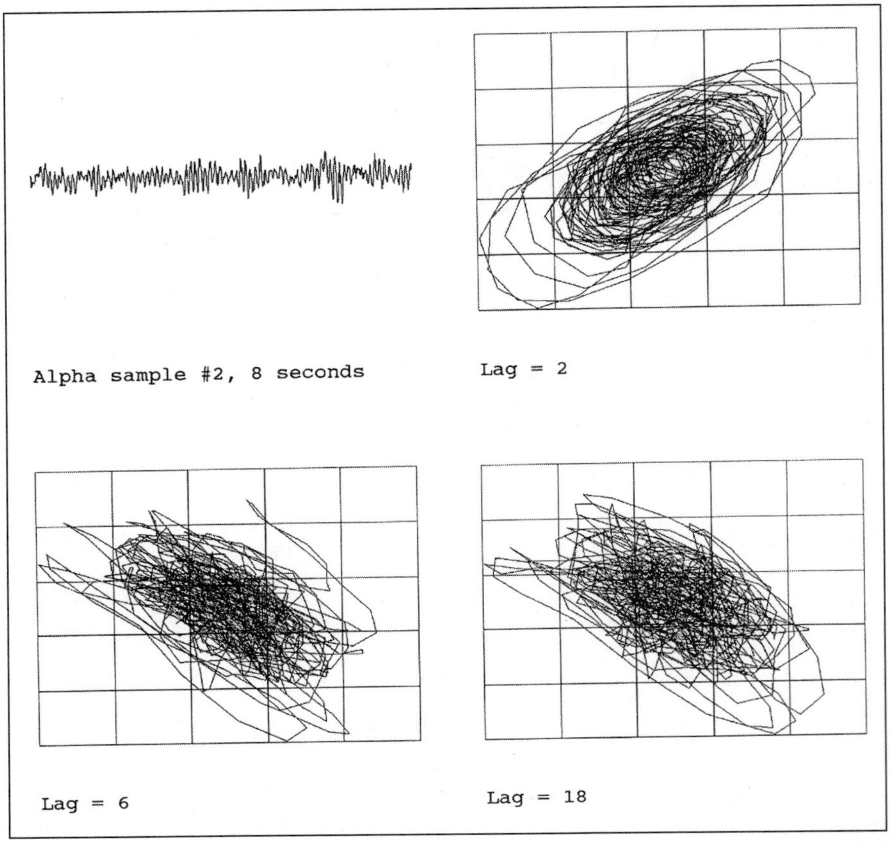

Fig. 1-8 Plots of the sequential dynamics of the voltage values of successive samples in the human electroencephalogram (EEG). The typical rhythmic activity (alpha) at an occipital electrode is shown in upper level. When a plot of these voltages is made between each successive voltage point and the next data point value two sample points in the future (lag 2), a clear pattern emerges, oriented at 45, with a center clustering that appears to have a hole in the middle. When lags are increased slightly, to 6 or 18 future data points, patterns orient in an opposite direction and lack the spiraling around the center. At still longer lags (not shown) the plots become randomly dispersed.

References

Babyloyantz, A. and Destexhe, A. 1986. Low-dimensional chaos in an instance of epilepsy. Proc. Nat. Acad. Sci. 83:3513-3517.

Duke, D., and Pritchard, W. (eds.) 1991. Measuring Chaos in the Human Brain. World Scientific Publishing Co., Singapore.

Hansen, B.H., and Brandt, M.E. 1993. Nonlinear Dynamical Analysis of the EEG. World Scientific Publishing Co., Singapore.

Lopes de Silva, F. 1991. Neural mechanisms underlying brain waves. From neural membranes to networks. Electroenceph. Clin. Neurophysiol. 79:81-93.

Pijn, J.P., Van Neerven, J., Noest, A., and Lopes da Silva, F.H. 1991. Chaos or noise in EEG signals; dependence on state and brain site.

Electroenceph. Clin. Neurophysiol. 79:371-381.
Skarda, A., and Freeman, W.J. 1987. How brains make chaos in order to make sense of the world. Behav. Brain Sci. 10:161-195.

Citation Classic

Freeman, W.J. 1975. Mass Action in the Nervous System. Academic Press, New York.

Circuit Design

Neural circuits are organized in certain basic ways: converging, diverging, parallel, and feedback. This provides an anatomical basis for distributed, parallel processing.

Explanation

Neural circuits have four basic design principles (Figure 1-9). The first three kinds of circuitry are feed-forward, where the flow of information is progressively distributed to sequentially located elements in the circuit. In converging circuits, activation of the input converges from many input cells to fewer cells in successive points (layers) in the circuitry. Diverging circuits have the opposite property. Parallel or "feed-forward" circuits are those that have many axonal branches in parallel. The parallel pathways may exhibit convergence or divergence. The fourth kind of circuit has feedback elements that permit some of the circuit's output to be led back into the input. This provides an anatomical basis for recurrent, rhythmic reverberation within the circuit.

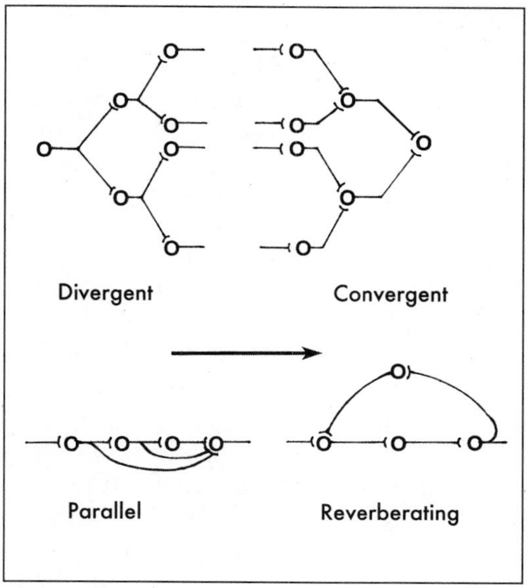

Fig. 1-9 Diagram of how neurons are organized into four basic types of circuitry. Direction of arrows shows the flow of impulses along the axons. Note that the feedback arrangement allows the output of the last neuron in the chain to feed back influence on the first neuron in the chain so that action potentials can continue to flow through the circuit long after the initial triggering condition.

Examples

Convergent Circuits – a biological example is that movement initiation and coordination signals can arise from many parts of the brain, yet converge to a single or a few movement-initiating neurons in the spinal cord. Such neurons are often referred to as the "final common pathway." Another example is the spinal sensory neuron, whose input processes receive sensory stimulation from various points on the body, including skin and viscera, and which then converge on a few neurons in the spinal cord. This arrangement can lead to "confusion," because other sensory neurons also converge at the same place. So-called referred pain, for example, the pain that one feels in the arm during a heart attack, occurs because the central nervous system refers the source of pain to the same spinal cord sensory neurons that are connected to both the heart and to the arm.

Divergent Circuits – an example is the input from a painful stimulus, which may begin at a single point (e.g., pin prick of the skin), and distribute widely to sensory cerebral cortex and multiple other parts of the cortex. Some neurons that control movement also are part of divergent circuits. Motor output from neurons in the motor cortex typically spreads in divergent circuits wherein some of the output goes to target muscles, while other output goes to coordinating centers in the basal forebrain, various regions of the brainstem, and to the cerebellum.

Parallel Circuits – one of the best studied neural network models is a multilayered feed-forward network model in which three or more layers of units interact in both converging and diverging fashion. Input units send divergent output to middle layers ("hidden units," "interneurons"), and these in turn send divergent outputs to the output units. The input layer of units is richly interconnected in parallel with units in successive layers and the connections between units are governed by nonlinear relations between the synaptic input and the output firing rate.

The study of neural circuits has led engineers and computational scientists to develop the field of "neural networks." The model system that they use, both hardware and software, provides opportunities for sophisticated computations and may prove to yield insights into how real neural networks operate (Figure 1-10).

A major biological example of feed-forward parallel circuitry is the primary body sensory system for touch, pressure, and thermal sensations. These pathways are routed to the cerebral cortex, but they also give off parallel collaterals to the brainstem, which in turn has its own projections to the cortex. The value of such a design is that target neurons in the cortex get two kinds of information simultaneously. The input from the brainstem goes to all of the cortex, not just the small zone of specific "sensory" cortex. That brainstem input has the effect of activating the entire cortex. It is a way for the brainstem to say, in effect, "wake up brain, here comes some sensory information that you need to attend to and process."

Feedback Circuits – This arrangement permits reverberation of excitation and is the basis for the tendency of many neuronal systems to oscil-

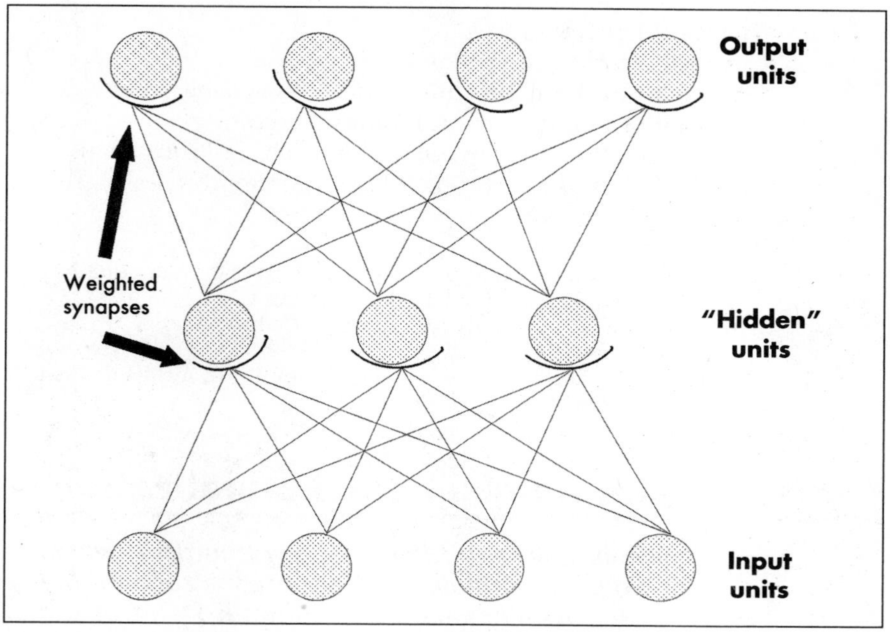

Fig. 1-10 Diagram of a common type of artificial neural network. A three-layered network features junctions ("synapses") that can be weighted in terms of connection strength and discharge threshold for yielding an output. The input and output can be considered as data arrays, with the middle layer serving as the primary processor. At each "synapse" a mathematical transfer function describes the current values of input signals, as well as values in local memory to produce the processing element's new output signal. The transfer function can also modify stored values stored in local memory, thus giving such circuits the capacity to "learn."

late. One well-studied example is a system involving the cerebral cortex, which connects with a major subcortical relay station known as the thalamus. Here, the reverberation becomes manifest in synchronous discharge of cortical cells in the visual cortex. This produces the rhythmic electroencephalographic waveform known as alpha rhythm.

TERMS

Feed-forward A circuit in which neuronal impulses are moving in the direction from input to output.

Feedback A circuit in which some of the neuronal impulses are moving back into the input.

Reverberation A circuit in which activity continues to re-cycle, producing rhythmic physiological processes and behavior.

Related Principles

Action Potentials (Cell Biology)
Feedback and Re-entry (Information Processing)
From Input to Output (Information Processing)
Parallel, Multi-level Processing (Information Processing)
Rhythmicity and Synchronicity (Information Processing)

References

Durbin, R., Miall, C., and Mitchison, G. 1989. The Computing Neuron. Addison-Wesley, Reading, Massachusetts.
Hecht-Nielsen, R. 1990. Neurocomputing. Addison-Wesley, Reading, Massachusetts.
Zornetzer, S.F., Davis, J.L., and Lau, C. (eds.) 1990. Academic Press, San Diego, California.

Symmetry and Hemispheric Lateralization

The brain is basically bilaterally symmetrical, which is a fundamental biological principle of vertebrate structure. Many functions in the brain are not bilaterally symmetrical, but rather are controlled by neuronal groups in one or the other hemisphere. These lateralizations seem to involve mostly cortical regions of the brain and higher nervous system.

Explanation

Bilateral symmetry is a fundamental principle of vertebrate anatomy. Despite the generally clear anatomical evidence for bilateral symmetry of brain structure and function, it is nonetheless true that certain functions are lateralized to one hemisphere or the other. These lateralizations seem to be primarily controlled by cortical regions of the brain. In most cases, the lateralizations that are known are functional, and anatomical asymmetries underlying lateralized function have not been detected. That may, however, just reflect the limitations of common anatomical methods for disclosing asymmetries in the microcircuitry of the brain.

However, no one place in the brain operates in isolation, and that is particularly true for higher brain functions that depend so greatly on parallel, distributed processes and interactive topographically mapped functions.

Examples

The first inkling of localization was with human speech, being first discovered by Broca in 1861. In most right-handed people, speech is controlled by a small cluster of neurons in the neocortex of the left hemisphere. One cluster, in the left temporal lobe of the cortex (Wernicke's area) acts as a language comprehension area. Input from both seen and heard words is first routed to Wernicke's area, and then this information is routed to another group of cells in the left frontal cortex (Broca's area). Broca's area projects to the lips, tongue, and mouth regions of the motor cortex and is necessary for appropriate articulation of speech sounds. These functions have been demonstrated in aphasic people, whose hemispheres have been disconnected by surgical cutting of the fiber tract joining the two hemispheres, and in conscious neurosurgical patients in response to electrical stimulation of the speech area. It has also been documented by positron emission tomography (PET).

In the young child, the right hemisphere also has a functional speech center that can assume command of speech should there be damage to the left hemisphere's speech center. There is a critical age, however, beyond which this plasticity does not exist.

Other lateralizations become evident from studying epileptic patients who have undergone surgical transection of the main fiber bundle connecting the two hemispheres to suppress the spread of epileptic seizure discharges. These "split brain" patients have essentially two independent brains, and each has been shown to specialize in certain functions. In most people, the left side controls, in addition to language processing, such functions as mathematical ability and complex movements. The right side of the brain specializes in spatial perception and complex pattern recognition.

Musical processing is a somewhat different case. The grammar and semantics of music have very little in common with verbal language, and thus it is not surprising to discover that musical functions are not performed by the speech centers in the left hemisphere. Research strategies are only now taking into account the important variables (for example, you need to study musicians, not just ordinary folks humming a tune). Topographical mapping of regional blood flow changes (PET scan) has shown that different regions are involved in different aspects of music processing, such as perception of sheet music or internally generated sounds in the "mind's ear" and the motor acts that translate music concepts to instrument manipulation. PET scans seem to show that several brain areas in *both* hemispheres participate in the translation of sight-read sheet music to instrument playing, after subtracting the control brain activations associated with the mechanics of playing scales.

Another indication that complex tasks engage different parts of both hemispheres is the observation that electroencephalographic (EEG) mapping reveals distinct EEG changes over many regions of both hemispheres when subjects performed a flight-landing simulation, after taking into account the brain activation during visual and motor control tasks.

For many years, localized functions were assumed to be absent in lower animals. Without the capacity for language, it is obviously not possible to demonstrate speech lateralization. However, recent studies in some bird species have shown that bird song is preferentially mediated from the left hemisphere. Lesions, for example, to the song area of the left hemisphere profoundly disrupted bird song, whereas homologous lesions to the right hemisphere had no such effects. Similarly, Japanese macaques have been shown to have a right ear advantage for processing species-specific vocalizations (indicating left hemisphere preference, because the sensory path crosses over to the opposite hemisphere).

Lateralizations to the right hemisphere for such functions as emotions and spatial perception have also been documented in animal species.

TERMS

Bilateral Symmetry Formal definitions abound in anatomy texts. A convenient working definition is that bilaterally symmetrical animals have a midline, and structures equidistant on each side of that midline are mirror images of each other. In other words, structures on the right side of the body (or brain) have corresponding structures on the left side.

Homologous Matching in structure, position, or function; that is, structures on the right side of the midline are homologous with those on the left.

Lateralization A difference of structure and/or function between the left and right side of the brain.

Positron Emission Tomography (PET) A technique for measuring cerebral blood flow, which increases in areas of the brain that are more active. The technique is based on location of a radioactive analog of glucose, which accumulates in the brain regions that are most metabolically active.

Related Principles

Cognition (States of Consciousness)
Emergent Properties (Information Processing)
Parallel, Multi-level Processing (Information Processing)
Selective Attention (States of Consciousness)
Rhythmicity and Synchronicity (Information Processing)
Topographical Mapping (Overview)

References

Deneberg, V.H. 1981. Hemispheric laterality in animals and the effects of early experience. Behav. Brain Sciences. 4:1-49

Geschwind, N. 1979. Specializations of the human brain. Sci. Amer. 241:180-199.

Glick, S.D. (ed.) 1985. Cerebral Lateralization in Non-human Species. Academic Press, Orlando.

Iaccino, J.F. 1993. Left Brain - Right Brain Differences: Inquiries, Evidence and New Approaches. Lawrence Erlbaum Assoc., Hillsdale, New Jersey.

Kitterle, F.L., (ed.) 1991. Cerebral Laterality: Theory and Research: the Toledo Symposium. Lawrence Erlbaum Assoc., Hillsdale, New Jersey.

Levy, J., Trevarthen, C., and Sperry, R.W. 1972. Perception of bilateral chimeric figures following hemispheric disconnection. Brain. 95:61-68.

Nottebohm, F. 1977. Neural lateralization of vocal control in a passerine bird. J. Exp. Neurol. 177:229-262.
Nottebohm, F., Stokes, T.M., and Leonard, C.M. 1976. Central control of song in the canary, *Serinus canarius*. J. Comp. Neurol. 165:457-486.
Sergent, J. 1993. Mapping the musician brain. Human Brain Mapping. 1:20-38.
Sperry, R.W. 1975. In search of the psyche, p. 425-434, In The Neurosciences: Paths of Discovery. Edited by F.G. Worden, J.P. Swazey, and G. Edelman. MIT Press. Cambridge, Massachusetts.
Springer, S.P. and Deutsch, G. 1989. Left Brain, Right Brain. 3rd ed. W. H. Freeman, New York.
Sterman, M. B., Mann, C.A., Kaiser, D.A., and Suyenobu, B.Y. 1994. Multiband topographic EEG analysis of a simulated visumotor aviation task. Int. J. Psychophysiol. 16:49-56.
Witelson, S.F. 1976. Sex and the single hemisphere: specialization of the right hemisphere for spatial processing. Science. 193:425-427.

Citation Classics

Sperry, R.W. 1961. Cerebral organization and behavior. Science. 133:1749-1757.
Sperry, R.W. 1968. Mental unity following surgical disconnection of the cerebral hemispheres. Harvey Lect. 62:293-323.
Sperry, R.W. 1970. Perception in the absence of the neocortical commissures. Res. Publ. Assoc. Res. Nerv. Ment. Dis. 48:123-138.

Modularity

The nervous system is organized as interacting subsystem assemblies of neuronal ensembles.

Explanation

Specific functions are subserved by specific neuronal systems. The neurons in a given system are not necessarily located all in the same place, but they do function collectively as a module. Both anatomical and functional boundaries between and among modules are usually fuzzy, and some of the neurons in one module may at times be recruited to participate in the functions of another module.

These modules are partially autonomous, but since they are interconnected they do influence and are influenced by other modules.

Examples

The simplest example of modularity is found in invertebrates, which often have repeated body segments, each with a cluster of neurons that controls the segment. Thus, each cluster of neurons and body segment is modular. This organizational principle is preserved in the vertebrate spinal cord, although the spinal segments and their corresponding body segments overlap, anatomically and functionally.

The spinal cord does not look modular from a distance, but actually it is composed of successive segments of neuronal groups that give rise to nerves that supply a restricted portion of the body. In principle, humans have a similar body segmentation as do earthworms. This only applies for the periphery, not the brain. Higher-order, but subconscious, operations are conducted by brainstem neurons and clusters of subcortical neurons (basal ganglia), which are grouped together at one end of the spinal cord. Highest-order, consciously operating functions are conducted by cerebral and cerebellar cortex, which have many of their connections with the brainstem and basal ganglia.

But even higher nervous functions have a degree of modularity. One module, the hypothalamus, largely controls neurohormones and nerves affecting viscera. Another module in the brainstem is a general activating system for the brain. The cerebellum exerts modular control over movement coordination. Sensation tends to be divided into modules for sound, sight, and other senses.

TERMS

Body Segments	Successive, bilaterally symmetrical regions of the body. Earthworms and insect larvae are the most obviously segmented.

Related Principles
Hierarchical Control
Neurohormonal Control
Readiness Response (States of Consciousness)
Topographical Mapping
Visceral Control (Motor Activity and Control)

Topographical Mapping

Major sensory and motor systems are topographically mapped. That is, the body, inside and out, is mapped by the nervous system. Major sensory systems map the external world within their own circuitry. Likewise, the nervous system contains a mapped control over the muscles of the body. Mapped regions may have different inputs or outputs or may share the same ones. Maps are interconnected so that projections from one map to another trigger a back projection to the first map.

Mapping can persist at all levels in a given pathway.

Explanation

The body, inside and out, is mapped by the nervous system. Mapped, interconnected neuron ensembles are the units of representation of perception and motor events. Locations in a three-dimensional sensory world are represented in the central nervous system neurons in such a way that neighboring locations in the sensory world also are represented in neighboring neurons in the nervous system. Likewise, in motor systems in the nervous system, neurons that activate certain muscles have neighboring neurons that activate neighboring muscles.

These topographical maps constitute an inner model of the body. To explain this model in terms of neural function, we can think of a model:

as an **Implementation** (nerve impulse patterns)
of a **Representation** (topographical maps)
of an **Abstraction** (sensory transduction/motor programs)
of a **Reality** (physical stimuli/?)

Note that it is not so obvious what the "reality" basis is for motor programs. The underlying reality must include some kind of combination of muscle and bone anatomy, neural circuitry, and various degrees of intentionality.

We know that topographical maps exist because with appropriate monitoring techniques, such as microelectrodes that can record responses at various points along a sensory or motor pathway, an observer can witness the point-to-point projections of activity. Conversely, if one knows, from electrical recordings for example, the anatomical locus of a projection, the information flow along the pathway can be mimicked by electrical stimulation or be abolished by a lesion that is strategically placed in the topographically mapped area.

Not all parts of the brain have clear topographical mapping. Such non-mapped areas include the hippocampus, hypothalamus, and basal ganglia. How these non-mapped regions interact with the mapped systems remains among the great enigmas of neuroscience.

Examples

The simplest example of mapping is found within the spinal cord. A restricted portion of the body is mapped by a few specific neurons in the part of the spinal cord that is part of the same body segment. The projections from the cord into the brain retain a topographic segregation in the first relation station in the subcortical part of the brain known as the thalamus. The bodily mapped representations in the thalamus are maintained in the projections to the sensory cortex. As mentioned earlier, the sensory information is also routed in a parallel non-mapped form via the central core of the brainstem. From there, projections go to diffuse areas of the cortex other than the sensory cortex.

This has substantial clinical application. From knowing the topography of a sensory projection, for example, one can predict the clinical effects of a lesion at a particular point in the spinal cord. Conversely, from careful observation of clinical signs, particularly spinal reflexes, one can predict the locus of a lesion (Figure 1-11).

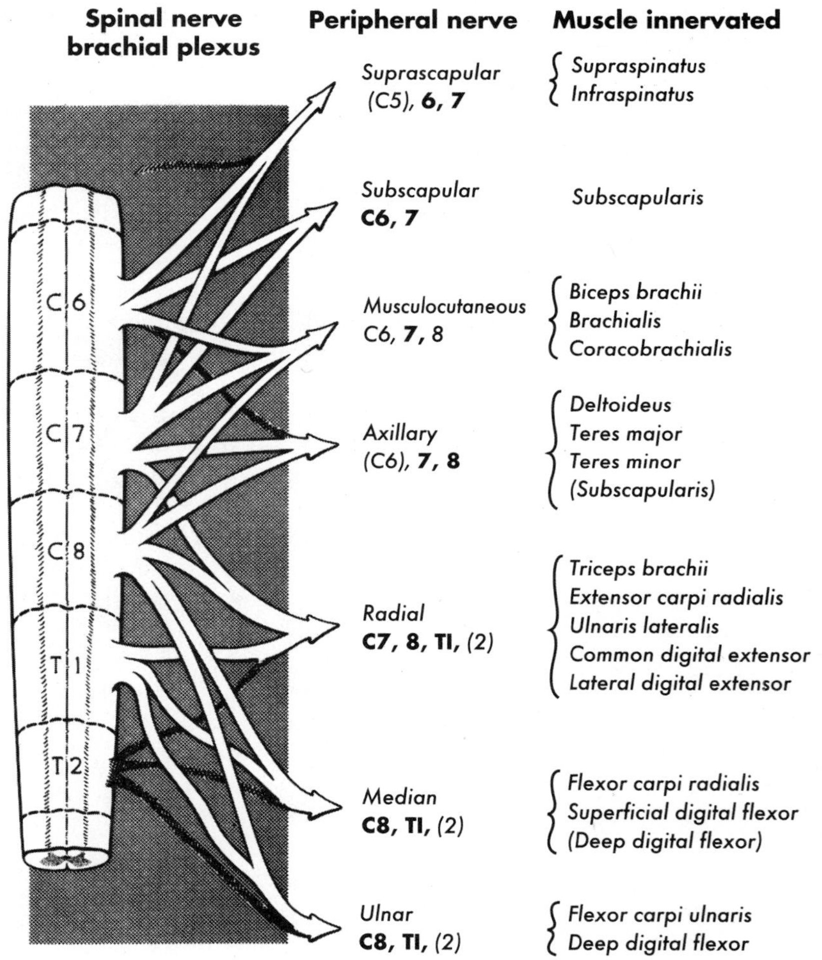

Flexor reflex:	Sensory: varies with area stimulated Motor: musculocutaneous, axillary, median, ulnar, radial
Biceps reflex:	Sensory and Motor: musculocutaneous
Triceps reflex:	Sensory and Motor: radial

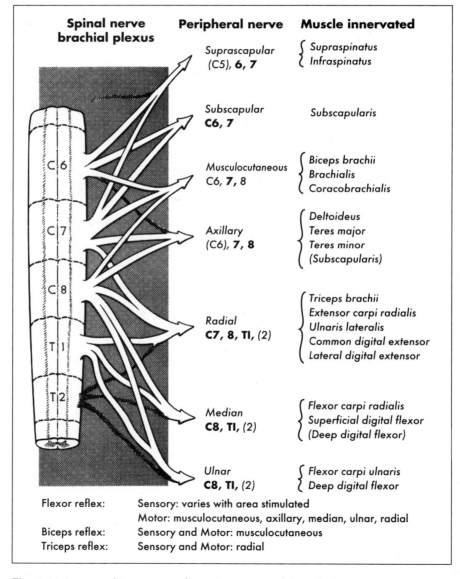

Fig. 1-11 Segmental innervation from the regions of the spinal cord that give rise to the brachial plexus in dogs. Note that specific regions of the cord specifically innervate different muscle groups. (From DeLahunta, A. 1983).

The projections from the cord into the brain retain a topographical segregation in the first relation station in the subcortical part of the brain known as the thalamus. The bodily mapped representations in the thalamus are maintained in the projections to the sensory cortex.

Topographic mapping is also a prominent part of the architecture of the higher parts of the nervous system, including much of the cerebral cortex. However, the clinical applications of this knowledge are not as straightforward as in the case of spinal pathways. In the cortex, there are regions where the body is mapped, both in terms of sensations and motor control. The precision of sensation or motor control is more or less proportional to the amount of map and number of neurons assigned to that function (Figure 1-12).

The persistence of mapping at all levels in a pathway can be illustrated by the pathways for vision. Visual input begins with a spatial mapping of the visual field in the retina of the eye. This mapping persists throughout the various fiber tracts and relay stations in the brainstem and thalamus

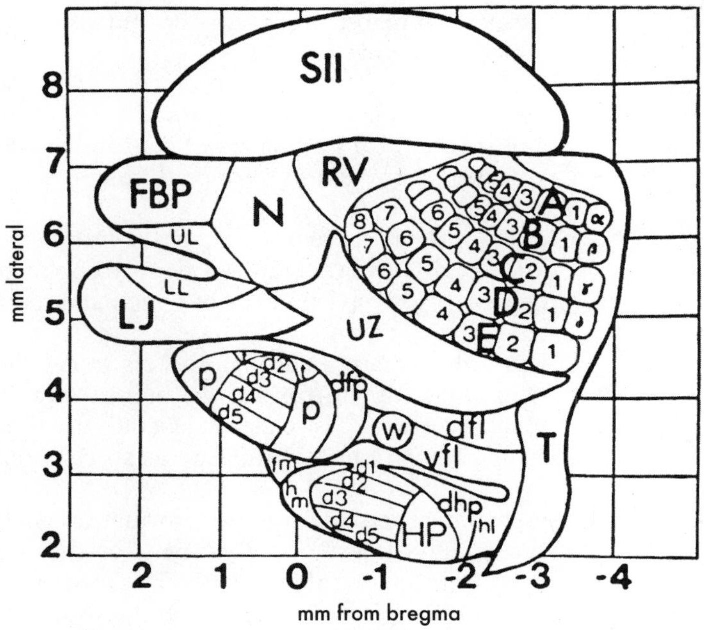

Fig. 1-12 Topographically detailed representation of the various parts of the skin of anesthetized rats in the primary sensory cortex. Map was constructed by microelectrode recording from neurons in the sensory cortex, while simultaneously stimulating different regions of skin. There is disproportionate representation of the vibrissae (A-E, 1-8) and the skin of the paws (dorsal hindpaws: dhp; dorsal forelimb: dfl; palm: P) and digits (d1-d5). Less well-represented areas include the trunk (T), nose (N), lips (UL,LL), lower jaw (LJ). Zone UZ was unresponsive in anesthetized rats. SII = secondary sensory cortex, which was unmapped. (From Chapin and Lin. 1984.)

on the way to the final termination in the visual cortex. The visual cortex of the monkey has over thirty different maps for coding different features of a visual scene, such as color, orientation, movement, dominant eye, etc. These maps are connected with each other by parallel pathways, and many of the mapped regions connect reciprocally.

TERMS

Basal Ganglia Several clusters of neurons that collectively modulate and help to coordinate body movements.

Body (Spinal) Segment The body can be mapped according to the innervation to and from specific zones of the spinal cord. For example, the innervation (both sensory and motor) of certain muscles resides in a specific region of the spinal cord; other cord areas innervate other body muscles.

Hippocampus A phylogenetically old kind of cortex, folded in under the neocortex in humans, that is involved in emotional behavior and in memory formation.

Hypothalamus A small zone on the ventral surface of the brain. This area contains several distinct clusters of neurons that are important for regulating visceral and hormonal functions.

Mapping Maps are neuronal representations (sometimes highly abstracted) of the real world. A map is a point-to-point representation of the real-world environment in which an animal lives.

Thalamus A group of many neuronal clusters along the midline of the brain, lying just in front of the brainstem and underneath the cerebral cortex. These clusters generally are topographically segregated for various sensations from specific parts of the body.

Topographical A spatial representation of information in the environment (can include internal environment of the body) that is projected onto neuronal circuitry. The representation of the environment cannot be "graphed" in the usual sense of the word, but can be revealed and studied by various physiological measures such as electrical recordings from various regions within the mapped areas.

Related Principles

Ensembles of Dynamic Neural Networks (Learning and Memory)
Modularity

Parallel, Multi-level Processing (Information Processing)
Emergent Properties (Information Processing)
Sensory Modalities and Channels (Senses)

References

Brodal, A. 1975. The "wiring patterns" of the brain: neuroanatomical experiences and their implications for general views of the organization of the brain, p. 123-140. In The Neurosciences: Paths of Discovery, edited by F.G. Worden, J.P. Swazey, and G. Adelman. MIT Press, Cambridge, Massachusetts.

Chapin, J.K. and Lin, C.-S. 1984. Mapping the body representation in the SI cortex of anesthetized and awake rats. J. Comp. Neurol. 229:199-213.

DeLahunta, A. 1983. Veterinary Neuroanatomy and Clinical Neurology. 2nd ed. W.B. Saunders Co., Philadelphia.

Edelman, G.M. 1992. Bright Air, Brilliant Fire. On the Matter of Mind. Basic Books.

Goldman-Rakic, P.S. 1988. Topography of cognition: parallel distributed networks in primate association cortex. Ann. Rev. Neurosci. 11:137-156.

Jenkins, T.W. 1978. Functional Mammalian Neuroanatomy, 2nd ed. Lea & Febiger, Philadelphia.

Knudsen, E.I., du Lac, S., and Esterly, S.D. 1987. Computational maps in the brain. Ann. Rev. Neurosci. 10:41-65.

Porter, R. ed. 1992. Exploring Brain Functional Anatomy with Positron Tomography: Ciba Foundation Symposium 163. Wiley, New York.

Udin, S.B. and Fawcett, J.W. 1988. Formation of topographic maps. Ann. Rev. Neurosci. 11:289-327.

Citation Classics

Edelman, G.M. 1988. Topobiology: An Introduction to Molecular Embryology. Basic Books, New York.

Hubel, D.H. and Wiesel, T.N. 1962. Receptive fields, binocular interaction and functional architecture in the cat's visual cortex. J. Physiol. London. 160:106-154.

Hubel, D.H. and Wiesel, T.N. 1968. Receptive fields and functional architecture of monkey striate cortex. J. Physiol. 195:215-243.

Neurohormonal Control

A major function of the nervous system is to release certain chemicals into the bloodstream that act as hormones to regulate various hormone-producing glands.

Explanation

Neurons are prominent neurosecretory cells. These secretions, when they act locally and directly on other neurons, are called neurotransmitters or neuromodulators. However, some neurons release their secretions into the bloodstream, where the target tissue is commonly an endocrine gland. The endocrine glands are thus part of an interacting neuroendocrine system, which brings the glands under the direct influence of the nervous system.

Neurosecretion is a prominent feature of invertebrate neural function and has been preserved in vertebrates for very important roles involving homeostatic regulation and reproductive fitness. In vertebrates, neuroendocrine cells occur mostly in the hypothalamus and the anterior pituitary. The pituitary gland can be thought of as an executor agent, being the proximate source of most neuroendocrine secretions. Endocrine neurons in the hypothalamus generally control the anterior pituitary through the release of hormones by releasing hormones into a venous system that drains directly into the anterior pituitary to stimulate or inhibit release of pituitary secretions.

Glandular secretions and muscle movements are the outputs of the nervous system, and collectively they govern what we view as behavior. Feedback information from the actions of glands and muscles, and from the consequences of behavior, affect the central nervous system's processing of sensory information and integration of thought processes and memory. Thus, neurohormonal controls can be seen to be one significant part of the overall function of the nervous system.

Examples

Several cell groups in the hypothalamus release various neurohormones into blood that drains into the anterior pituitary. Most of these compounds stimulate the release of specific pituitary hormones. There are "releasing factors" for adrenocortical stimulating hormone, follicle stimulating hormone, luteinizing hormone, and thyroid stimulating hormone. In addition, there is a release-inhibiting factor for luteinizing hormone.

Two clusters of cells in the hypothalamus send their terminals into the posterior pituitary, where they release two neurohormones: oxytocin, which contracts smooth muscle of the uterus, and vasopressin, which contracts smooth muscle of blood vessels and additionally stimulates water reabsorption in the kidney.

Sex hormones provide a good example of feedback influences on the central nervous system. When blood levels of estrogens rise suddenly at parturition, they are found to bind on specific neurons in the medial preoptic region of the hypothalamus. This binding activates the cells, as demonstrated by increased c-fos (see Terms on p. 31) and electrical activity. As a consequence, a whole series of maternal behaviors are triggered, in large part because the olfactory system becomes sensitized to detect the odor of the babies. This triggers approach behavior and licking and touching of the babies.

Another example of biasing in an opposite direction is the effect of endogenous opiates on pain. Here, the hormones have pain-alleviating properties akin to morphine and other opiates. The hormones are released under stressful and painful conditions.

TERMS

C-fos — This is a proto-oncogene in neurons that produces a nuclear protein, call fos. Fos is believed to promote the translation of synaptic activity into further changes of gene expression. Basal levels are typically quite low, but intense stimulation of neurons causes detectable increases in the fos protein, which is typically monitored immunohistochemically. Thus, c-fos expression is an index of heightened neural activity. If such activity persists, c-fos expression can also be a molecular index of learning.

Hormone — Chemical that circulates in the blood to alter the function of specific target organs.

Hypothalamus — An area at the base of the brain that is symmetrically divided about the midline. It lies directly over the pituitary gland. It is bounded dorsally by the thalamus, posteriorly by the midbrain, and anteriorly by the septum and certain other forebrain structures.

Neurohormone — A chemical released from neurons that is released into the blood and acts like a hormone.

Pituitary Gland — The "master gland" of the body. Its secretions control the activities of many of the other glands.

Releasing Factor — A neurohormone that causes other neurons to release their hormones.

References

Cohen, A.H. et al. 1988. Neural Control of Rhythm and Movement in Vertebrates. Wiley. New York.

Gurney, M.E., and Konishi, M. 1980. Hormone-induced sexual differentiation of brain and behavior in zebra finches. Science. 208:1380-1383.

Lincoln, D.W., and Wakerly, J.B. 1974. Electrophysiological evidence for the activation of supraoptic neurons during the release of oxytocin. J. Physiol. 242:533-554.

Pierpaoli, W., Regelson, W., and Fabris, N. 1994. The Aging Clock: the Pineal Gland and Other Pacemakers in the Progression of Aging and Carcinogenesis. N.Y. Acad. Sci., New York.

Scharrar, B. 1987. Neurosecretion: beginnings and new directions in neuropeptide research. Ann. Rev. Neurosci. 10:1-18.

Swanson, L.W., and Sawchenko, P.E. 1983. Hypothalamic integration: organization of the paraventricular and supraoptic nuclei. Ann. Rev. Neurosci. 6:269-324.

Verney, E. B. 1974. The anti-diuretic hormone and the factors that determine its release. Proc. Roy. Soc. London B. 135:25-106.

Hierarchical Control

The nervous system functions as a hierarchy of semiautonomous subsystems whose rank order is variable. There is no permanent "supervisor" neuron or population of neurons.

Any subsystem may take part in many types of interrelationships. Whichever subsystem happens to dominate a situation, each subsystem is independent only to a certain extent, being subordinate to the unit above it and modulated by the inputs from its own subordinate subsystems and by other subsystems whose position in the hierarchy is ill-determined. This design feature of the mammalian nervous system provides maximum flexibility and is probably the basis for the brain's marvelous effectiveness.

Explanation

Invertebrates such as lobsters, gastropods, and crayfish have specific groupings of neurons that reflexly elicit escape swimming. These circuits have identifiable sensory neurons, interneurons, and motor neurons, as well as "command" neurons, that activate the entire network's output and achieve a stereotyped and relatively complete repertoire of behavior.

The extent to which such "preprogrammed," command-neuron circuitry exists in mammals is not known, but such circuitry probably does exist in "masked" form because of the extensive interaction with other, less stereotypically organized neurons and circuits. The best-known examples of such command circuits in mammals involve systems that regulate hand and eye movements.

These neurons are high in the hierarchy because of their executor action. Nonetheless, other neural systems can determine whether the executor neurons are free to issue their motor commands.

Example

When considering the gross organization of more advanced nervous systems, it is tempting to view them as a hierarchy of subsystems that the cerebral cortex "supervises." In humans, for instance, neurons in the spinal cord can carry out mundane control functions for their respective body segment while the cortical neurons are "free to think higher thoughts." But the assignment of rank order to the spinal cord and subsystems in the brain is not as obvious as it may seem. Although the part of the brain that provides intelligence, the neocortex, ranks above the reflex systems in the brainstem and spinal cord, there are practical limits on the degree of control exerted by the neocortex. If control were absolute, for example, people would not succumb to the dizziness and ataxia associated with motion sickness and the vestibular system of the brainstem. People would be able to suppress the pain that is mediated in the thalamus. They could stave off sleep indefinitely by keeping the reticular activating system active.

TERMS

Neocortex	The cell-dense, outer mantle of the brain, more particularly, the part that appears most prominent in mammals.
Reticular Activating System	A large pool of neurons in the brainstem that is essential for the maintenance of attentiveness and consciousness.

Related Principles
Conscious Awareness (States of Consciousness)
Homeostasis
Modularity
Reflex Action (Information Processing)

Homeostasis

The brain regulates the bodily internal milieu through coordinated control over hormones and the nerves that supply viscera.

Explanation

The nervous system is the body's chief coordinating agency, exerting control over almost all functions. It does this through a combination of control over numerous hormone systems and by direct neural control of visceral systems, such as those controlling respiration, digestion, circulation, and urine and genital functions.

Examples

The endocrine system, which regulates metabolism and growth, is not only controlled by the nervous system, but also, to a large extent, is part of it. One part of the brain, the hypothalamus, releases secretions that regulate the master endocrine gland, the pituitary. Hormones from the endocrine system also act on the nervous system, suggesting that it is a control system with feedback not only from the impulse conducting nerve pathways but also from hormones. Although it is not well understood, some chemical secretions from neurons that normally act locally on adjacent neurons can escape into the larger bulk of extracellular fluid. Thus, they can act over large distances, in the manner of circulating hormones from endocrine glands. Then there is certain neural tissue whose predominant function is to produce neurohormones. These neural secretions are dumped more or less directly into the blood. Examples include certain neurons in the hypothalamus whose secretions control the pituitary gland. The posterior pituitary itself contains neural tissue that releases vasopressin hormone into the blood. Finally, the medulla of the adrenal gland is modified neural tissue, and its secretion, adrenalin, is released directly into the blood.

Figure 1-13 shows the interrelations among the nervous system and the two main mechanisms it has for maintaining homeostasis: control over hormone release and direct nervous system control. The brain, particularly the hypothalamus and associated limbic system structures, integrates internal and external stimuli and regulates the pituitary. The pituitary in turn governs function of other major endocrine glands via tropic hormones. The hypothalamus also has direct actions governing such functions as growth and fluid balance. Feedback information from glands and target tissues occurs in the form of biochemical and neuronal messages: these messages provide the internal stimuli that influence the brain and the hypothalamus in their regulatory control.

Hormone-feedback effects are often mediated in the hypothalamus by circulating hormones. Hormones have specific receptor-binding sites in the hypothalamus, and such binding produces both physiological and

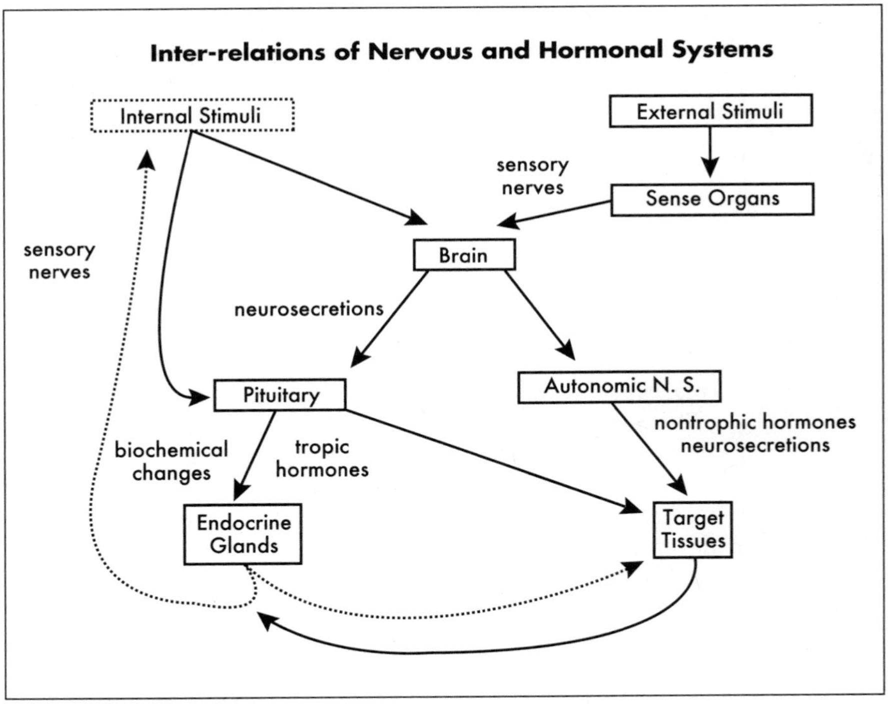

Fig. 1-13 Diagrammatic illustration of the various components of the brain and hormonal systems that collectively provide homeostasis for the body's organ systems.

behavioral consequences. For example, injection of estrogen directly into the hypothalamus of ovariectomized cats can restore sexual behavior and even induce nymphomania. (Because this treatment does not immediately affect the ovary or uterus, it indicates a direct central effect.) Another principle that this example illustrates is that the level of hormone may alter the probability and intensity of the associated behavior but does not necessarily alter its form.

Besides affecting pituitary and adrenal function, adrenal glucocorticoids have important direct effects on neural processes that affect behavior. Treatment of sick animals with glucocorticoids often makes them seem to feel better. Adrenocorticotropic hormone (ACTH), glucocorticoids, and even certain peptide fragments of ACTH that have no endocrine function have distinct but poorly understood influences on learning and various behaviors.

The brain also provides direct nervous system control over visceral functions through a coordinating and control system in the hypothalamus, collectively called the "autonomic nervous system."

TERMS

Endocrine Gland	Gland that expels its secretions into the bloodstream.
Homeostasis	The set of physiological regulatory processes that serves to restore and maintain the normal state.
Mesenteric Ganglia	Clusters of nerve cell bodies located outside of the spinal cord and brain and in the delicate tissues (mesentery that holds and surrounds viscera in the abdominal cavity.)
Parasympathetic Nervous System	That part of the nervous system that generally deactivates visceral organs—promotes such bodily maintenance activities as digestion, reproduction, and deactivation of the cardiovascular system.
Sympathetic Nervous System	That part of the nervous system that generally activates visceral organs for "fight or flight" kinds of responses.

Related Principles

Hierarchical Control
Feedback and Re-entry (Information Processing)
From Input to Output (Information Processing)
Reciprocal Action (Information Processing)
System Modulation (Information Processing)
Neurohormonal Control
Visceral Control (Motor Activity and Control)

References

Brown, R. E. 1993. Introduction to Neuroendocrinology. Cambridge U. Press, New York.
Cohen, A. H. et al. 1988. Neural Control of Rhythm and Movement in Vertebrates. Wiley. New York.
Jasmin, G. and Cantin, M. (eds.) 1991. Methods and Achievement in Experimental Pathology Series. Vols. 14 and 15. S. Karger, Farmington, Connecticut.
Muller, Enginio and Nistico, Giuseppe. 1989. Brain Messengers and the Pituitary. Academic Press, New York.
Motta, M. 1991. Brain Endocrinology, 2nd ed., Comprehensive Endocrinology, Revised Series, L. Martini, Ed. Raven Press, New York.
Schulkin, Jay, (ed.) 1993. Hormonally Induced Changes in Mind and Brain. Academic Press, New York.

Citation Classic

Cannon, W. B. 1932. The Wisdom of the Body. Norton, New York.

Behavior

Behavior is what emerges from the nervous system's output to glands and muscles, particularly muscles.

Explanation

Behavior is an emergent property of nervous system function. It cannot be readily explained from the properties of individual neurons or even selected neuronal populations. Behavior is the expressed sum of individual muscle contractions and hormonal secretions. The order, timing, and relative amount of muscle contractions and secretions determine what kind of behavior will occur.

Motor neurons (and glands) are the "final common pathway" through which behavioral patterns are initiated and made overt. The interaction of sensory input and central nervous system (CNS)-initiated input to muscle and gland output neurons causes behavior to be displayed against the backdrop of the existing state of muscle tone, of primitive spinal reflexes, and of current glandular activity. Feedback modulation of behavior occurs directly on the CNS, on neurohumoral secretory cells and proprioceptors, and indirectly by the behavior itself, which may well alter both the external situation and the internal state. Accordingly, behavior is the result of the way the various nervous modules and subsystems interact with each other and with the internal and external world (Figure 1-14).

Important to the concept of behavior is not only what behavior *is* but also what behavior *does*. Because behavior alters an animal's relationship to its external world, one of the principle consequences of behavior is to adjust the sensory input.

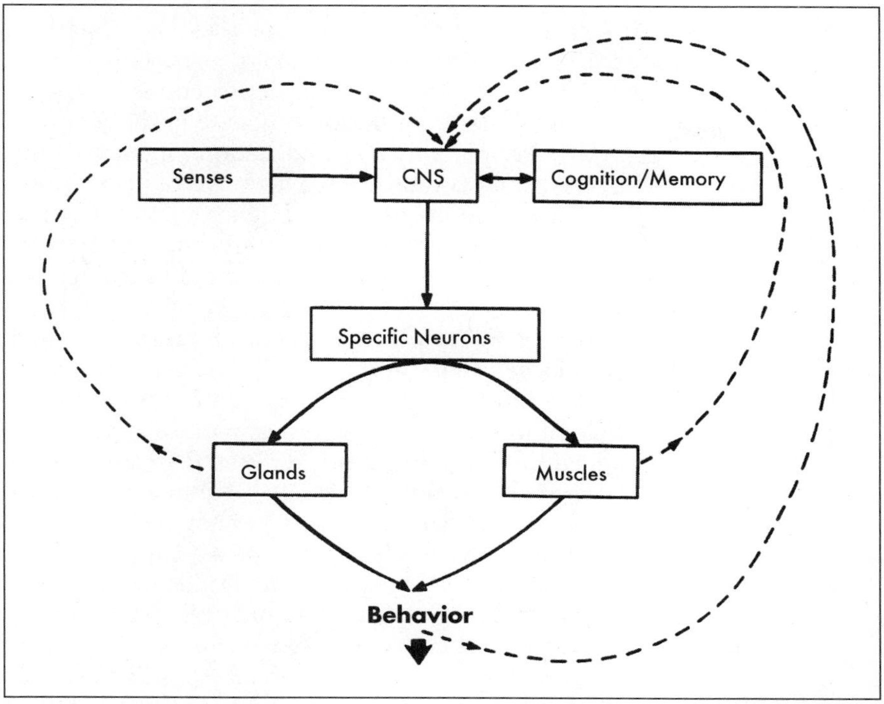

Fig. 1-14 Diagram of a physiological concept of behavior. Behavior emerges from the specific pattern of motoneuron and muscle activity, which in turn is governed by sensory and other inputs, including feedback information from the muscular activity and the behavior itself. Note that behavior can be either self-initiated or driven from various external or internal contingencies. CNS: central nervous system. (From Melvin J. Swenson and William O. Reece, editors: Dukes' Physiology of Domestic Animals, 11th ed. Copyright 1993 by Cornell University. Used by permission of the publisher, Cornell University Press.)

Certain behaviors are relatively inflexible (stereotyped), and these consist of subunits of so-called fixed-action patterns that occur in a predictable sequence. The stimulus that initiates the sequence of subunits acts only as a trigger and exerts little or no effect on the behavior once it is under way. Such patterns are generally genetically programmed and species specific.

Most obvious animal behaviors result from serially ordered patterns of motor acts. Though the sequence is not always predictable, many types of behavior, particularly in response to a specific stimulus, are probabilistically organized; that is, one can predict the probability that a given subset of behavior will occur.

Examples

When muscles contract in a certain pattern and sequence, the resulting movement creates an observable behavior. Walking is a good example. When a cat arches its back, the hair over the back and neck becomes erect, and spitting occurs; we recognize these muscular and glandular actions as fear behavior. Similarly, when dogs bare their teeth and growl, we recognize the underlying muscular contractions as aggression.

Patterns of muscular contraction can arise in the periphery or be generated centrally. Peripherally induced actions include stimulus-induced reflex behavior. Many complex behaviors are built on sequential chains of simple stereotyped individual acts. Higher animals can construct complex behaviors that are not stereotyped, but rather determined by flexible adaptation to environmental conditions, prior learning, and even conscious intent.

Many complex behaviors are triggered centrally rather than as a peripherally induced reflex. In the cat, for instance, various forms of sensory deprivation have no effect on the motor discharge that underlies purring. All domestic animals have central pattern generators in the brainstem and spinal cord for locomotion. Several regions of the brain and spinal cord, when electrically or chemically stimulated, will elicit near-normal locomotor activity. The movements are stereotyped and undirected, thus suggesting that the stimulation activates a command circuit for locomotion, rather than activating some motivational or drive mechanism.

When the sequence and pattern of muscular contractions occur predictably, they cause stereotypical or fixed-action behaviors. Simple spinal reflexes are examples of fixed-action behaviors. Another example at a higher level of nervous system integration is the courting behavior of bulls. Most fixed action patterns in domestic animals are not as fixed as they seem. Most chains of reflex actions are modifiable by other stimuli and assorted CNS influences.

TERMS

Chained Behaviors A series of behavioral components, linked together in specific sequential order, that produces a complex, global behavioral repertoire.

Reflex A specific, reproducible pattern of response to specific sensory stimuli.

Related Principles

Emergent Properties (Information Processing)
Fixed-action Patterns (Motor Activity and Control)
Modularity
Reflex Action (Information Processing)

Reference

Klemm, W. R., and Vertes, R. 1990. Brainstem Mechanisms of Behavior. John Wiley & Sons, New York, N.Y.

Overview: Study Questions

1. List and explain each of the principles in this category.
2. For each of the principles, provide an example *that is not mentioned in this text.*
3. What are the consequences of neurons being polarized in their input/output connections?
4. What are the consequences of neurons being electrically polarized?
5. Why is it useful for neurons to have such diversity in structure and the chemicals they release?
6. Compare and contrast neuronal computing with computing in man-made digital systems.
7. What besides brain size determines intrinsic mental and behavioral capability?
8. Explain the paradox of the brain being both stochastic and deterministic.
9. Give some examples, other than those mentioned, to illustrate the dual stochastic and deterministic nature of the nervous system.
10. Are neuronal circuits anatomical or functional?
11. What are the advantages of the various kinds of neuronal circuits?
12. What are the advantages of the brain being symmetrical, yet having cerebral lateralization?
13. What kinds of movements are asymmetrical? Could these be performed as well without hemispheric lateralization?
14. Why should functions such as speech arise from lateralized cortical tissue?
15. Why is topographical mapping necessary?
16. Speculate on the relationships between mapped and non-mapped regions of brain.
17. How do the nervous system and endocrine system interact to produce homeostasis?
18. Why do we say that the rank order of nervous system components is variable?
19. What is the adaptive value of a system that has a shifting hierarchy?
20. From a phylogenetic and evolutionary perspective, explain why a central nervous system is necessary and inevitable.
21. What is the value of a segmental body plan with modular nervous system control?
22. Why is behavior defined in terms of glandular secretion and muscle contraction?

CHAPTER 2

Cellular Biology

Like all but the simplest of organisms, we have a nervous system. And for the same reason: a nervous system permits the discriminative guidance of behavior. But a nervous system is just an active matrix of cells, and a cell is just an active matrix of molecules

— Paul M. Churchland

The basic unit of the nervous system is the neuron, although it has important biochemical interactions with the more numerous "glial" cells that support neuronal function. Our focus here will be on neurons, because as far as we know now, they are the proximate driving and integrating force of the nervous system. Glial cells are, however, quite important and necessary for neuronal function. They create the physical matrix in which neurons are embedded, form insulation around many of the neurons, exchange nutrients and metabolites with neurons, and interact electrically.

The most obvious functional property of neurons is that they generate electrical pulses, called **Action Potentials,** that propagate down the cytoplasmic extensions of neurons to reach other target cells (muscles, glands, or other neurons). A succession of action potentials carries biologically meaningful messages, expressed in the form of the rate of impulse discharge and in the pattern of intervals between successive discharges (Figure 2-1).

In addition to the direct action of action potentials on target cells, the extracellular electrical fields associated with action potentials can bias neuronal membranes. Regions of an axon, for example, that are close to an action potential source may be nearby regions of that same neuron or perhaps even other neurons within the field. Such **Electrotonus** is reduced in neurons that have an insulated wrapping of glial cells around the neuron and its processes, but gaps in this insulation underlie a rapid propagation of impulses down the axons of such partially insulated neurons. Some degree of electrical "cross-talk" may exist in closely packed axons

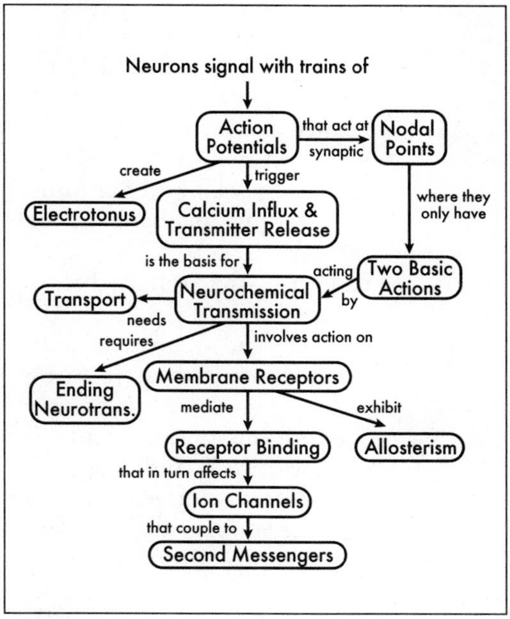

Fig. 2-1 Concept map for the cell-level actions of the nervous system.

in a fiber bundle, but such cross talk is more likely in regions where large numbers of neurons discharge action potentials synchronously, so that their extracellular fields summate.

The action of action potentials on specific target cells is commonly mediated through specialized membrane regions of target cells known as synapses. These act as go or no-go **Nodal Points**. A given neuron may have a thousand or more of these synapses, coming from the axon terminals of as much as a hundred or more other neurons. Synapses constitute decision points at which input from one neuron may be blocked, facilitated, or modified before action on a target neuron. The collective impact of all the synapses on a target cell determines the net state of activity of the target and governs, for example, the kind of output discharge of action potentials that occurs in the target neurons.

Actions at a synapse can have only **Two Basic Actions:** excitation or inhibition. These influences are graded, rather than all or none, affecting the bias or probability that a target neuron will initiate its own action potential response to input.

Although synaptic interactions can be electrical (i.e., due to electrotonus), a major class of interaction is neurochemical. That is, action potentials in a pre-synaptic axonal terminal trigger **Calcium Influx and Transmitter Release.** These actions are the basis for **Neurochemical Transmission,** which is characterized by the release of specific transmitter chemicals that communicate chemically with synaptic targets.

The ability to sustain a continuous supply of neurochemical transmitter to axon terminals requires an extensive **Transport** system to get nutrients and precursors into the terminals and to remove metabolic wastes from the terminals. Transport occurs by shuttling precursor chemicals and metabolites back and forth between the RNA sites of production in the cell body and the synaptic terminals.

Various mechanisms exist for **Ending Neurotransmission.** These include enzymatic destruction of the transmitter and re-uptake into the terminals that released it. In some cases, glial cells take up the released transmitter.

Released neurotransmitter diffuses across the synaptic cleft whereupon it interacts with postsynaptic **Membrane Receptors.** Receptors are special signal recognition proteins, found on both presynaptic and postsynaptic membrane, and they recognize not only neurotransmitters but also hormones and drugs. The tertiary and quaternary structure of receptor proteins confers upon them special recognition capability for specific neurochemical transmitters, hormones, and drugs.

Receptor Binding is temporary, because the bonds are not covalent, but rather involve a combination of hydrogen bonds, electrostatic attractions, and Van der Waals forces. The amount of neurotransmitter bound and its consequent biological effect are proportional to the number of stereospecific receptors sites on the target cells and on the binding affinity, which is a function of the conformation of the receptor protein.

Modulation of receptor protein binding sites is called **Allosterism.** There may be multiple binding sites that, when occupied, alter the overall conformation in ways that bias the receptor's capability for binding its stereospecific neurochemical transmitter. Certain normally occurring ligands act in this way as neuromodulators.

Neurotransmitter binding alters the permeability to flow of selected ions through **Ion Channels,** so that the asymmetric distribution of such ions in the resting state converges to equilibrium as the ions are free to move across the membrane in response to existing concentration and electrostatic gradients.

Activation of ion channels may create a cascade of secondary biochemical reactions, mediated by so-called Second Messengers. These second messenger systems are responsible for changing the metabolic machinery of the target neurons. Commonly, this involves phosphorylation of intracellular proteins.

List of Principles

Action Potentials A resting neuron maintains a resting electrical charge, relative to the outside fluids. When a neuronal membrane is destabilized, either by an appropriate electrical field or reaction with neurotransmitter chemical, the resting membrane potential reverses polarity. This polarity reversal propagates from point to point along the membrane surface.

Electrotonus The action potential electrical currents flowing across the membranes of adjacent neurons create extracellular voltage fields that bias the probability of generating other action potentials. This so-called electrotonus (or "cross-talk") influences not only the neuron that generates action potentials but perhaps other nearby neurons as well. In neurons whose axons are partially insulated by adjacent glial cells, the electrotonus is responsible for re-initiation of the action potential at successive gaps in the insulation.

Nodal Point The synapse is a nodal point between two neurons at which input from one neuron may be blocked, facilitated, or modified. Interactions of multiple inputs at a common target neuron are summed algebraically.

Two Basic Actions Neurons have direct actions and indirect actions. Direct actions take only one of two forms: to excite or to inhibit. Indirect actions can occur in several ways, but the important consequence is that they modulate excitatory or inhibitory activity.

Calcium Influx and Transmitter Release Calcium entry couples action potentials to neurotransmitter release. In many synapses, depolarization of presynaptic terminals opens membrane gates for calcium channel influx, which in turn triggers the release of certain transmitters. These channels are voltage sensitive, unlike other classes of membrane channels that are ligand sensitive.

Neurochemical Transmission Neurons release "transmitter" chemical secretions from their terminals into synaptic junctions, which serve to excite or inhibit postsynaptic cells. Synaptic transmission is ultimately associated with altered conductance across the membrane of specific ions. The conformational changes and channel opening are regulated by membrane potential changes in some ion channel systems and by specific neurotransmitter-receptor interactions in other systems (acetylcholine and gamma-aminobutyric acid [GABA], for example).

Neurotransmitters affect virtually all physiological processes in the brain and all behaviors and any given process or behavior is generally influenced by more than one neurotransmitter or neuromodulator.

Transport Neurons transport metabolic products, sometimes over relatively great distances, through small cytoplasmic extensions of the cell body.

Ending Neurotransmission The action of released transmitters is terminated by several mechanisms: enzymatic destruction, diffusion away from receptors, and active transport processes that take transmitter or its precursors back into nerve terminals or into adjacent glia cells.

Membrane Receptors Neural membranes, both presynaptic and postsynaptic, contain embedded proteins that act as receptors for specific ligands. These ligands are normally neurotransmitters and hormones, but they can also be drugs. The ligands must "fit" the conformation and electric charge properties of their receptors. This means that the ordering of amino acids in the receptor protein is very important, along with the ordering of constituent atoms in the transmitter and the arrangement of the atoms in space (i.e., the "handedness" of certain key atoms).

Receptor Binding Neurotransmitters bind receptors temporarily, because the interactions do not lead to covalent bonds but rather are a combination of weaker forces involving hydrogen bonding, electrostatic attraction, and Van der Waals forces. Such binding is stereospecific and confers a particular conformation on the membrane receptor. This receptor conformation determines whether the aqueous channels within the protein are open or closed to ionic flux.

Allosterism Binding of ligand to receptor occurs at specific sites within the receptor. Membrane receptors are responsive to conformational alterations produced by various endogenous and exogenous ligands, and thus the accessibility to binding sites is likewise altered. These changes, called allosteric, alter the binding properties of the receptor, making it more or less sensitive to particular ligands. Ligands that cause allosteric effects are sometimes referred to as neuromodulators, as opposed to neurotransmitters, because they do not "transmit" information per se but rather alter the bias for transmission of information by specific neurotransmitters.

Ion Channels Neuronal membranes have coiled proteins that are embedded in the hydrophobic lipid bilayer. Some of the amino acids surround a variable-sized pore, which constitutes a channel through which ions can flow according to the electrostatic and concentration gradients.

The degree to which these ion channels are open can be regulated by voltage fields (voltage-gated channels) or by specific receptor ligands (ligand-gated channels). Voltage-gated channels, at least those for sodium in mammalian brain, are also regulated by phosphorylation of protein kinase C and cAMP-dependent protein kinase.

Second Messengers Activation (opening) of an ion channel creates a cascade of intracellular biochemical reactions that leads ultimately to phosphorylation of intracellular proteins. This cascade of reactions constitutes a biochemical signal transduction mechanism that converts biochemical signals at the membrane surface to biochemical signals inside the cell.

Action Potentials

A resting neuron maintains a resting electrical charge, relative to the outside fluids. When a neuronal membrane is destabilized, either by an appropriate electrical field or reaction with neurotransmitter chemical, the resting membrane potential reverses polarity. This polarity reversal propagates from point to point along the membrane surface.

Explanation

In the resting state, certain ions are asymmetrically distributed across the cell membrane (Figure 2-2). For example, potassium is concentrated inside the cell, while sodium and chloride are concentrated outside the cell. Because proteins, which have a net electronegative charge, are trapped inside the cell, the net charge is negative, relative to the outside.

The asymmetrical distribution of ions requires energy, being maintained by protein transport molecules in the membrane that act like pumps. These "pumps" move ions against their concentration and electrostatic gradients so that asymmetry is maintained.

If anything disturbs the resting membrane permeability to ionic flow, sodium runs down its concentration and electrostatic gradients to enter the neuron, resulting in a temporary reversal of the membrane potential. Likewise, potassium runs down its concentration gradient to leave the neuron. These changes do not require energy because the needed energy was stored by the membrane pump proteins in creating the concentration

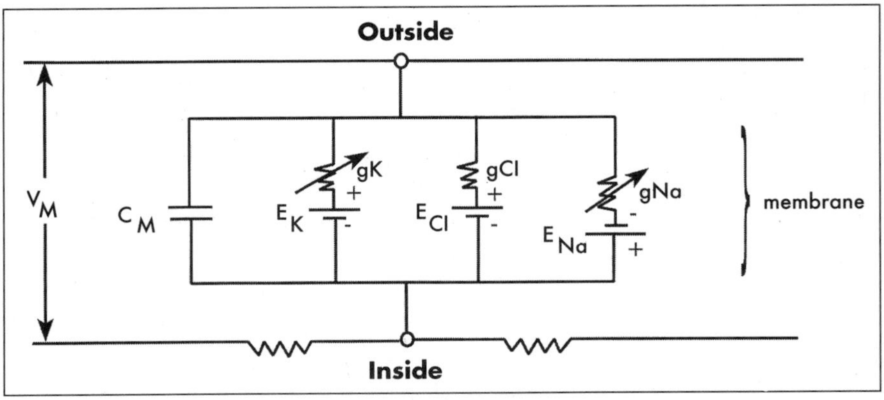

Fig. 2-2 Diagram of an electrical circuit that is equivalent to the electrical properties of nerve-cell membrane. The battery symbols reflect the membrane voltage (Vm) differential for each of the respective ions, potassium (K), chloride (Cl) and sodium (Na). That is, the voltage gradient for K^+ is inside negative, which means that K^+ ions flow toward the outside. The gradient for NA^+ is inside positive, which means that Na^+ currents flow toward the inside. Conductances (g) for the various ions are represented as variable resistors, meaning that flow of a given ion is regulated by a variable resistance imposed by whether or not the membrane pores (ion channels) for that particular ion are open to allow the ion to move down its concentration and electrostatic gradient.

and electrostatic gradients. This is exactly what happens when a neuronal receptor organ is excited by its environmental stimulus, or when a neuron is excited by electrical stimulation or by chemicals released from other neurons. Sodium enters neurons via pores (ion channels) in specific receptor proteins that traverse the plasma membrane. As sodium enters the cell, the positive charge that it carries cancels the internal negativity, and actually reverses the neuron's polarity so that the membrane voltage temporarily becomes about 60 mV, inside positive. This change is sudden and is the most conspicuous sign of activity; it therefore is called an action potential (Figure 2-3). Action potentials generate from the summation of small voltage changes in dendrites, spread down the axon, and may even back-propagate into dendrites to affect their responsivity to input.

The process is also called "depolarization" because the electrical polarization of the resting membrane is eliminated; actually, it is reversed. The size of the action potential is limited by concentration and electrical gradients. As sodium rushes in, it is erasing both the concentration and electrical gradients that enable the influx; at about 60 mV positive an "equilibrium potential" is reached where the forces compelling influx are offset by the forces that develop to stop influx. For a given axon, the size of the action potential is generally the same. This is the so-called "all or none" principle. However, a few neuronal types that fire in rapid bursts may produce action potentials in the burst that show a progressive decrease in amplitude.

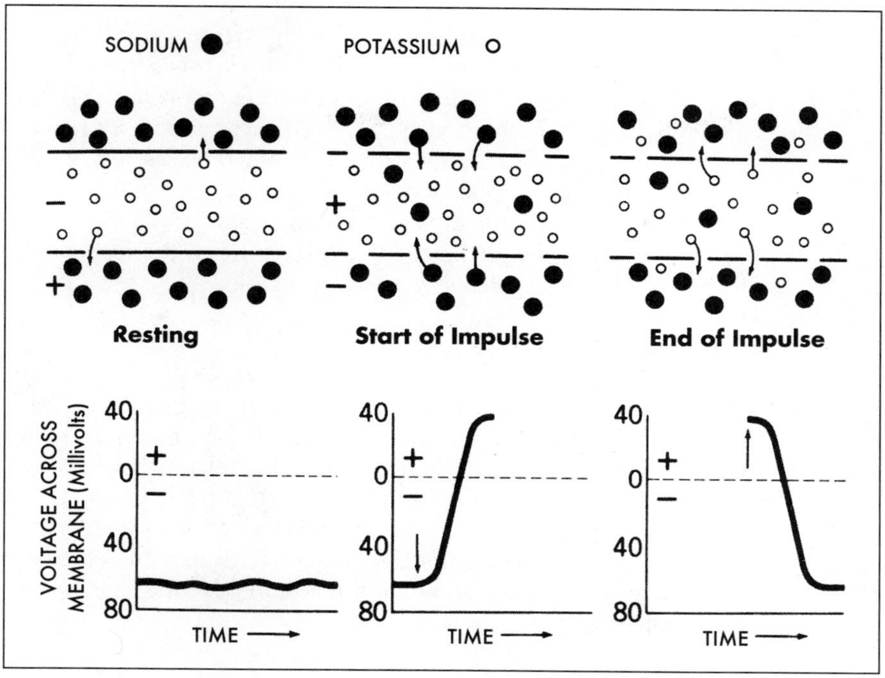

Fig. 2-3 Diagrams of movements of charged elements (ions) across neuronal membranes. The graphs of the corresponding membrane voltage change during rest, start of action potential, and end of action potential. Small circles represent potassium ions, some of which gradually leak out in the resting state. Large closed circles represent sodium ions, which, when neuronal membrane is stimulated, are suddenly able to rush into the cell, carrying their positive charge inside. The action potential is ended by a sudden outrush of potassium, which, because it moves positive ions out of the cell, helps to return the net membrane voltage to the initial inside negative state. Slow acting "ion pump" mechanisms act over a much longer time course to achieve redistribution of the ions that have moved during an action potential. (From Klemm, 1972.)

The all-or-none principle also does not apply to action potentials as they are recorded from a peripheral nerve, which contains axons of many different neurons.

We have made the point that sodium influx accounts for the action potential. What is the significance of the potassium efflux? It too continues until its equilibrium potential is reached, which occurs somewhat later than is the case for sodium influx. Since potassium is leaving the neuron, it carries electropositivity with it; i.e., the efflux tends to make the inside of the neuron electronegative, thus tending to cancel the electropositivity that was created by sodium influx. In short, potassium efflux terminates the action potential.

The sodium pump is the best-understood mechanism for maintaining asymmetric distribution of ions. The pump is a transport protein called

Na^+, K^+, ATPase. ATPase has differing affinity for both K^+ and Na^+, depending on its conformation, which in turn is affected by such things as the surrounding electrical field and access to water. When the enzyme has high affinity for Na^+, it has low affinity for K^+, and vice versa. Thus in one conformation, it picks up Na^+ and releases K^+, while in another conformation the opposite occurs. The enzyme apparently changes the locus of its ion binding sites as the conformation changes, so that an ion can be picked up from the extracellular fluid and released into the inside of the cell, for example.

Example

The change in membrane potential during an action potential is actually a plot of volts vs. time, and, as could be expected, the wave shape features a rapid rise and fall in potential. The absolute values obtained depend on the recording environment. For example, if two extracellular electrodes are placed outside an axon, the peak depolarization will be only a few millivolts, seen as an upward deflection of an oscilloscope beam when the action potential is directly under one electrode, and seen as a downward deflection when the action potential has moved under the second electrode.

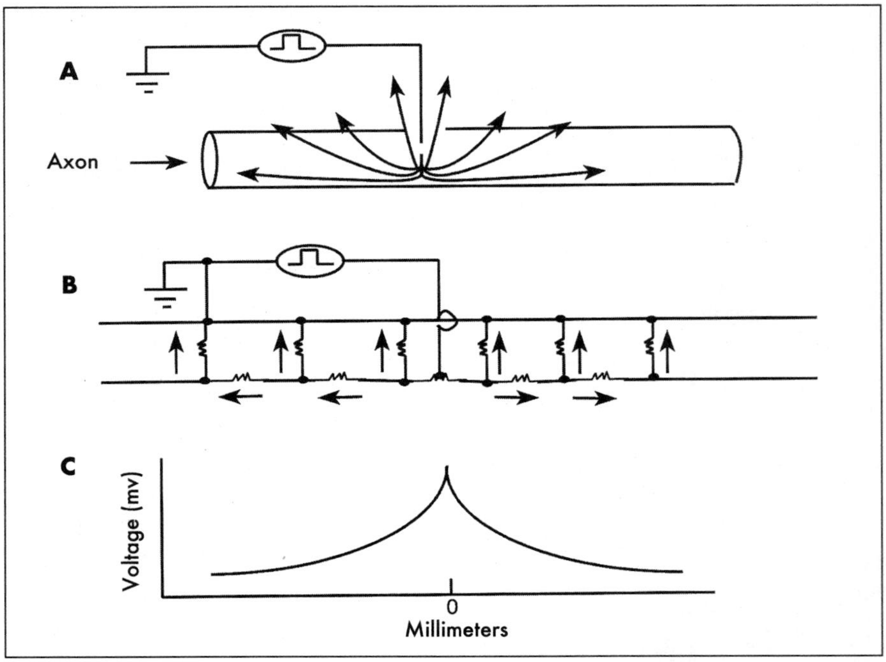

Fig. 2-4 Electrical equivalent diagrams of the axon during the passage of an action potential. Paths of ionic current flow are indicated by arrows, (**A**) Equivalent resistances for intracellular and transmembrane flow are indicated in (**B**) The dissipation over distance of the voltage field associated with an action potential is shown in C.

As seen in Figure 2-4, the plot of the action potential begins at the resting voltage, around -70 mV, followed by a sudden rise in external positivity, followed by a quick return to the resting inside negativity. Normally these plots are displayed electronically on oscilloscopes, because the high speed of the process is too fast. The whole action potential typically occurs in less than a millisecond.

TERMS

Concentration (Chemical) Gradient A differential distribution of chemicals over space, in this case ions such as Na^+, K^+, and Cl^-. There is a spatial gradient between areas where a given ion concentration is large to areas where it is low. The laws of diffusion normally obliterate concentration gradients, UNLESS something else, such as a specific ion pump and/or selective membrane permeabilities, acts to offset diffusion.

Electrostatic Gradient A voltage difference between two locations, in this case between the inside of a neuron and the outside. Because the charge density decreases over distance from the electrical source, it is said to create a gradient.

Related Principles

Electrotonus
Ion Channels
Neuron: The Operational Unit

References

Brazier, M.A.B. 1968. Electrical Activity of the Nervous System, 3rd ed. Williams & Wilkins, Baltimore, Maryland.

Klemm, W.R. 1972. Science, The Brain, and Our Future. Bobbs-Merrill Co., New York.

Lauger, P. 1992. Electrogenic Ion Pumps. Sinauer, Sunderland, Massachusetts.

Magee, J.C. and Johnston, D. 1995. Synaptic activation of voltage-gated channels in the dendrites of hippocampal pyramidal neurons. Science 268:301-304.

Citation Classics

Goldman D.E. 1943. Potential, impedance, and rectification in membranes. J. Gen. Physiol. 27:37-60.

Hodgkin A.L. and Katz B. 1949. The effect of sodium ions on the electrical activity of the giant axon of the squid. J. Physiol. 108:37-77.

Hodgkin A.L. and Huxley A.F. 1952. A quantitative description of membrane current and its application to conduction and excitation in nerve. J. Physiol. Lon. 117:500-544.

Hubel D.H. 1957. Tungsten microelectrode for recording from single units. Science 125:549-550.

Narahashi T., Moore J.W., and Scott W.R. 1964. Tetrodotoxin blockage of sodium conductance increase in lobster giant axons. J. Gen. Physiol. 47:965-974.

Skou J. C. 1960. Further investigations on a Mg^{++} + Na^{+}-activated adenosintriphosphatase, possibly related to the active, linked transport of Na^{+} and K^{+} across the nerve membrane. Biochim. Biophys. Acta 42:6-23.

Electrotonus

The action potential electrical currents flowing across the membranes of adjacent neurons create extracellular voltage fields that bias the probability of generating other action potentials. This so-called electrotonus (or "cross-talk") influences not only the neuron that generates action potentials but perhaps other nearby neurons as well. In neurons whose axons are partially insulated by adjacent glial cells, the electrotonus is responsible for re-initiation of the action potential at successive gaps in the insulation.

Explanation

Consider two axons from different neurons that lie parallel and close to each other. The impulse discharge in one axon could well serve as an electrical stimulus to initiate excitation in the neighboring axon. This is usually undesirable, inasmuch as many axons that have different targets are nonetheless bundled together in the same nerve.

Nature's way of solving this problem is to provide an insulating wrapping of axons. Another cell type in the nervous system, called glia, sends out membranous extensions that wrap around an axon, often several times, to provide electrical insulation. Currents cannot flow through all those membranes; the electrical resistance is just too high.

But now, we must ask, how then can any insulated axon generate impulses if electrical current cannot flow across its membranes? Nature solved this problem with a compromise: every few millimeters along the length of an axon, there is a gap of non-insulated region. The gap occurs because a succession of glia cells is needed to provide wrapping along the complete length of an axon.

The gaps (called "nodes of Ranvier") serve another purpose. They speed up the propagation of nerve impulses. Impulses no longer have to progress, micron by micron, down the axon, but they can "leap" from node to node because the electrical field at one node is strong enough to act as an electrical stimulus at the next adjacent node. Thus, if one experimentally tries to stimulate axons, the typical observation is that the current strength needed to reach excitation threshold varies at different points along the axon. That is, the threshold is lowest where the uninsulated nodes are and highest at the midpoint between nodes (Figure 2-5).

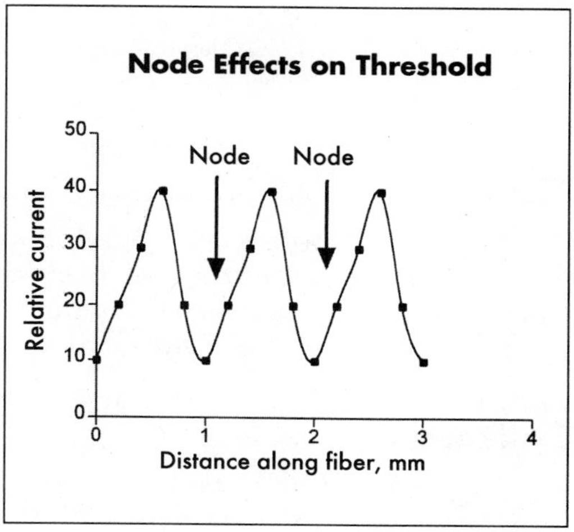

Fig. 2-5 Illustration of the change in excitatory threshold at successive points along an axon. Lowest threshold occurs in the regions of the uninsulated nodes.

Examples

Certain diseases cause demyelination. The best known demyelination disease in animals is canine distemper, and the best known in people is multiple sclerosis. In canine distemper, a virus attacks and kills glial cells, among other effects, thus creating axons that lack the glial insulation. The resulting symptoms of nervous dysfunction depend on whether sensory or motor systems suffer the damage.

In humans, there are several demyelinating diseases, but multiple sclerosis is the most common type. Typical targets are the cerebellum and brainstem, and motor dysfunction is a common symptom. Rarely, demyelination can occur from certain viral infections, such as measles, or as an abnormal reaction to certain vaccines, such as smallpox or rabies. Demyelination may also occur secondary to other problems, such as a deficiency of vitamin B_{12} or vascular insufficiency.

Related Principles

Action Potentials
Neuron: The Operational Unit (Overview)
Rhythmicity and Synchronicity (Information Processing)

References

Adey, W.R. 1988. The cellular microenvironment and signaling through cell membranes., pp. 81-106. In Electromagnetic Fields and Neurobehavioral Function, Progress in Clinical and Biological Research. Vol. 27. Edited by M.E. O'Connor and R.H. Lovely. Alan R. Liss, New York.

Blackman, C.F. 1988. Stimulation of brain tissue in vitro by extremely low frequency, low intensity, sinusoidal electromagnetic fields, pp. 107-118. In Electromagnetic Fields and Neurobehavioral Function, Progress in Clinical and Biological Research. Vol. 27. Edited by M.E. O'Connor and R.H. Lovely. Alan R . Liss, New York.

Citation Classic

Tasaki, I. 1953. Nervous Transmission. Charles C Thomas, Springfield, Illinois.

Nodal Point

The synapse is a nodal point between two neurons at which input to a neuron may be blocked, facilitated, or modified before action on a target neuron. Interactions of multiple inputs at a common target neuron are summed algebraically.

Explanation

Synaptic regions have morphological specialization of both pre- and post-synaptic membranes. Transmission across synapses is of two kinds, electrical and chemical. In an electrical synapse, there is direct electrical coupling between a presynaptic neuron and its postsynaptic target. Little-to-no modification of signal occurs in electrical synapses. Their purpose seems to be the mediation and spread of activity throughout circuits that subserve automated behaviors that need to occur reliably and quickly. Chemical synapses, on the other hand, couple presynaptic and post-synaptic neurons via the release of chemical signals in which a "transmitter" chemical is released from a presynaptic neuron, whereupon it interacts with specific receptor molecules on its postsynaptic neuronal target.

In chemical synapses, once information about stimuli enters the CNS, chemical and electrical changes in the synapses provide enormous capacity for processing that information as it passes from neuron to neuron. Release of the chemical transmitter can produce a graded change in the membrane voltage of the postsynaptic neuron. If the polarity of that potential is in the right direction, the cell becomes excited and begins to discharge impulses; this change is called an excitatory postsynaptic potential (EPSP). Such excitatory systems cause sodium channels to open and allow sodium ions to enter the cell. The net response is graded: a small depolarization occurs if a few channels open, whereas a large depolarization and impulse discharge occurs if many channels are open. The associated ion flow gradually declines to zero as it runs down the electrochemical gradients; the duration is usually on the order of one to a few milliseconds.

Another key feature of chemical neurotransmission is that as long as the membrane potentials are below threshold for firing impulses, the membrane potential can summate inputs. That is, if the neurotransmitter at one synapse causes a small depolarization, a simultaneous release of transmitter at another synapse located elsewhere on the same cell body will summate to cause a larger depolarization. This so-called "spatial" summation mechanism is complemented by "temporal" summation, wherein successive releases of transmitter from one synapse will cause progressive polarization change as long as the presynaptic changes occur faster than the decay rate of the membrane potential changes in the postsynaptic neuron. Neurotransmitter effects last several times longer than presynaptic impulses, and thereby allow summation of effect. Thus, the EPSP differs from action potentials in a fundamental way: it summates inputs and expresses a graded response, as opposed to the "all-or-none" response of impulse discharge. The summation process is an integral feature of the nervous system's ability to "process" information.

At the same time that a given postsynaptic neuron is receiving and summating excitatory neurotransmitter, it may also be receiving "conflicting" messages that are telling it to shut down firing. These inhibitory influences (inhibitory postsynaptic potentials, IPSPs) are mediated by inhibitory neurotransmitter systems that cause postsynaptic membranes to hyperpolarize. Such effects are generally attributed to the opening of selective ion channels that allow either intracellular potassium to leave the postsynaptic cell or to allow extracellular chloride to enter. In either case, the net effect is to add to the intracellular negativity and move the membrane potential farther away from the threshold for generating impulses.

When EPSPs and IPSPs are generated simultaneously in the same cell, the output response will be determined by the relative strengths of the excitatory and inhibitory inputs. Output "instructions," in the form of impulse generation is thus determined by this "algebraic" processing of information. Because the discharge threshold across a synapse is a function of the presynaptic volleys that act upon it, and because a given neuron may receive branches from many axons, the passage of impulses in a network of such synapses can be highly varied. The versatility of the synapse arises from its ability to modify information by algebraically summing input signals. The subsequent change in stimulation threshold of the postsynaptic membrane can be enhanced or inhibited, depending on the transmitter chemical involved and the ion permeabilities. Thus the synapse acts as a decision point at which information converges, and it is modified by algebraic processing of EPSPs and IPSPs. In addition to the IPSP inhibitory mechanism, there is a presynaptic kind of inhibition that involves either a hyperpolarization on the inhibited axon or a persistent depolarization; whether it is the former or the latter depends on the specific neurons involved.

Examples

Several kinds of synaptic junctions are known: axosomatic, between axons and soma or cell body (for example, between axons and cell bodies in autonomic ganglia); axoaxonic, between two axons (for example, the junction between axons of spinal interneurons and primary afferent fibers); axodendritic, between axons and dendrites (for example, as in the dorsal horn of the spinal cord); and dendrodendritic, between dendrites (for example, in cerebellum cortex).

Experimentally, the most obvious way to distinguish chemical synapses from electrical synapses is to monitor the time delay between activation

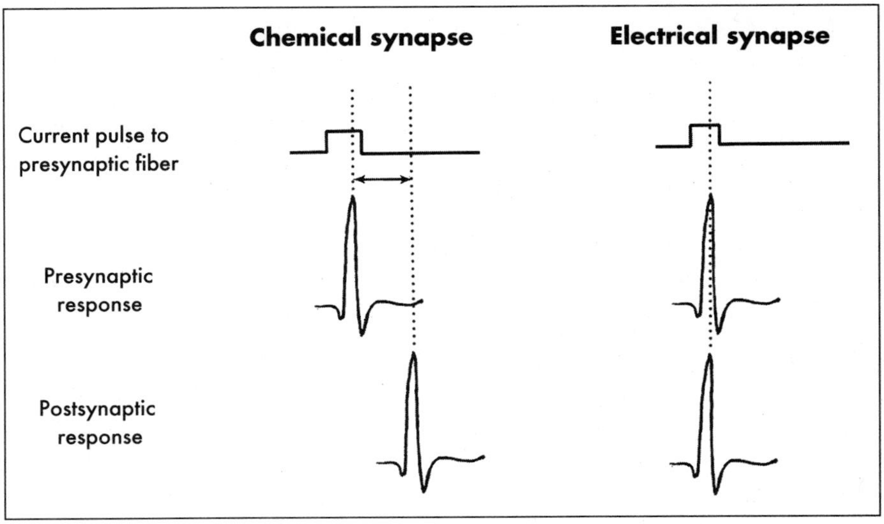

Fig. 2-6 Illustration of how one can determine whether a given synapse is chemical or electrical. LEFT: an excitatory electrical stimulation (pulse in upper trace) delivered to a presynaptic fiber excites an immediate action potential (middle trace). RIGHT: any response in a postsynaptic neuron will exhibit a measurable delay of a few milliseconds (bottom trace). In an electrical synapse, the pre- and post-synaptic responses to stimulation are essentially immediate, with no readily detectable postsynaptic delay.

of a presynaptic neuron and the response of its postsynaptic target neuron. In electrical synapses, there is virtually no synaptic delay, inasmuch as the coupling occurs via the electrical field. In chemical synapses, there are usually one to a few milliseconds of delay, associated with the release of transmitter, diffusion across the synaptic cleft, binding to receptor molecules, and alteration of ion channels and the ionic flux that causes the postsynaptic response (Figure 2-6).

Electrical synapses are most useful to organisms where movements need to be uniform, reliable, and rapid. Simple invertebrates, such as Hydra, that have nervous systems composed of relatively unorganized networks of neurons, use electrical synapses. Stimulation of any one point of a tentacle, for example, will cause an immediate response throughout the whole organism. Even higher organisms may have some isolated, electrically coupled circuits. The network that controls tail movement in fish, for example, uses some electrical synapses.

Chemical synapses come in several excitatory and inhibitory types. Excitation in a chemical synapse tends to accomplish the same things it does in electrical synapses: spread of excitation throughout a distributed network. However, because the coupling is chemical, there are abundant opportunities for the signal to get augmented, diminished, or even blocked at strategic points in the overall circuitry.

With inhibitory chemical synapses, when inhibition occurs at specific points in an interactive circuit of neurons, it can act as a gate to route information flow. In circuits where the inhibition is fed back into the "input" neurons, a synchronization of the activity of many neurons can occur; this process is believed to underlie the production of large, rhythmic EEG waves such as alpha-waves (8-10/sec) from the visual cortex of humans and 0-waves (4-7/sec) from the hippocampus of mammals.

Synaptic function can be modeled mathematically by a transfer function that accounts for the potency of the input (expressed as a weight for likelihood of causing discharge of the output neuron) and the intrinsic threshold for activation of the output neuron. The transfer function is commonly viewed as taking the form of:

$$f_x(\Sigma w_{ik} n_{ik} - \Phi_i)$$

where f(x) is either a discontinuous state function or a smoothly increasing sigmoidal function, w_{ik} is a real-valued weighting value assigned to the links i and k between two neurons, Φ_i is a real-valued bias (threshold of activation) for the synaptic junction, and n_{ik} is a state variable for the junction.

TERMS

Alpha Waves Electroencephalographic activity in which the duration of individual waves is rhythmic, with about 8 to 12 waves per second.

State Variable In this case, a biochemical or physiological state that imposes a constraint on function.

Transfer Function In this case, an equation that describes the transformation between two states.

Related Principles

Action Potentials
Calcium and Transmitter Release
Ending Neurotransmission
Ion Channels
Membrane Receptors
Receptor Binding
Two Basic Actions

References

Baudry, M., Thompson, R.F., and Davis, J.L., (eds.) 1993. Synaptic Plasticity: Molecular, Cellular, and Functional Aspects. MIT Press, Cambridge, Massachusetts.

Brown, T.H., Kairiss, E.W., and Keenan, C.L. 1990. Hebbian synapses: biophysical mechanisms and algorithms. Ann. Rev. Neurosci. 13:475-511.

Eccles, J.C. 1982. The synapse: from electrical to chemical transmission. Ann. Rev. Neurosci. 5:325-339.

Reinhardt, B.M.J. 1990. Neural Networks. Springer-Verlag, Berlin.

Citation Classics

Eccles, J.C. 1982. The synapse: from electrical to chemical transmission. Ann. Rev. Neurosci. 5:325-339.

Gray E.G. 1956. Axo-somatic and axo-dendritic synapses of the cerebral cortex: an electron microscope study. J. Anat. 93:420-433.

Gray E.G. and Whittaker V.P. 1962. The isolation of nerve endings from brain: an electron microscopic study of cell fragments derived by homogenization and centrifugation. J. Anat. 96:79-87.

Two Basic Actions

Neurons have direct actions and indirect actions. Direct actions take only one of two forms: to excite or to inhibit. Indirect actions can occur in several ways, but the important consequence is that they modulate excitatory or inhibitory activity.

Explanation

The nervous system's electrical activity is the most conspicuous aspect of nervous tissue function; much of what we know about the nervous system derives from study of its electrical activity. Electrical activity arises from the fact that ions distribute differentially on either side of the cell membrane. Flow of ions across neuronal membrane is regulated by so-called "ion channels" that are proteins that undergo conformational changes to open and close a pathway selectively for a specific ion to flow down its electrochemical gradient.

Electrophysiological data are usually expressed as voltage, or the potential force generated by the separation of electrically charged ions. The most fundamental biological potential is the steady cell voltage that exists between the inside and the outside of the resting, or unstimulated, neuron.

Activity in adjacent neurons alters the direction and magnitude of this polarization of postsynaptic neurons. If the membrane polarity tends to reverse to the extent of achieving a critical threshold level of depolarization, a nerve impulse is discharged. Such membrane potential changes are called postsynaptic potentials. A neuron receives not only excitatory postsynaptic input but often also inhibitory (hyperpolarizing) input, which can be added to determine the net level of polarization and excitability.

Typically, a given neuron exerts its influence on many target cells. These cells may be muscles, glands, or other neurons. The traditional view is that a given neuron can only excite or inhibit a given target. This was attributed to the fact that neuronal actions on their targets are usually due to release of a chemical transmitter that can produce only one kind of effect on the target cell. However, we now know that many neurons have more than one transmitter and these may have dual and even opposite actions on target cells. Moreover, the action may be one of modulation rather than direct excitation or inhibition.

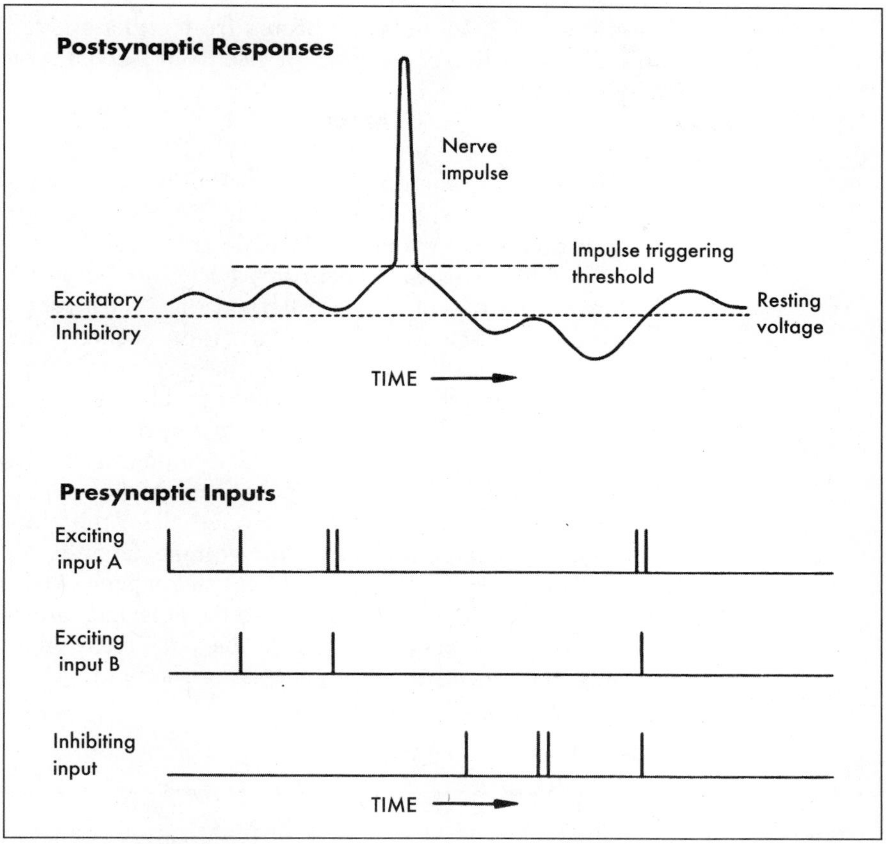

Fig. 2-7 Schematic diagram of the summation of changes in postsynaptic membrane that can occur in excitatory neurotransmitter systems. When excitatory inputs are received simultaneously at several synapses on the same neuron, the postsynaptic potentials summate ("spatial summation"), bringing the threshold (dashed line - Impulse Triggering Threshold) for firing an action potential. When inhibitory input is received at a synapse, it produces an opposite polarization change ("hyperpolarization"). If inhibitory input occurs at the same time as excitatory input from another source, it may offset the excitatory influence. (From Klemm, WR: Science, the Brain, and Our Future, 1972, Bobbs-Merrill, New York.)

Excitatory effects are mediated by depolarizations of postsynaptic membrane. When depolarization of sufficient magnitude is reached ("threshold"), the membrane triggers action potentials; thus, it has been excited. In other situations, synaptic membrane can be hyperpolarized, making it less likely to discharge action potentials. Thus, it is inhibited (Figure 2-7).

Modulation in the living nervous system is achieved in several ways, both electrical and chemical. The action may occur on either side of the junction ("synapse"), acting on the presynaptic neuron or on the postsynaptic (target) neuron. The net effect is that the probability for synaptic transmission is altered. Modulation occurs from either a given presynaptic neuron's ability to deliver an effect or else from a given target neuron's responsivity to input.

Example

This basic principle was disclosed in Nobel-prize winning studies of Eccles on neurons in the spinal cord of cats. By placing microelectrodes into motor neurons of the spinal cord, while at the same time electrically stimulating input fibers to those neurons, Eccles saw only two kinds of changes: depolarization or hyperpolarization. With depolarization, Eccles observed that when enough depolarization occurred (i.e., a certain threshold was reached), the cell would discharge action potentials.

In the original experiments, these voltage changes were generally elicited by electrical stimulation of the postsynaptic neurons or of their input fibers. Since that time, we know that under natural conditions, most of these depolarization or hyperpolarization changes are mediated largely by chemical neurotransmitters. Nonetheless, the electrical fields surrounding synapses can alter membrane voltages in either depolarizing or hyperpolarizing directions. The electrical fields generated by neurons thus affect the neurons that generate them. This has practical consequence when large numbers of neurons in the same location and with the same geometry fire coherently. The neurons of the pyramidal layer of the hippocampus are perhaps the best examples.

TERMS

Coherent Firing When neurons have membrane polarization changes at the same time, the extracellular fields may summate to create voltage fields large enough to influence the excitability of the neurons that generated the fields.

Polarization A state of electrical charge separation, as in a battery. In the case of neurons, the unperturbed cell maintains a more or less steady voltage difference across the plasma membrane of about 70 mV, inside being negative relative to the outside.

Depolarization A relative shift in polarization tending toward cancellation of charge differential across the plasma membrane. In the case of neurons, when depolarization reaches a certain threshold (about 60 mV inside negative) the neurons become very unstable, discharging a complete charge reversal ("nerve impulse").

Hyperpolarization A relative shift in polarization toward making the charge differential across the plasma membrane even greater. In the case of neurons, hyperpolarization can occur only up to a certain maximum (about 90 mV, inside negative).

Related Principles

Action Potentials
Electrotonus
Information Carriers (Information Processing)
Ion Channels
Neuron: Operational Unit (Overview)
Rhythmicity and Synchronicity (Information Processing)

References

Brock, L.G., Coombs, J.S., and Eccles, J.C. 1952. The nature of the monosynaptic excitatory and inhibitory processes in the spinal cord. Proc. R. Soc. B. 140:170-176.

Eccles, J.C. 1975. Under the spell of the synapse, pp. 159-179. In The Neurosciences: Paths of Discovery, edited by F.G. Worden, J.P. Swazey, and Edelman, G. M.I.T. Press, Cambridge, Massachusetts.

Eccles, J.C., Fatt, P., and Landgren, S. 1956. Central pathway for direct inhibitory action of impulses in largest afferent nerve fibres to muscle. J. Neurophysiol. 19:75-98.

Citation Classics

Brock, L.G., Coombs, J.S., and Eccles, J.C. 1952. The recording of potentials from motoneurons with an intracellular electrode. J. Physiol. 117:431-460.

Eccles, J.C. 1964. The Physiology of Synapses. Springer-Verlag. Berlin.

Calcium and Transmitter Release

Calcium entry couples action potentials to neurotransmitter release. In many synapses, depolarization of presynaptic terminals opens membrane gates for calcium channel influx, which in turn triggers the release of certain transmitters.

These channels are voltage sensitive, unlike other classes of membrane channels that are ligand sensitive.

Explanation

Transmitter can be released slowly and in small amounts without calcium influx, but in many synapses the probability of rapid release of large amounts of transmitter is greatly increased by calcium influx into presynaptic terminals. The influx occurs through voltage-sensitive calcium channels in the membrane. Note that the electrostatic and concentration gradients favor the entry of calcium under these conditions. Calcium is normally concentrated outside the cell about 10,000 times more than inside the cell (1.8 mM vs. 0.1 to 0.2 μM). Moreover, most of this internal calcium is bound to such proteins as calmodulin and parvalbumin. Thus, if the calcium channel proteins are perturbed by depolarization

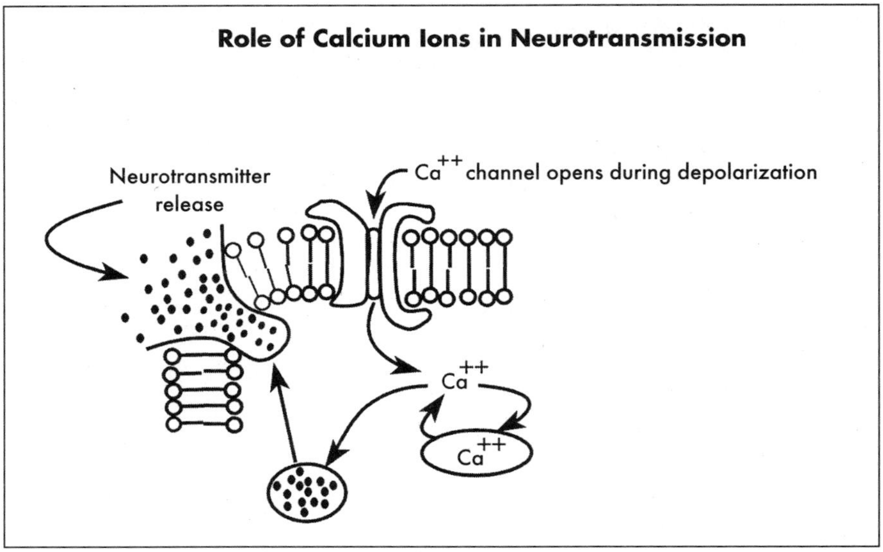

Fig. 2-8 Diagram illustrating role of calcium ions in neurotransmitter release. In normal, resting state, calcium channels are closed and the extracellular calcium concentration is greater than the intracellular concentration. When a calcium channel opens, in response to specific ligand binding or to a depolarizing electrical field, calcium ions rush into the cell. Some calcium becomes bound to protein reservoirs (circled calcium), and some acts on synaptic vesicles to promote their rupture and release of neurotransmitter at the cell membrane surface.

enough to accommodate calcium ions, they will rush in. As with sodium influx, the inside negativity also pulls positive ions in, until the membrane potential reverses (Figure 2-8).

In such voltage-gated calcium channels, membrane depolarization itself opens the calcium channels, and the influx of calcium promotes the release of intracellularly bound calcium. However, some neurons have calcium channels that are modulated by neurotransmitter action. In such ligand-gated channels, the neurotransmitter binding activates membrane enzymes that generate inositol phosphate (IP3) from membrane phospholipids. IP3 diffuses and binds specific receptors on calcium-storage proteins. This sends a signal to membrane calcium channel proteins, causing them to open to let in more free calcium (which typically has a higher concentration outside the cell than within.

Examples

One way that this principle was demonstrated was with an *in vitro* preparation of a neuromuscular junction. Omitting calcium ions from the bathing medium blocked neuromuscular transmission. At the same time, there was no effect on the magnitude of presynaptic depolarization or the action of exogenously applied acetycholine on the junction. Elevated extracellular magnesium ions can apparently antagonize the calcium effect, because it reduces neurotransmitter release.

Other ways to demonstrate this effect is with calcium-sensitive dyes, such as FURA-2, which chelate calcium and acquire an altered fluorescence spectrum as a result. For example, unbound FURA-2 fluoresces maximally when excited at a wavelength of 385 nm, but the most effective wavelength shifts to 345 nm when calcium becomes bound. When such dyes are injected into presynaptic terminals, calcium entry occurs very quickly after depolarization. This suggests that the ions are entering through voltage-sensitive calcium channels. When squid axons are voltage clamped, applied step currents of depolarization activate calcium influx, but the influx currents do not deactivate rapidly as they do with sodium influx during depolarization.

TERMS

Receptor-mediated Channels Ions that are opened or closed as a response to neurotransmitter or hormonal action on postsynaptic receptor molecules. Also called ligand-gated channels.

Step Currents Electrical current that has near-zero rise and fall times (i.e., a square wave shape of onset and offset).

Voltage Clamping A technique whereby the membrane voltage created by applying a steady electrical current, either depolarizing or hyperpolarizing, is monitored and coupled to an electronic feedback circuit that adjusts applied current to sustain a steady membrane voltage, "clamped" in a depolarized or hyperpolarized state. Other electronic circuitry can then monitor the ionic current flows that are associated with different levels of clamped membrane voltage.

Voltage-mediated Channels Ion channels that are opened or closed by changes in membrane voltage, i.e., strictly an electrical response.

Related Principles

Action Potentials
Neurochemical Transmission
Ion Channels
Membrane Receptors
Nodal Point
Second Messengers

References

Augustine, G.J., Charlton, M.P., and Smith, S.J. 1985. Calcium entry and transmitter release at voltage-clamped nerve terminals of squid. J. Physiol. 367:163-181.

Augustine, G.J., Charlton, M.P., and Smith, S.J. 1987. Calcium action in synaptic transmitter release. Ann. Rev. Neurosci. 10:633-693.

Hagiwara, S. and Byerly, L. 1981. Calcium channel. Ann. Rev. Neurosci. 4:69-125.
Hess, P. 1990. Calcium channels in vertebrate cells. Ann. Rev. Neurosci. 13:337-356.
Hofmann, F., Biel, M., and Flockerzi, V. 1994. Molecular basis for CA^{2+} channel diversity. Ann. Rev. Neuroscience. 17:399-418.
Kelly, R.B., Deutsch, J.W., Carlson, S.S., and Wagner, J.A. 1979. Biochemistry of neurotransmitter release. Ann. Rev. Neurosci. 2:399-446.
Miller, R.J. 1987. Multiple calcium channels and neuronal function. Science. 235:46-52.
Putney, J.W. Jr. 1993. Excitement about calcium signaling in inexcitable cells. Science. 262:676-678.
Smith, S.J. and Augustine, G.J. 1988. Calcium ions, active zones and synaptic transmitter release. Trends Neurosci. 11:458-464.
Sobel, E.C. and Tank, D.W. 1994. In vivo Ca^{2+} dynamics in a cricket auditory neuron: an example of chemical computation. Science. 263:823-825.

Citation Classics

Baker P.F., Hodgkin A.L. and Ridgway E.B. 1971. Depolarization and calcium entry in squid giant axons. J. Physiol.-London 218:709-755.
Frankenhaeuser B. and Hodgkin A.L. 1957. The action of calcium on the electrical properties of squid axons. J. Physiol.-London 137:218-244.
Rubin R.P. 1970. The role of calcium in the release of neurotransmitter substances and hormones. Pharmacol. Rev. 22:389-428 (32/83/LS).

Neurochemical Transmission

Neurons release "transmitter" chemical secretions from their terminals into synaptic junctions, which serve to excite or inhibit postsynaptic cells. Synaptic transmission is ultimately associated with altered conductance across the membrane of specific ions. The conformational changes and channel opening are regulated by membrane potential changes in some ion channel systems and by specific neurotransmitter-receptor interactions in other systems (acetylcholine and GABA, for example).

Neurotransmitters affect virtually all physiological processes in the brain and all behaviors, and any given process or behavior is generally influenced by more than one neurotransmitter or neuromodulator.

Explanation

Synaptic contacts are made between axons and the postsynaptic soma or cell body and even between axons and other axons, and the neurotransmitter-receptor interactions may depolarize or hyperpolarize the postsynaptic target. The density of synaptic junctions is extraordinary. As much as 65% of the surface of a neuron's soma and dendrites may be covered with synapses from afferent terminals.

Neurotransmitter release is quantal, involving the release in packets of equal number of molecules. Neurotransmitter is packaged in the presynaptic terminals in small vesicles, about 50 nm in diameter. Each packet of transmitter is released upon arrival of action potentials in the nerve terminal and associated calcium ion influx. The release occurs as vesicles move to the terminal membrane, fuse with it, and dump their contents. The release process seems to involve second messengers that activate protein kinases, which in turn phosphorylate vesicle-associated proteins.

After release from their storage sites in presynaptic vesicles, neurotransmitters diffuse across the synaptic gap to bind with specific receptor proteins. Neurotransmitters affect postsynaptic membranes by binding reversibly with certain sites on receptors, which they "recognize" on the basis of compatible size, three-dimensional shape, and electric-charge distribution. Similar to the "lock and key" reactions of enzymes and substrates, the transmitter is thought to react only with receptors, destructive enzymes, and drugs that have comparable physical and electrochemical characteristics of the binding site of the molecule. The binding is transient, so that transmitter in the gap can also be destroyed enzymatically, can be taken back up into the presynaptic terminal, or may sometimes even bind to presynaptic "autoreceptors." Postsynaptically, neural membranes are modified by incorporation into the lipid bilayer of specific receptor proteins and their associated complex lipids, called gangliosides.

Examples

The standard criteria for deciding which neurochemicals actually have a transmitter function are that the candidate transmitter must be normally present, synthesized, stored, released, reacted with postsynaptic membranes to alter polarization, and destroyed or otherwise removed from the site of action. There are now over 50 recognized neurotransmitters. The reason for the variety of these transmitters is not at all clear, but such "communication" systems certainly provide an enormous amount of highly specific synaptic "messages" (Figure 2-9).

Many transmitters have multiple receptors and multiple functions. Dopamine, for example, has at least two receptors with such diverse actions as: (1) inhibition of neurons in the caudate nucleus of the brain, (2) inhibition of postganglionic neurons in the superior cervical ganglion, (3) activation of adenylate cyclase in postsynaptic neurons, (4) stimulation of parathyroid hormone release, (5) inhibition of prolactin release from the anterior pituitary, and (6) regulation of ß-endorphin release from the intermediate pituitary.

For many decades a central doctrine of neurophysiology was that a given neuron contained only one kind of neurotransmitter. We now know this is not true for many neurons. For example, the "classical" neurotransmitters

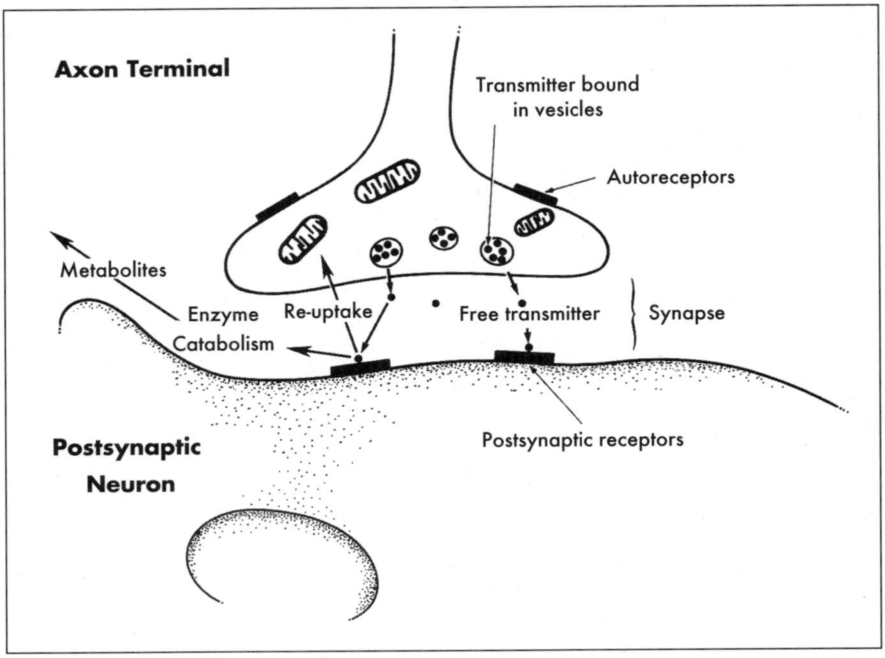

Fig. 2-9 Diagram of the basic principles of neurochemical transmission in the nervous system. As action potentials move into presynaptic terminals, neurotransmitter that is stored in molecular packets in vesicles is released into the synapse. Released transmitter participates in many reactions: 1) binding to stereospecific receptors on postsynaptic membrane, whereupon excitation or inhibition is promoted; 2) enzymatic binding and destruction to produce metabolites of the neurotransmitter; 3) re-uptake into the presynaptic terminal for later re-use. Some neurotransmitters can also act on presynaptic membrane receptors (autoreceptors) on the axon terminals. Each of the three reactions is vulnerable to drug treatment, and many drugs exist to exploit these mechanisms. (From Melvin J. Swenson and William O. Reece, editors: Dukes' Physiology of Domestic Animals, 11th ed., 1993, Cornell University Press.)

(GABA, acetylcholine, dopamine, norepinephrine, and serotonin) co-exist in certain neurons with a peptide (commonly cholecystokinin, enkephalin, endorphin, substance P, neuropeptide Y, or somatostatin). The functional significance of dual transmitters seems to be that one member of a pair modulates the postsynaptic effect of the other or its presynaptic release.

Historically, as the list of neurotransmitters grew, concepts of transmitter actions expanded to include a variety of second messenger systems that regulate active and passive ionic conductances across membranes. Not much is known about how transmitters interact to regulate common synaptic targets.

Some neural secretions do not directly excite or inhibit other cells, but instead modulate the release of transmitters indirectly . Some compounds, such as norepinephrine, may be a neurotransmitter in one synapse system and be a neuromodulator in another. The neuromodulator that is currently causing the most excitement in neuroscience research is nitric oxide, because it seems to have ubiquitous actions.

What happens to released neurotransmitter? Some is taken back up into the terminals that release it. Some may be taken up and stored in nearby glial cells. Other transmitter molecules may be destroyed by enzymes either presynaptically or postsynaptically. And some other molecules may actually diffuse in extracellular space to act at a distance from their site of release. This is especially likely with membrane-permeable gas transmitters, such as nitric oxide.

Transmission from axons to skeletal muscle

When an impulse in a motor axon supplying a skeletal muscle arrives at the neuromuscular junction, it normally evokes a membrane potential depolarization and action potential in the target muscle. One of the first clues that this transmission process did not involve direct electrical coupling came from the observation that neuromuscular transmission could be blocked by curare (a plant extract used by South American Indians on the tips of the blowpipe arrows that they use to capture monkeys and other prey). We now know that neuromuscular transmission depends on release of acetylcholine from an axon terminal, which reacts with a specific class of receptors (nicotinic) on the postjunctional membrane of muscle cells; curare is a specific blocker of nicotinic receptors (the commonly used muscle relaxant, d-tubocurarine, is a purified form of curare).

TERMS

Adenylate Cyclase A "second messenger" (see the Principle, "Second Messenger").

Depolarize To reduce the degree of electrical polarization across the neuronal membrane. That is, to move the membrane potential from its resting-state electronegativity toward 0 voltage difference across the membrane.

Hyperpolarize To increase the degree of voltage difference across the cell membrane. That is, to make the inside of the neuron more electronegative than it normally is.

Pituitary The "master gland" of the body. It makes many chemical products (hormones) that influence the activity of glands that secrete other hormones.

Prolactin A hormone with several effects, the most notable of which is the promotion of milk formation.

ß endorphin An endogenous opiate peptide that has many of the properties of opiate narcotic drugs.

Related Principles

Action Potentials
Calcium and Neurotransmitter Release
Information Carriers (Information Processing)
Ion Channels
Long-term Post-tetanic Synaptic Potentiation (Learning and Memory)
Membrane Receptors
Nodal Point
Receptor Binding Affinities
Second Messengers

References

Agnati, L.F., Bjelke, B., and Fuxe, K. 1992. Volume transmission in the brain. Amer. Scientist. 80:362-373.

Akil, H., Watson, S.J., Young, E., Lewis, M.E., Khachaturian, H., and Walker, J.M. 1984. Endogenous opioids: biology and function. Ann. Rev. Neurosci. 7:223-255.

Augustine, G.J., Charlton, M.P., and Smith, S.J. 1987. Calcium action in synaptic transmitter release. Ann. Rev. Neurosci. 10:633-693.

Björklund, A., Hökfelt, T. 1984. Classical Transmitters in the CNS. Part 1. Elsevier, Amsterdam.

Björklund, A., Hökfelt, T., and Kuhar, M.J. 1984. Classical Transmitters and Transmitter Receptors in the CNS, Part II. 1984. Handbook of Chemical neuroanatomy. Volume 3. Elsevier, Amsterdam.

Black, I.R. 1991. Information in the Brain: A Molecular Perspective. MIT Press, Cambridge, Massachusetts.

Carlsson, A. 1987. Perspectives on the discovery of central monoaminergic neurotransmission. Ann. Rev. Neurosci. 10:19-40.

Cotman, C.W., Monaghan D.T., and Ganong, A.H. 1988. Excitatory amino acid neurotransmission: NMDA receptors and Hebb-type synaptic plasticity. Ann. Rev. Neurosci. 11:61-80.

Eccles, J.C. 1982. The synapse: from electrical to chemical transmission. Ann. Rev. Neurosci. 5:325-339.

Greengard, P., Valtorta, F ., Czernik, A.J., and Benfenati, F. 1994. Synaptic vesicle phosphoproteins and regulation of synaptic function. Science 259:780-785.

Harris, K. and Kater, S.B. 1994. Dendritic spines: cellular specializations imparting both stability and flexibility to synaptic function. Ann. Rev. Neuroscience. 17:341-372.

Jahn, R. and Studhof, T.C. 1994. Synaptic vesicles and exocytosis. 1994. Ann. Rev. Neuroscience. 17:219-246.

Kelly, R.B., Deutsch, J.W., Carlson, S.S., and Wagner, J.A. 1979. Biochemistry of neurotransmitter release. Ann. Rev. Neurosci. 2:399-446.

Kupfermann, I. 1979. Modulatory actions of neurotransmitters. Ann. Rev. Neurosci. 2:447-465.

Montague, P.R. Gancayco, C.D. Winn, M.J., Marchase, R.B., and Friedlander, M.J. 1994. Role of NO production in NMDA receptor-mediated neurotransmitter release in cerebral cortex. Science 263:973-977.

Moore, R.Y., and Bloom, F.E. 1979. Central catecholamine neuron systems: anatomy and physiology of the norepinephrine and epinephrine Systems. Ann. Rev. Neurosci. 2:113-168.

Scharrer, B. 1987. Neurosecretion: Beginnings and new directions in neuropeptide research. Ann. Rev. Neurosci. 10:1-17.

Schuman, E.M. and Madison, D.V. 1994. Nitric oxide and synaptic function. Ann. Rev. Neurosci. 17:153-184.

Tallman, J.F. and Gallager, D.W. 1985. The GABA-ergic system: a locus of benzodiazepine action. Ann. Rev. Neurosci. 8:21-44.

Citation Classics

Amin A.H., Crawford T.B.B. and Gaddum J.H. 1954. The distribution of substance P and 5-hydroxytryptamine in the central nervous system of the dog. J. Physiol. 126:596-618.

Anden N-E, Dahlstrom A., Fuxe K., Larsson K., Olson L. and Ungerstedt U. 1966. Ascending monoamine neurons to the telencephalon and diencephalon. Acta Physiol. Scand. 67:313-326.

Axelrod J. and Tomchick R. 1958. Enzymatic O-methylation of epinephrine and other catechols. J. Biol. Chem. 233:702-705.

Bertler A., Carlsson A. and Rosengren D. 1958. A method for the fluorimetric determination of adrenaline and noradrenaline in tissues. Acta Physiol. Scand. 44:273-292.

Bogdanski D.F., Pletscher A., Brodie B. and Udenfried S. 1956. Identification and assay of serotonin in brain. J. Pharmacol. Exp. Ther. 117:82-88.

Burn J.H. and Rand M.J. 1958. The action of sympathomimetic amines in animals treated with reserpine. J. Physiol.-London 144:314-336.

Burnstock G. 1972. Purinergic nerves. Pharmacol. Rev. 24:509-581. (3/85/LS)

Carlsson A. and Waldeck S. 1958. A fluorometric method for the determination of dopamine (3-hydroxytyramine). Acta Physiol. Scand. 44:293-298.

Curzon G. and Green A.R. 1970. Rapid method for the determination of 5-hydroxytryptamine and 5-hydroxyindolacetic acid in small regions of rat brain. Brit. J. Pharmacol. 39:653-655.

Dahlstrom A. and Fuxe K. 1964. Evidence for the existence of monoamine-containing neurons in the central nervous system. Acta Physiol. Scand. 62(Supp. 232):1-55.

De Robertis E., Pellegrino de Iraldi A., Rodriquez de Lores Arnaiz G., and Salganicoff L. 1962. Cholinergic and non-cholinergic nerve endings in rat brain-I. Isolation and subcellular distribution of acetylcholine and acetylcholinesterase. J. Neurochemistry 9:23-35.

Falck B. 1962. Observations on the possibilities of the cellular localization of monoamines by a fluorescence method. Acta Physiol. Scan. 56(Suppl. 197):6-25.

Falck B., Hillarp N-A., Thieme G., and Torp A. 1962. Fluorescence of catecholamines and related compounds condensed with formaldehyde. J. Histochem. Cytochem. 10:348-354.

Fernstrom J.D. and Wurtman R.J. 1971. Brain serotonin content: physiological dependence on plasma tryptophan levels. Science 173:149-152.

Fernstrom J.D. and Wurtman R.J. 1972. Brain serotonin content: physiological regulation by plasma neutral amino acids. Science 178:414-416.

Fuxe K. 1965. Distribution of monoamine nerve terminals in the central nervous system. Acta. Physiol. Scand. 64:37-85.

Glowinski J. and Iversen L.L. 1966. Regional studies of catecholamines in the rat brain. I. The disposition of [^{3}H]norepinephrine, [^{3}H]dopamine, and [^{3}H]dopa in various regions of the brain. J. Neurochem. 13:655-669.

Haggendal J. 1963. An improved method for fluorimetric determination of small amounts of adrenaline and noradrenaline in plasma and tissues. Acta Physiol. Scand. 59:242-254.

Heuser J.E. and Reese T.S. 1973. Evidence for recycling of synaptic vesicle membrane during transmitter release at the frog neuromuscular junction. J. Cell Biol. 57:315-344.

Hökfelt T., Kellerth J.O., Nilsson G., and Pernow B. 1975. Substance P: localization in central nervous system and in some primary sensory neurons. Science 190:889-890.

Hökfelt T., Johanssom O., Ljungdahl A., Lundberg J.M., and Schultzberg M. 1980. Peptidergic neurons. Nature 284:515-521.

Hughes J., Smith T.W., Kosterlitz H.W., Fothergill L.A., Morhan B.A., and Morris H.R. 1975. Identification of two related pentapeptides from the brain with potent opiate agonist activity. Nature 258:577-579. [Unit Res. Addictive Drugs and Dept. Biochem., Univ. Aberdeen, Scotland; Pharmaceutical Div., Reckitt and Colman Ltd., Hull; and Dept. Biochem., Imperial Coll., London, England]

Koe B.K. and Weissman A. 1966. p-Chlorophenylalanine: a specific depletor of brain serotonin. J. Pharmacol. Exp. Ther. 154:499-516.

Koelle G.B. and Friedenwald J. S. 1949. A histochemical method for localizing cholinesterase activity. Proc. Soc. Exp. Biol. Med. 70:617-622 (11/84/LS).

Krnjevic K. 1974. Chemical nature of synaptic transmission in vertebrates. Physiol. Rev. 54:418-540 (22/84/LS).

Krnjevic K. and Phillis J.W. 1963. Iontophoretic studies of neurons in the mammalian cerebral cortex. J. Physiol. 165:274-304.

Krnjevic K. and Schwartz S. 1967. The action of gamma-aminobutyric acid on cortical neurones. Exp. Brain Res. 3:320-336.

Laverty R. and Taylor K.M. 1968. The fluorometric assay of catecholamines and related compounds: improvements and extensions to the hydroxyindole technique. Anal. Biochem. 22:269-279.

Maickel R.P., Cox R.H., Jr., Saillant J., and Miller F.P. 1968. A method for the determination of serotonin and norepinephrine in discrete areas of rat brain. Int. J. Neuropharmacol. 7:275-282.

Nagatsu T., Levitt M., and Udenfriend S. 1964. Tyroxine hydroxylase: the initial step in norepinephrine biosynthesis. J. Biol. Chem. 239:2910-2917.

Palkovits M. and Jacobowitz D.M. 1974. Topographic atlas of catecholamine and acetylcholin esterase-containing neurons in rat brain. J. Comp. Neurol. 157:29-42.

Phillis J.W. and Wu P.H. 1981. The role of adenosine and its nucleotides in central synaptic transmission. Prog. Neurobiol. 16:187-239.

Spector C., Sjoerdsma A., and Udenfriend S. 1965. Blockade of endogenous norepinephrine synthesis by alpha-methyl-tyrosine, and inhibitor of tyrosine hydroxylase. J. Pharmacol. Exp. Ther. 147:86-95.

Snyder S.H., Axelrod J., and Zweig M. 1965. A sensitive and specific fluorescence assay for tissue serotonin. Biochem. Pharmacol. 14:831-835.

Soubrie P. 1986. Reconciling the role of central serotonin neurons in human and animal behavior. Behav. Brain Sci. 9:319-335.

Starke K. 1977. Regulation of noradrenaline release by presynaptic receptor systems. Rev. Physiol. Biochem. Pharmacol. 77:1-124.

Swanson L.W. and Hartman B.K. 1975. The central adrenergic system. An immunofluorescence study of the location of cell bodies and their

efferent connections in the rat utilizing dopamine-B-hydroxylase as a marker. J. Comp. Neurol. 163:467-506.
Thoenen H. and Tranzer J.P. 1968. Chemical sympathectomy by selective destruction of adrenergic nerve endings with 6-hydroxydopamine. Naunyn-Schmied. Arch. Pharmakol. Exp. Pathol. 261:271-288.
Ungerstedt U. 1971. Stereotaxic mapping of the monoamine pathways in the rat brain. Acta Physiol. Scand. (Suppl. 367):1-48.
Vogt M. 1954. The concentration of sympathin in different parts of the central nervous system under normal conditions and after the administration of drugs. J. Physiol. 123:451-481.

Transport

Neurons transport metabolic products, sometimes over relatively great distances, through small cytoplasmic extensions of the cell body.

Explanation

Most of the metabolic machinery of the neuron is located in the nucleus and cytoplasm of the cell body. This includes the DNA and RNA of the cell, as well as most enzyme systems. Because neurons have many cytoplasmic extensions (dendrites and axons) that require energy and biochemicals for the continual maintenance and reconstruction of their membranes and other structures, neurons must have an effective transport system for moving materials from the cell body to the dendrites and axons. Moreover, since neurons are actually secretory cells that synthesize, store, and release chemicals from their terminals, an effective transport system for precursor chemicals is also needed.

Examples

Neuronal processes have a system of tubules (neurotubules) and parallel fibrils (neurofibrils) that enable transport. Transport is predominantly, though not exclusively, from the cell body to the periphery, which was first demonstrated in an experiment where an axon was ligated. In a few days, a bulge was seen immediately before the ligature, on the cell body side (Figure 2-10). Some retrograde transport also occurs wherein materials in the terminals are returned to the cell body for degradation or re-use.

The distances over which transport must occur in neurons can range from centimeters to many meters (as in the case of the sciatic nerve of an elephant). In such long nerves, the volume of cytoplasm in the total axon length may be as much as 10,000 times that of the cell body.

Two systems accomplish neuronal transport (Figure 2-11). One is a "fast" transport system, consisting of neurotubules, which carry radiolabeled materials injected into the cell body down the axon at rates of about 200-400 mm/day. There are also slow transport systems (0.2-1 mm/day) that use neurofibrils as substrates for guiding the flow of materials. Peristaltic waves of cytoplasmic movement can be seen in time-lapse photography as the propulsive force behind both kinds of transport.

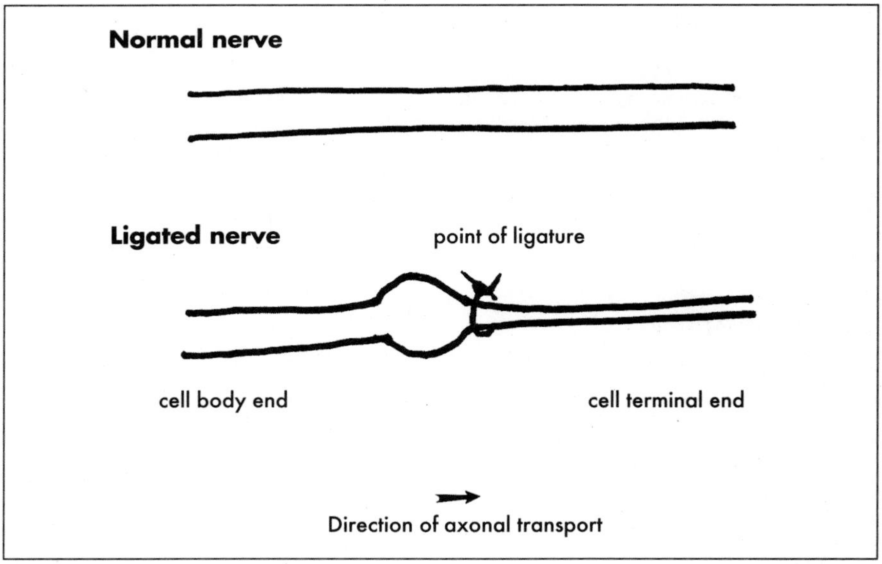

Fig. 2-10 Diagram of the original experiments that suggested axonal transport. When a nerve was ligated by a thread, a bulging of the nerve was evident on one side of the ligature. Since that bulge was on the cell-body side of the ligature, it suggested that most of the transport was anterograde, from cell body to terminal end of the axons contained within that nerve.

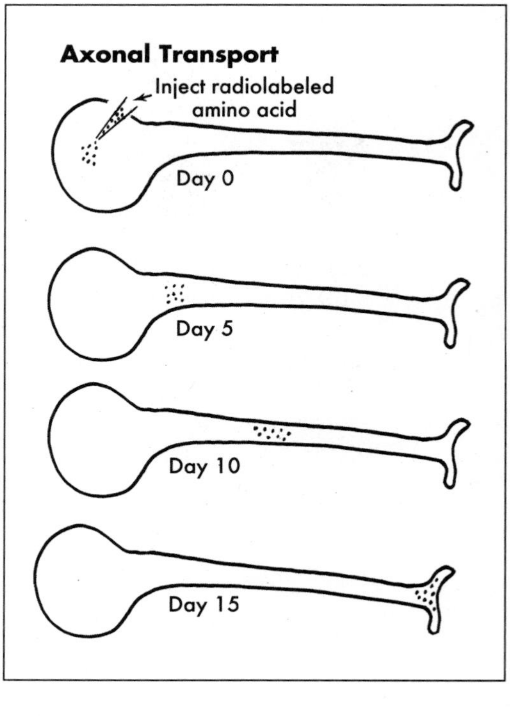

Fig. 2-11 Diagram of the experimental methodology for proving axonal transport and for quantifying the rate for various substances. The fundamental feature of the method is to inject radiolabeled precursor compounds into the cell body and periodically monitor the amount of radioactivity along the axon at various times post injection. For example, if radiolabeled amino acids are injected on Day 0, by day 5 the radioactivity may show up in the proximal third of the fiber. By day 10, the radioactivity may be in the middle third of the axon, and by day 15, the radioactivity may be in the terminal third. (From Schwartz, JH: Axonal transport: components, mechanisms, and specificity, Ann Rev Neurosci 2: 467-504, 1979.)

Among the substances transported into neuron terminals are mitochondria, which move at 50 mm/day. This fast transport occurs along neurofibrils, being powered by force-generating motors, such as kinesin and dynein. One can only speculate about how proteins and an energy source can move molecules, but it seems certain that energy allows conformational changes in protein and thus alters the binding properties of the protein. If protein is anchored, such as along neurofibrils, its process of binding at one location and release at another location, because of conformational change, will effectively move the chemical being bound and released.

Proteins synthesized in the cell body seem to be transported at slow rates as intact polymers inside the neurotubules. Finally, there is evidence that glial cells transport proteins into axons and thus facilitate transport, especially in invertebrates and lower vertebrates. In such animals, even a transected axon can remain viable for many weeks and even months.

Related Principles

Allosterism (Cell Biology)
Neuronal Growth (Development)

References

Hall, Zach W. 1992. An Introduction to Molecular Neurobiology. Sinauer. Sunderland, Massachusetts.

Hoffman, P.N. and Lasek, R.J. 1975. The slow component of axonal transport: identification of major structural polypeptides of the axon and their generality among mammalian neurons. J. Cell. Biol. 66:351-366.

Mitchison, T. and Kirschner, M. 1988. Cytoskeletal dynamics and nerve growth. Neuron. 1:761-772.

Mori, H., Komiya, Y., and Kurokawa, M. 1979. Slowly migrating axonal polypeptides: inequalities in their rate and amount of transport between two branches of bifurcating axons. J. Cell. Biol. 82:174-184.

Poo, M. 1985. Mobility and localization of proteins in excitable membranes. Ann. Rev. Neurosci. 8:369-406.

Schwartz, J.H. 1979. Axonal transport: components, mechanisms, and specificity. Ann. Rev. Neurosci. 2:467-504.

Vallee, R.B. and Bloom, G.S. 1991. Mechanisms of fast and slow axonal transport. Ann. Rev. Neurosci. 14:59-92.

Citation Classic

Ochs, S. 1972. Fast transport of materials in mammalian nerve fibers. Science. 176:252-260.

Ending Neurotransmission

The action of released transmitters is terminated by several mechanisms: enzymatic destruction, diffusion away from receptors, and active transport processes that take the transmitter or its precursors back into nerve terminals or into adjacent glia cells.

Explanation

If neurotransmitter action were persistent, then there would be no way for the nervous system to be continually responsive to new stimuli and contingencies. In short, a sensory message needs to be read and acted upon, and then "forgotten" so that new sensory messages can be received and acted upon. Thus, it is imperative that mechanisms be present to terminate the action of neurotransmitters (Figure 2-12).

The simplest mechanism is simply diffusion away from the synapse. In fact, this diffusion process has led many investigators to argue that neurotransmitters have effects over great expanses of tissue, in addition to the localized action in the synapse in which release occurred. However, diffusion dilutes the transmitter and it is absorbed into the bloodstream where it becomes vulnerable to metabolic destruction in such organs as the liver.

Re-uptake of certain neurotransmitters has been demonstrated in both axon terminals and in nearby glia cells. In either case, this process not only terminates the action but also conserves the neurotransmitter, making it available for re-use.

Finally, some neurotransmitters are enzymatically destroyed in or near the synapse or axon terminal.

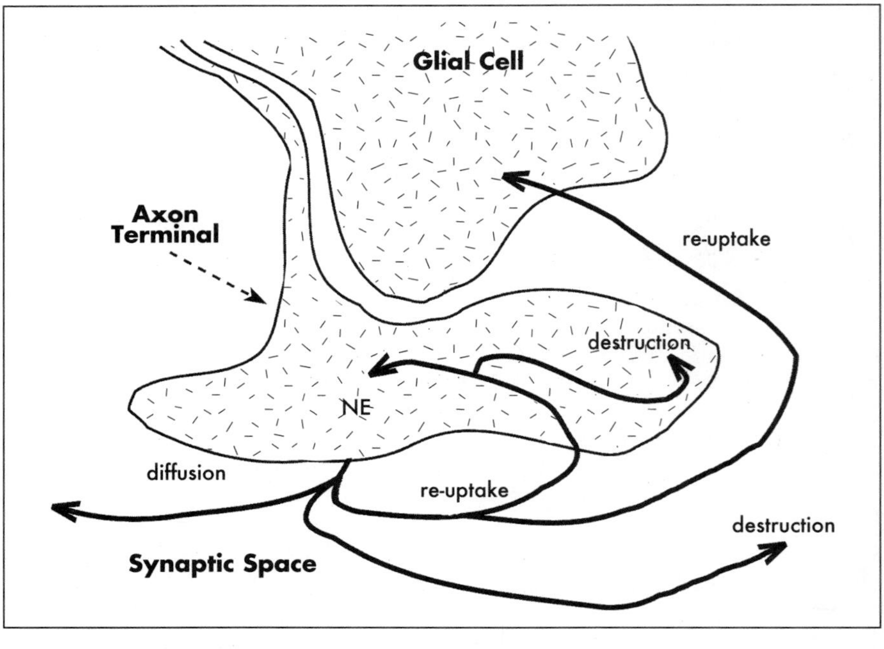

Fig. 2-12 Diagram of the fate of norepinephrine (NE) after its release into the synaptic gap. Molecules that are not bound to postsynaptic receptors can do several things: diffuse into extracellular fluid, be destroyed by enzymes in extracellular fluid or in neurons, or be taken back up into neurons or adjacent glial cells.

Example

Many of these modes of termination of neurotransmitter action can be illustrated with norepinephrine (NE). Once released into the synaptic gap, some NE diffuses into extracellular fluid. Other NE molecules are destroyed by one of two enzymes, monoamine oxidase or catechol-O-methyl transferase. These enzymes act in extracellular fluid as well as within neurons. Other NE molecules get taken back up, intact, into either neurons or adjacent glial cells.

References

Axelrod, J. 1971. Noradrenaline: fate and control of its biosynthesis. Science. 173:598-606.

Cooper, J.R., Bloom, R.E., and Roth, R.H. 1991. The Biochemical Basis of Neuropharmacology, 6th ed. Oxford U. Press, New York.

Iversen, L.L. 1967. The Uptake and Storage of Noradrenaline in Sympathetic Nerves. Cambridge U. Press, Cambridge.

McGeer, P.L., Eccles, J.C., and McGeer, E.G. 1987. Molecular Neurobiology of the Mammalian Brain, 2nd ed. Plenum, New York.

Citation Classics

Axelrod, J. and Tomchek, R. 1958. Enzymatic O-methylation of epinephrine and other catechols. J. Biol. Chem. 233:702-705.

Dale, H. 1935. Pharmacology of nerve-endings. Proc. R. Soc. Med. (Lond) 28:319-332.

Membrane Receptors

Neural membranes, both presynaptic and postsynaptic, contain embedded proteins that act as receptors for specific ligands. These ligands are normally neurotransmitters and hormones, but they can also be drugs. The ligands must "fit" the conformation and electric charge properties of their receptors. This means that the ordering of amino acids in the receptor protein is very important, along with the ordering of constituent atoms in the transmitter and the arrangement of the atoms in space (i.e., the "handedness" of certain key atoms). When a ligand binds a receptor, the receptor molecule changes conformation, which may in turn change the "channels" in receptor protein that allow ions to flow across the cell membrane.

Explanation

Receptor proteins are embedded in cell membranes. The polypeptide chain of these receptors spans the membrane from one to more than a dozen times in the form of long alpha helixes that contain hydrophilic channels. These water-filled channels allow ions to flow across the otherwise impermeable membrane. Whether or not the channels are open

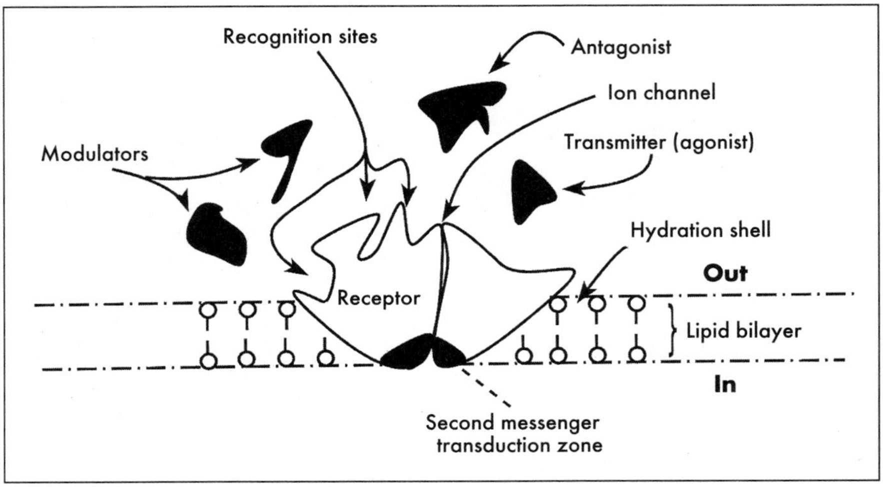

Fig. 2-13 Diagram of the relationship of a protein receptor that is embedded in cell membrane. The receptor is seen to have multiple sites on its extracellular surface that can bind various compounds in very specific ways. Some ligands, called agonists, produce the normal neurotransmitter responses at that synapse. Other ligands, antagonists, produce opposite effects. Other ligands (modulators) alter the sensitivity of agonist and /or antagonist binding sites to their ligands. The conformational changes produced by ligand binding determine whether the pores (ion channels) are open or closed to regulate ionic flow across the membrane.

depend on the three-dimensional conformation of the receptor protein. This conformation is altered by the presence of ligands that bind to specific molecular recognition sites on the membrane.

Because proteins have tertiary and quaternary structure (i.e., coiling and multiple folding) and because some of the amino acids are electrically charged, they present a microdomain that can electrostatically bind specific ligands in very selective ways.

Examples

A finger-and-glove analogy illustrates the point. When ligand binding occurs, the conformation of the protein may well change, altering for example the aqueous-filled pores in the protein. Such pores can constitute channels for flux of certain ions that are of the appropriate hydrated size and electrical charge. Thus, one can imagine how ligand binding can open or close specific ion channels (Figure 2-13).

Receptor recognition may occur at different sites of the same protein. For example, one site may bind a neurotransmitter molecule that produces the typical postsynaptic response for that particular cell, while another site may bind a compound that has an opposite or modulatory action. The chemical that causes direct effects upon binding is called an "agonist," and it will alter ion flow across the membrane to cause either excitation or inhibition, whichever is the typical response to that transmitter. Other chemicals may bind other recognition sites on the membrane and cause allosteric changes that antagonize or modulate the action of the normal transmitter.

On the cytoplasmic side of a membrane receptor, there are typically long strings of the protein that have reactive sites for coupling to intracellular "second messenger" systems to activate changes in functional proteins within the cell.

Molecular recognition sites are so specific that they usually recognize only one isomeric form of the molecule. For some unknown (perhaps chance) reason, most biologically active receptor ligands are of the "left-handed" versions of the molecule (Figure 2-14).

For a given neurotransmitter, there may be a family of receptor proteins that bind it, yet they differ in subtle ways in their affinity for the transmitter. This phenomenon has made possible the development of drugs that act preferentially on one receptor subtype, thus producing a more selective effect than is produced by the natural transmitter. For example, the receptors for acetylcholine include two major subtypes. One type is muscarinic, which preferentially binds muscarine and certain other agonist drugs and other antagonist drugs (atropine, scopolamine). So-called nicotinic receptors bind not only acetylcholine but also nicotine and certain agonist drugs and other antagonist drugs (curare). Because receptors may be distributed differentially in the body, drugs that are selective for one or another receptor subtype can produce selective actions. In the case of acetylcholine receptors, the nicotinic form is the only form found at neuromuscular junctions and in ganglia of the autonomic nervous system. Thus, atropine, an antagonist for muscarinic sites, can be used medically without fear of causing muscle paralysis (such as failure to breathe).

Of special interest is the question of how certain drugs are able to act on membrane receptors. In many cases, the drug used in a human-made replica of the natural transmitters epinephrine and norepinephrine are typical examples. In other cases, the drug molecule just coincidentally fits a receptor. Strychnine, a poison that causes fatal convulsions, acts as an antagonist on receptors of certain inhibitory cells, which has the effect of releasing motor neurons from inhibition. But strychnine does not occur as a normal neurotransmitter. It just happens to bind to and block the receptor for an inhibitory neurotransmitter, glycine. Some other drugs have caused neuroscientists to look for previously undiscovered neurotransmitters and their receptors. The opiates are a good example. Opiates, such as morphine and heroin, do not occur naturally in the body. Yet, we know they are able to act because they bind certain membrane receptors. This suggests that there are naturally occurring opiate-like transmitters that act on receptors that coincidentally can also bind opiate drugs. Indeed, that is the case, and it has led to the discovery of several endogenous opiates (endorphins and enkephalins) and several opiate receptor subtypes (such as delta, mu, sigma, etc.).

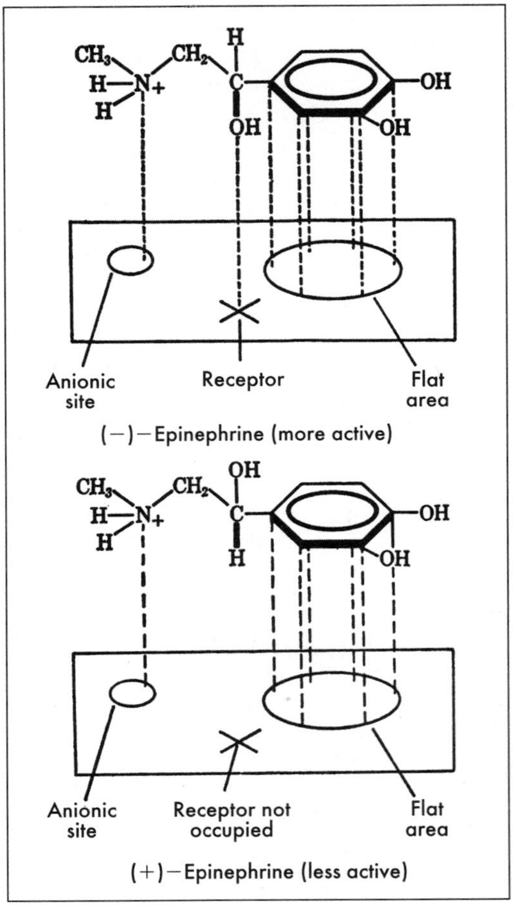

Fig. 2-14 Illustration of the stereospecific "handedness" of ligand binding to receptor. In this example, "left-handed" (-) epinephrine is shown to be capable of binding the receptor, whereas the right-handed (+) compound cannot. Binding apparently occurs only with the (-) form because the receptor has the right shape and atomic composition at that region to accommodate and attract the OH group (above), but cannot accommodate the H of the + form of epinephrine (below). We assume that the entire molecule is attracted to the receptor because the receptor has a strong electronegative domain that attracts and accommodates the positively charged nitrogen domain of epinephrine. The OH group presumably is attracted to hydrogen bond to protons that are situated on the receptor midway between its anionic site and its recess that holds the ring structure. But this proton region cannot form hydrogen bonds with the proton presented to it by the (+) form of epinephrine.

TERMS

Allosterism A transformation in the three-dimensional conformation of a protein that is produced by ligand binding.

Autonomic Nervous System That part of the nervous system that controls the viscera. It consists of some pools of control neurons in the hypothalamic part of the brain and peripheral nerves that supply blood vessels and viscera in the thorax and abdomen.

Disinhibition Releasing from inhibition; thus, this can produce an excitatory-like effect.

Second Messenger One or more series of molecular reactions that occur secondarily inside the cell to a preceding ligand binding action on a surface receptor.

Related Principles

Allosterism
Information Carries (Information Processing)
Ion Channels
Neurochemical Transmission
Receptor Binding
Second Messengers

References

Cotman, C.W., Monaghan, D.T., and Ganong, A.H. 1988. Excitatory amino acid neurotransmission: NMDA receptors and Hebb-type synaptic plasticity. Ann. Rev. Neurosci. 11:61-80.

Cowan, S.W. and Rosenbusch, J.P. 1994. Folding pattern diversity of integral membrane proteins. Science. 264:914-916.

Creese, I., Sibley, D. R., Hamblin, M.W., and Leff, S.E. 1983. The classification of dopamine receptors. Ann. Rev. Neurosci. 6:43-71.

Daw, N.W., Stein, P.S.G., and Fox, K. 1993. The role of NMDA receptors in information processing. Ann. Rev. Neurosci. 16:207-222.

Dunwiddie, Thomas V., and Lovinger, David M. (eds.) 1993. Presynaptic Receptors in the Mammalian Brain. Birkhauser, Boston.

Gennaro, A.R. 1985. Remington's Pharmaceutical Science. 17th ed. Mack Publishing Co., Easton, Pennsylvania.

Hucho, F. 1993. Neurotransmitter Receptors. New Comprehensive Biochemistry Volume 24. Elsevier, Amsterdam.

Kawai, N., Nakajima, T., and Barnard, E. 1992. Neuroreceptors, Ion Channels and Brain. Elsevier, Amsterdam.

Kupfermann, I. 1979. Modulatory actions of neurotransmitters. Ann. Rev. Neurosci. 2:447-465.

Poo, M. 1985. Mobility and locations of proteins in excitable membranes. Ann. Rev. Neurosci. 8:369-406.

Macdonald, R.L. and Olsen, R.W. 1994. $GABA_A$ receptor channels. Ann. Rev. Neuroscience. 17:569-602.

McCarthy, M.P., Earnest, J.P., Young, E.F., Choe, S., and Stroud R.M. 1986. The molecular neurobiology of the acetycholine receptor. Ann. Rev. Neurosci. 9:383-413.

O'Dowd, B.F., Lefkowitz, R.J., and Caron, M.G. 1989. Structure of the adrenergic and related receptors. Ann. Rev. Neurosci. 12:67-83.

Peroutka, S.J. 1988. 5-Hydroxytryptamine receptor subtypes. Ann. Rev. Neurosci. 11:45-60.

Snyder, S.H. and Childers, S.R. 1979. Opiate receptors and opioid peptides. Ann. Rev. Neurosci. 2:35-64.

Citation Classics

Ahlquist R.P. 1948. A study of the adrenotropic receptors. Amer. J. Physiol. 153:586-600 (45/78).

Anden N-E., Butcher S.G., Corrodi H., Fuxe K., and Ungerstedt U. 1970. Receptor activity and turnover of dopamine and noradrenaline after neuroleptics. Eur. J. Pharmacol. 11:303-314.

Braestrup C. and Squires R.F. 1977. Specific benzodiazepine receptors in rat brain characterized by high affinity [^{3}H]diazepam binding. Proc. Nat. Acad. Sci. USA 74:3805-3809.

Folch J., Ascoli I., Lees M., Meath J.A., and LeBaron F.N. 1951. Preparation of lipid extracts from brain tissue. J. Biol Chem. 191:833-841.

Iversen L.L. 1975. Dopamine receptors in the brain. Science 188:1084-1089.
Kebabian J.W. and Calne D.B. 1979. Multiple receptors for dopamine. Nature 277:93-96.
Kleckner, N.W. and Dingledeine, R., 1988. Requirement for glycine in activation of NMDA receptors expressed in Xenopus oocytes. Science. 241:835-837.
Langer S.Z. 1977. Presynaptic receptors and their role in the regulation of transmitter release. Brit. J. Pharmacol. 60:481-497.
Levitan, E.S., Schofield, P.R., Burt, D.R. et al. 1988. Structural and functional basis for $GABA_A$ receptor heterogeneity. Nature. 335:76-79.
Leysen J.E., Gommeren W., and Laduron P.M. 1978. Spiperone: a ligand of choice of neuroleptic receptors. 1. Kinetics and characteristics of in vitro binding. Biochem. Pharmacol. 27:307-316.
Lisk R.D. 1960. Estrogen-sensitive centers in the hypothalamus of the rat. J. Exp. Zool. 145:197-207.
Mohler H. 1977. Benzodiazepine receptor: demonstration in the central nervous system. Science 198:849-851.
Monod J., Wyman J., and Changeux J-P. 1965. On the nature of allosteric transitions: a plausible model. J. Mol. Biol. 12:88-118.
Yamamura H.I. and Snyder S.H. 1974. Muscarinic cholinergic binding in rat brain. Proc. Nat. Acad. Sci. USA 71:1725-1729; 86-1783.
Zukin R.S. and Zukin S.R. 1981. Multiple opiate receptors: emerging concepts. Life Sci. 29:2681-2690.

Receptor Binding

Neurotransmitters bind receptors temporarily, because the interactions do not lead to covalent bonds but rather are a combination of weaker forces involving hydrogen bonding, electrostatic attraction, and Van der Waals forces. Such binding is stereospecific and confers a particular conformation on the membrane receptor. This receptor conformation determines whether the aqueous channels within the protein are open or closed to ionic flux.

Explanation

The binding forces (electrostatic, Van der Waals, etc.) that hold neurotransmitter and receptor protein molecules together are only temporary. Moreover, conformational change in the protein requires that the receptor fit, in hand-in-glove fashion, into specific receptor sites on the proteins that have complementary conformation and electrostatic charge. Binding properties depend on the nature of the lipid environment, and, in some cases, on lipids that are covalently bonded to receptor proteins.

Net binding effectiveness is a function of the number of receptors and the affinity they have for their ligands. By carefully extracting receptor protein from neuronal membranes, one can test *in vitro*, the relative potency of various agonists, antagonists, and modulators. The drug industry does this routinely in attempts to find drugs that can act very specifically on certain receptor subtypes, so that discrete and specific actions are elicited, with minimal side-effects.

There is also nonspecific binding of various compounds to receptor proteins. This kind of binding is characterized as being of low affinity (i.e., large concentrations of ligand must be present), as opposed to highly stereospecific binding of ligands that bind with high affinity (i.e., in low concentrations). This phenomenon is the basis for a common methodological approach in studying receptor binding phenomena. The experimenter will deliberately saturate a receptor protein preparation with nonspecific ligands and then test for binding affinities of putative high-affinity ligands.

Transmitter molecules can bind their respective receptors according to the simple relation:

$$[T] + [R] \underset{k_2}{\overset{k_1}{\rightleftharpoons}} [TR]$$

where T = transmitter molecule, R = receptor molecular complex, TR = transmitter-receptor complex, and k_1/k_2 = constants for the rate at which the association takes place.

The ratio of dissociation to association, k_2/k_1, is a function of the binding affinity, called K_d, of the ligand and its receptor. Certain drugs and neuromodulators can alter the K_d through allosteric action.

Binding of different molecular species occurs all the time, but usually the binding is nonspecific. The criteria for stereospecific binding includes reversibility and saturability. That is, receptor proteins have a finite number of stereospecific receptor sites, and once they are all occupied by neurotransmitter, the system is saturated. Adding more neurotransmitter will not produce more conformational change nor physiological consequence.

Examples

Experimentally, when an investigator wants to measure the binding properties of a neurotransmitter, or a drug that acts on neurotransmitter receptors, the goal is to distinguish how much ligand is stereospecifically bound (to "high affinity" receptor sites), as opposed to how much is nonspecifically bound (to "low affinity" sites, that can even include binding to experimental apparatus such as filters and glassware).

These experiments always include the testing of a range of concentrations of the ligand, on the assumption that at low concentrations most of the ligand will be bound to high-affinity, stereospecific sites, while at high concentrations, saturation of the high-affinity sites has occurred and most of the binding occurs at nonspecific sites. In practice, the experiment is based on a set of pairs of test tubes in which the tissue homogenate is incubated with increasing concentrations of radiolabeled ligand, with and without added excess of the same ligand that is not radiolabeled. When only radioligand is present, one is measuring

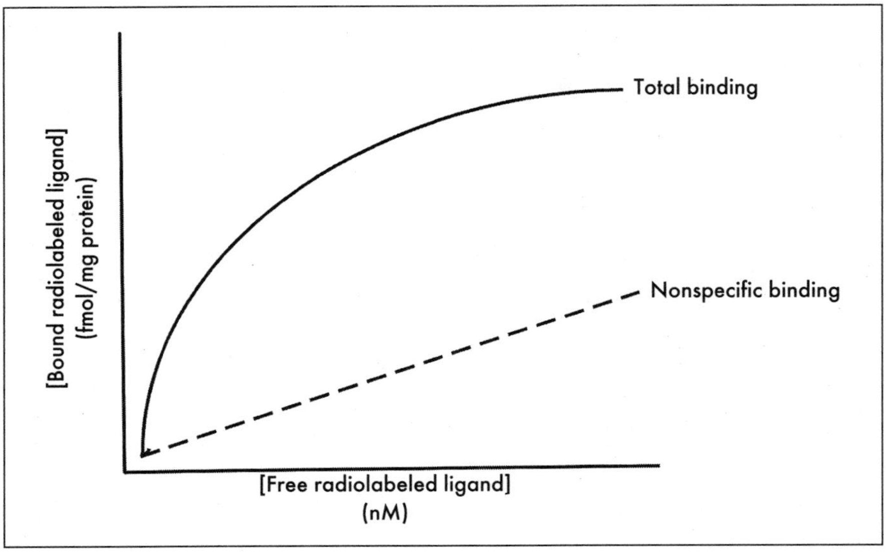

Fig. 2-15 Typical curves obtained from a direct binding assay of neurotransmitter (or drug) and receptors in extracts of synapses. The portion of the total radioactivity bound that is due to nonspecific interactions is determined by measuring total binding in another sample tube in which a surplus of non-radioactive ligand (of the same or related compound), which will saturate all the nonspecific binding sites. The stereospecific binding at each concentration is determined by subtracting the nonspecific binding value from the total binding value.

specific binding at low concentrations, but specific plus nonspecific binding at high concentrations. In either case, this is TOTAL binding. In the presence of an excess of nonlabeled ligand, the radiolabeled ligand can displace the fraction of the total binding that is attributable to stereospecific binding sites. The specific binding at each concentration is calculated as the difference between total and nonspecific binding. Note from Figure 2-15 that the nonspecific binding curve is linear, but that total binding reaches asymptote, as does the fraction of the total that is stereospecific.

If these data are plotted in a certain way (Scatchard plot) (Figure 2-16), it is possible to determine the total number of specific binding sites (B_{max}) and the binding affinity (K_d). One can plot the ratio of bound and free ligand (B/F) against the amount of bound ligand (B); the slope of the regression line of this plot is equal to $-1/K_d$, which allows calculation of the K_d. The maximum number of specific binding sites is indicated by the intercept of the line on the x-axis. When such plots yield a nonlinear curve, it suggests that the ligand being tested binds to more than one site with different affinity or that there is cooperative interaction between two or more sites with different affinity or cooperative interaction. The presence of cooperativity is evaluated by other analytical schemes, such as use of the Hill Plot.

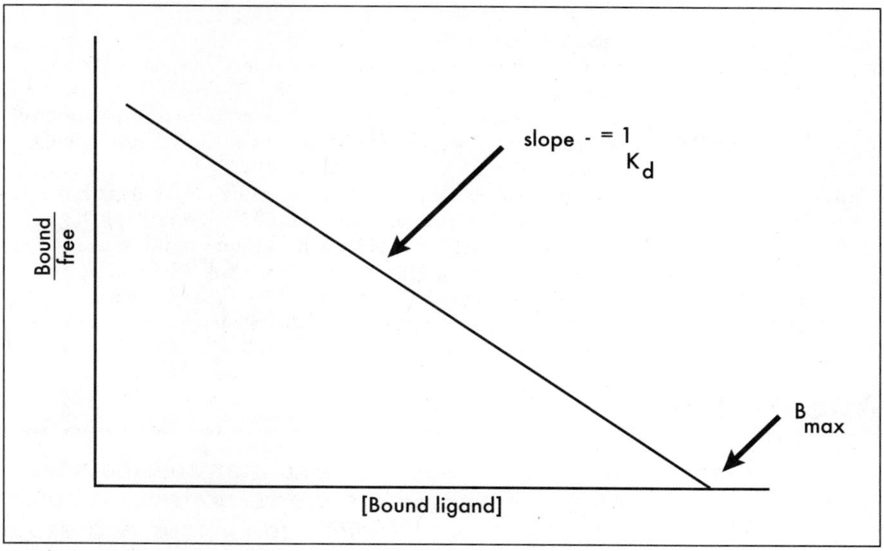

Fig. 2-16 Typical curve used to illustrate binding affinity (K_d) and number of binding sites (B_{max}) (so-called Scatchard plot of the bound: free ratio versus the bound). K_d is calculated from the slope of the curve and B_{max} is read from the point of intercept on the x axis. Units are usually in fmol/mg protein.

TERMS

Ligand Molecule that binds to another molecule; typically a small molecule that binds relatively large molecules.

Related Principles
Allosterism
Membrane Receptors
Neurochemical Transmission

References

Boulton, A.A., Baker, G.B., and Hrdina, P.D. 1986. Neuromethods. 4. Receptor Binding. Humana Press. Clifton, New Jersey.

Casey, P.J. 1995. Protein lipidation in cell signalling. Science 268:221-225.

Lamble, J.W. 1981. Towards Understanding Receptors. Elsevier, Amsterdam.

Lamble, J.W. 1982. More About Receptors. Elsevier, Amsterdam.

Venter, J.C., and Harrison, L.C . (eds.) 1984. Membranes, Detergents, and Receptor Solubilization. Receptor Biochemistry and methodology, Vol. 1. Liss, New York.

Citation Classics

Pert C.B. and Snyder S.H. 1973. Opiate receptor: demonstration in nervous tissue. Science 179:1011-1014.
Simon E.J., Hiller J.M. and Edelman I. 1973. Stereospecific binding of the potent narcotic analgesic [3H]etorphone to rat-brain homogenate. Proc. Nat. Acad. Sci. USA 70:1947-1949. [19/78] 86-0612.
Yamamura H.I. and Snyder S.H. 1974. Muscarinic cholinergic binding in rat brain. Proc. Nat. Acad. Sci. USA 71:1725-1729. 86-1783
Zukin S.R. and Zukin R.S. 1979. [3H]Phencyclidine binding in rat central nervous system. Proc. Nat. Acad. Sci. USA 76:5372-5376.
Zukin R.S. and Zukin S.R. 1981. Demonstration of [3H]cyclazocine binding to multiple opiate receptor sites. Mol. Pharmacol. 20:246-254.

Allosterism

Binding of ligand to receptor occurs at specific sites within the receptor. Membrane receptors are responsive to conformational alterations produced by various endogenous and exogenous ligands, and thus the accessibility to binding sites is likewise altered. These changes, called allosteric, alter the binding properties of the receptor, making it more or less sensitive to particular ligands. Ligands that cause allosteric effects are sometimes referred to as neuromodulators, as opposed to neurotransmitters, because they do not "transmit" information per se but rather alter the bias for transmission of information by specific neurotransmitters.

Explanation

Whether or not a ligand-specific receptor binding site is occupied depends not only on the presence of appropriate ligand, but also on the accessibility of that ligand. Depending on the neurotransmitter system, ligands may bind on the receptor moieties that project into extracellular space, on the surface, deep within a pocket in the receptor protein, or even at a site on the surface that is adjacent to the receptor. Wherever binding occurs, it may alter the shape and electric charge orientation of binding sites within the receptor. These sites may become more or less accessible to their ligand. Thus, we can see how binding of one compound onto a large receptor protein can alter the binding of other ligands that normally bind at other sites on that same protein.

Examples

Because receptor-ligand interactions typically catalyze key intracellular chemical reactions, allosteric effects translate into effects on this catalysis. For example, the mechanism whereby G-protein coupled receptors catalyze adenosine nucleotide reactions involves an allosteric effect of two ligands binding at two separate sites of the receptor, with the binding of one ligand decreasing the binding of the other. Not only do these allosteric effects alter ligand affinity, but another major consequence is on the speed of reaction. Allosteric ligands often occur in low concentrations, and thus receptor binding sites must have high affinity for them. But high affinity and speed are often incompatible. Allosteric interactions may be negative, whereby the binding of one ligand increases the rate of dissociation of another ligand that otherwise would dissociate slowly. To generate a rapid response in the face of high ligand affinity, it helps to have a doubly liganded protein. If both ligands have high dissociation constants, then when both are bound one or the other ligand will dissociate rapidly. Guanine nucleotides are exchanged by this mechanism. Normally, the dissociation of nucleotide directly from guanine nucleotide is slow, but the binding of the associated ligand-bound receptor protein facilitates dissociation of nucleotide.

A good example of a multiple-site receptor is a protein that binds a methylated amino acid, known as *N*-methyl-D-aspartate (NMDA). This NMDA receptor is a membrane protein that contains an excitatory ion channel that is opened under typical conditions by binding to glutamate. The channel can also be activated by binding to the drug, PCP. Binding to magnesium can block the NMDA receptor.

Another example is the benzodiazepine receptor, whose ion channels mediate inhibition. Agonist sites on this protein have been found for benzodiazepine tranquilizers, gamma amino butyric acid (a major inhibitory neurotransmitter), barbiturate anesthetics, and perhaps a yet-to-be discovered endogenous tranquilizer neurotransmitter.

Related Principles

Ion Channels
Membrane Receptors
Neurochemical Transmission
Receptor Binding

References

Cotman, C.W., Iversen, L.L., Watkins, J.C. et al. 1987. Special issue: excitatory amino acids in the brain—focus on NMDA receptors. Trends Neurosci. 10:263-301.

Hall, Z.W. 1992. An Introduction to Molecular Neurobiology. Sinauer Associates. Sunderland, Massachusetts.

Hirsch, J.D., Garrett, K.M., and Beer, B. 1985. Heterogeneity of benzodiazepine binding: a review of recent research. Pharmacol. Biochem. Behav. 23:681-685.

Olsen, R.W. and Leeb-Lundberg, F. 1981. Convulsant and anticonvulsant drug binding sites related to GABA-regulated chloride ion channels. Adv. Biochem. Psychopharmacol. 26:93-102.

Paul, S.M., Crawley, J.N., and Skolnick, P. 1986. The neurobiology of anxiety: the role of the GABA/benzodiazepine receptor complex, p. 581-596. In American Handbook of Psychiatry, Vol. 8, edited by P. A. Berger and K. H. Brodie. Basic Books, New York.

Citation Classics

Fonnum, F. 1984. Glutamate: a neurotransmitter in mammalian brain. J. Neurochem. 42:1-11.

Mohler, H. and Okada, T. 1977. Benzodiazepine receptor: demonstration in the central nervous system. Science. 198:849-851.

Squire, R.F. and Braestrup, C. 1977. Benzodiazepine receptors in rat brain. Nature. 266:732-734.

Ion Channels

Neuronal membranes have coiled proteins that are embedded in the hydrophobic lipid bilayer. Some of the amino acids surround a variable-sized pore, which constitutes a channel through which ions can flow according to the electrostatic and concentration gradients.

The degree to which these ion channels are open can be regulated by voltage fields (voltage-gated channels) or by specific receptor ligands (ligand-gated channels). Voltage-gated channels, at least those for sodium in mammalian brain, are also regulated by phosphorylation of protein kinase C and cAMP-dependent protein kinase.

Explanation

When neurons discharge action potentials, the propagation of these potentials from one point along the membrane to another is accomplished by voltage-gated ion channels. That is, the electrical field around a point of membrane depolarization serves as a trigger for nearby ion channels to open. The resulting change in membrane potential from resting polarization to depolarization is actually the summated response of sodium and potassium ion channels, as they open and then close to allow the flux of ions.

Postsynaptic activity is ultimately governed by the flow of ions through specialized ion channels. These are of two types: (1) voltage gated, i.e., regulated by membrane voltage *per se,* as is seen in axonal propagation of impulses, and (2) chemically gated, i.e., regulated by neurotransmitter, as is typical of much synaptic transmission. These two types of channels are differentially affected by poisons; tetrodotoxin, for example, blocks voltage-gated but not chemically gated Na^+ channels, while cobra venom does just the opposite. There are other kinds of junctions between cells, called "gap junctions, in which the cytoplasm of one cell is in direct contact with that of an adjacent cell; in these, there is no ion channel capable of opening and closing.

Microelectrode techniques ("patch clamp") involving recording of current flow through only one or a few ion channels reveal that the channels open in an all-or-none fashion. The graded over-all changes in postsynaptic potentials arise because the channels of the whole neuron open in staggered sequence.

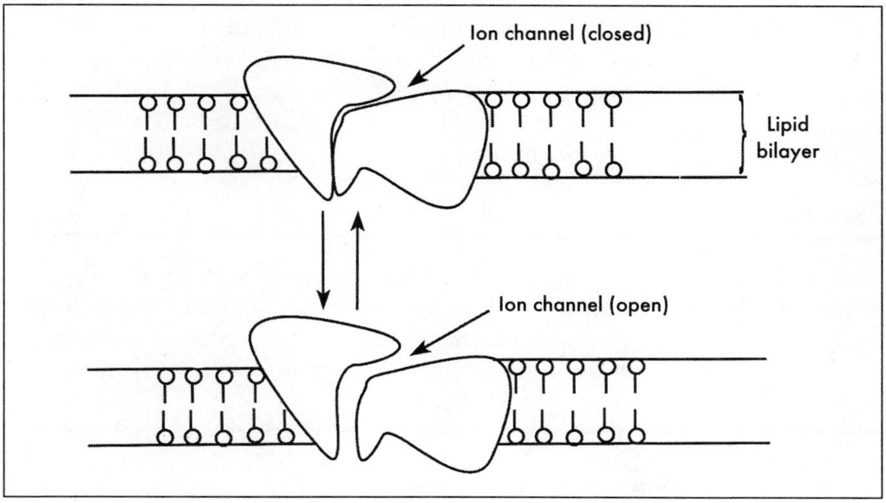

Fig. 2-17 Diagram of a membrane-bound protein that creates a channel for flux of specific ions when the protein is in certain conformations that create an opening in the membrane.

Examples

During the generation of an action potential, the nerve cell membrane becomes permeable to certain ions, because their ion channels open. Because the main diffusible ions in neurons, sodium and potassium, are asymmetrically across the membrane, there will be a net flux if the ion-specific channels are open. In a resting, nonfiring, neuron these channels are generally closed (Figure 2-17).

An "equilibrium potential," for example, is membrane voltage at which the electrostatic and concentration gradients offset each other so that there is no *net* flex for a given ion. The equilibrium potential for sodium, for example, is about +60 mV. This means that during the generation of an action potential, which is created by sodium influx through open sodium channels, the net influx of sodium stops at +60 mV because the level of inside positivity starts to repel sodium influx and because the high concentration of sodium outside the cell has been reduced by the influx. As the voltage field builds up around this region of membrane, the sodium channels close, followed by an opening of potassium channels. Intracellular potassium, which has a higher concentration inside the cell than outside the cell, then is able to diffuse along its concentration gradient until the equilibrium potential for potassium is reached. The efflux of positively charged potassium ions, reduces the relative positivity of the interior, thus "pulling down" the action potential to the normal resting level, which in absolute terms is about 70 mV, inside negative.

Besides sodium and potassium channels, calcium channels are important in the release of neurotransmitter chemicals. In addition, there are dozens of subtypes of channels for the same ion, particularly for potassium and calcium channels. It is not clear why these subtypes exist. Perhaps they indicate undiscovered functions. Or perhaps they simply create a redundancy that could have the advantage of offering some resistance to channel-blocking poisons.

TERMS

Protein Kinase and Protein Kinase C	Enzymes that contain catalytic and regulatory (allosteric) subunits. These become activated when they are bound by specific neurotransmitters, second messengers, hormones, or growth factors. Their general function is to activate intracellular proteins by phosphorylation.

Related Principles

Allosterism
Membrane Receptors
Neurochemical Transmission
Receptor Binding
Second Messengers

References

Hille, B. 1992. Ionic Channels of Excitable Membranes. 2nd ed. Sinauer, Sunderland, Massachusetts.

Levitan, I.B. 1988. Modulation of ion channels in neurons and other cells. Ann. Rev. Neurosci. 11:119-136. Levitan, I. B.

Li, M., West, J.W., Numann, R., Murphy, B.J., Scheuer, T., and Catterall, W.A. 1993. Convergent regulation of sodium channels by protein kinase C and cAMP-dependent protein kinase. Science. 261:1439-1442.

Spitzer, N.C. 1979. Ion channels in development. Ann. Rev. Neurosci. 2:363-397.

Citation Classics

Hirning, L.D., Fox, A.P., McClesky, E.W. et al. 1988. Dominant role of N-type Ca^{2+} channels in evoked release of norepinephrine from sympathetic neurons. Science. 239:57-61.

Plummer, M.R., Logothetis, D.E., and Hess, P. 1989. Elementary properties and pharmacological sensitivities of calcium channels in mammalian peripheral neurons. Neuron. 2:1453-1463.

Second Messengers

Activation (opening) of an ion channel creates a cascade of intracellular biochemical reactions that leads ultimately to phosphorylation of intracellular proteins. This cascade of reactions constitutes a biochemical signal transduction mechanism that converts biochemical signals at the membrane surface to biochemical signals inside the cell.

Explanation

Opening of an ion channel has the immediate effect of allowing the flow of specific ions along their concentration and electrostatic gradients. But if this were all that happened, it would be insufficient to alter the biochemical and genetic machinery inside the cell. Some kind of transduction process must operate to alter the functional proteins in the cell. Moreover, it would be helpful if the transduction process could magnify the effect of opening a few ion channels.

Ion-channel actions are amplified indirectly by a chain of biochemical events, a cascade of biochemical reactions. The cascade begins with an initial formation of "second messenger" molecules, such as calcium, cyclic adenosine monophosphate (cAMP), cyclic guanosine monophosphate (cGMP), or phosphoinositides (PI). These are commonly coupled to a series of postsynaptic biochemical reactions that act in cascade fashion to produce activated, phosphorylated, proteins that cause the ultimate change in cellular activity (Figure 2-18).

Although there are several distinct "second messenger" systems, all seem to share common principles. In particular, high-energy phosphate bonds are formed on proteins, such as enzymes, membrane channel proteins, and nuclear histones, which activates them to catalyze cellular responses. The addition of charged phosphate groups to protein can alter conformation and thereby change the function of an enzyme, a regulatory protein, or an ion channel subunit. Sometimes, phosphorylation is inhibitory. Finally, phosphatases cleave the activated phosphate groups, terminate the second messenger action, and thus allow regeneration of the system.

Just as nature has found it expedient to use a cascade of reactions to couple and transfer energy in the metabolism of glucose, a series of coupled reactions at and near the cytoplasmic side of a neuronal membrane produces the intracellular response to neurotransmitter action at the membrane surface and magnifies it in the process.

Example

Two well-studied transducer systems are those involving cAMP and inositol phosphate.

Many neurotransmitters act on receptors that are coupled to cAMP systems. Some of the receptor types that operate in this mode are the

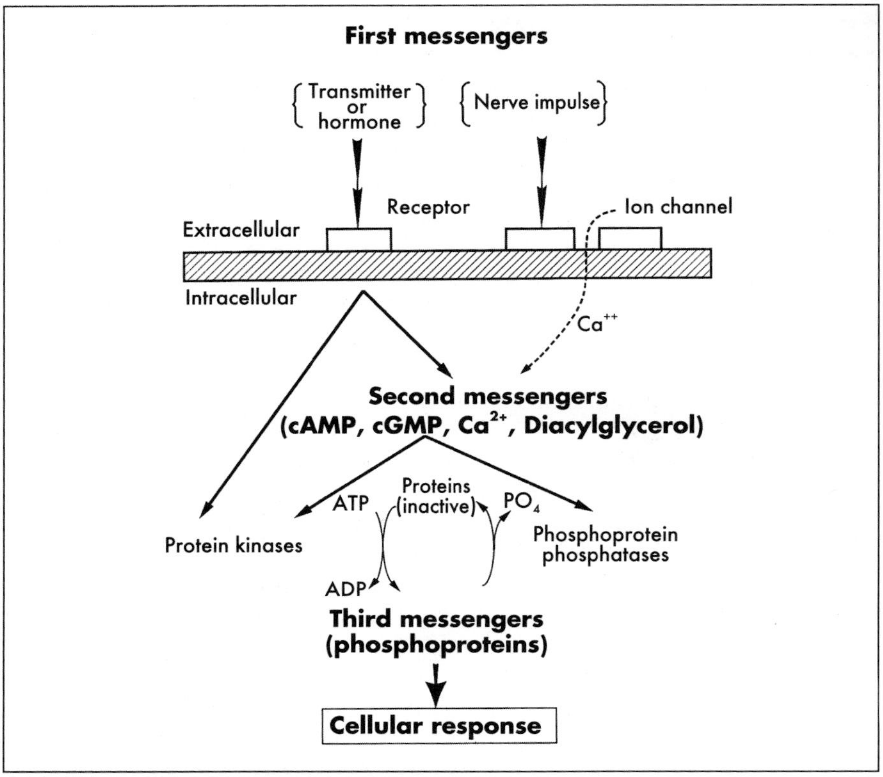

Fig. 2-18 Ion gate actions are amplified indirectly by a chain of biochemical events, a cascade of biochemical reactions that serve to amplify the input. In some synaptic systems, a presynaptic nerve impulse opens voltage-gated ion channels (top right), whereas in others a transmitter-receptor interaction is involved (top left). The cascade begins with an initial formation of "second messenger" molecules, such as cyclic adenosine monophosphate (cAMP), cyclic guanosine monophosphate (cGMP), or phosphoinositides. These are commonly coupled to a series of postsynaptic biochemical reactions that act in cascade fashion to produce activated, phosphorylated, proteins ("third messengers") that cause the ultimate change in cellular activity. The addition of charged phosphate groups to protein can alter conformation and thereby change the function of an enzyme, a regulatory protein, or an ion channel subunit.

$alpha_2$, $beta_1$ and $beta_2$ adrenergic receptors, and receptors for adenosine, opiates, and some muscarinic receptors.

In the case of beta adrenergic receptors, for example, there are three membrane proteins of interest: the adrenergic receptor protein, G protein, and the enzyme, adenylate cyclase. When a neurotransmitter (norepinephrine) binds to its receptor protein, the induced conformational change makes it attract and attach to G protein. This induces a conformational change in G protein, which makes it release one of its protein subunits. The freed subunit then releases a guanine diphosphate (GDP) component in exchange for guanine triphosphosphate (GTP). Now, the GTP-activated protein subunit acts as a catalyst for the enzyme, adenylate cyclase, which is then able to cleave high-energy phosphate bonds of adenosine triphosphate (ATP) to form cyclic adenosine monophosphate (cAMP), which is the second messenger. In the process, the G protein's GTP-linked subunit becomes dephosphorylated, allowing it to become reconstituted with the other G protein subunits, which regenerates the system to allow it to respond again to new neurotransmitter.

The benefits of this complex chain of reactions lie in amplification. A single norepinephrine molecule can induce formation of several hundred cAMP second messengers. Note also that the system operates only as long as the transmitter is bound to the receptor. Thus, the system can be shut down rapidly by the removal of transmitter, either through its diffusion, re-uptake, or enzymatic destruction.

The phosphoinositide (PI) system can be activated in certain neurons by norepinephrine (alpha$_1$ receptors only), serotonin, and some muscarinic receptors. Here, membrane-bound phosphoinositides, which have multiple phosphate groups facing into the cytoplasm, respond to neurotransmitter binding by releasing some of their phosphate to form diacylglycerol (DG) and inositol triphosphate (IP_3). DG can act as a second messenger to activate protein kinase C, a major activator of intracellular proteins. IP_3 releases bound intracellular Ca^{++}, which has its own actions. In some cases, calcium ions can act as second messengers without the participation of protein phosphorylation.

TERMS

Cascade (Biochemical) A chain of linked biochemical reactions in which each depends on successful completion of a preceding reaction.

High-energy Phosphate Bonds Covalent bonds involving phosphate that act as storage sites for large amounts of energy that can be used to drive chemical reactions.

Transduction The conversion of one form of energy or information to another; in this case, the conversion of neurochemical transmitter signals at the postsynaptic membrane surface to intracellular chemical messengers.

Related Principles

Calcium and Transmitter Release
Membrane Receptors
Neurochemical Transmission
Ion Channels
Transduction (Senses)

References

Ghosh, A. and Greenberg, M.E. 1995. Calcium signalling in neurons: molecular mechanisms and cellular consequences. Science 268:239-247.

Morinorbu, S., Kuwayama, N., Kawanami, T., Okuyama, N., Takahashi, M., Totsuka, S., and Endoh, M. 1992. Influence of the acute stress on agonist-stimulated phosphoinositide hydrolysis in the rat cerebral cortex. Prog. Neuro-Psychopharm. Biol. Psychiat. 16:561-570.

Schwartz, J.H. and Greenberg, S.M. 1987. Molecular mechanisms for memory: second-messenger induced modifications of protein kinases in nerve cells. Ann. Rev. Neurosci. 10:459-476.

Siegel, G.J., Agranoff, B.W., Albers, R.W., and Molinoff, P.B. 1989. Basic Neurochemistry. Molecular, Cellular, and Medical Aspects. 4th Ed. Raven Press, New York.

Smith, C.U.M. 1989. Elements of Molecular Neurobiology. 1989.

Tanaka, C. and Nishizuka, Y. 1994. The protein kinase C family for neuronal signaling. Ann. Rev. Neurosci. 17:551-568.

Citation Classics

Berridge M.J. and Irvine R.F. 1984. Inositol trisphosphate, a novel second messenger in cellular signal transduction. Nature 312:315-321.

Berridge M.J. 1984. Inositol trisphosphate and diacylglycerol as second messengers. Biochem. J. 220:345-360.

Clement-Cormier Y.C., Kebabian J.W., Petzold G.L., and Greengard P. 1974. Dopamine-sensitive adenylate cyclase in mammalian brain: a possible site of action of antipsychotic drugs. Proc. Nat. Acad. Sci. US 71:1113-1117.

Hirata F. and Axelrod J. 1980. Phospholipid methylation and biological signal transmission. Science 209:1082-1090.

Kebabian J.W., Petzlod G. L., and Greengard P. 1972. Dopamine-sensitive adenylate cyclase in caudate nucleus of rat brain, and its similarity to the "dopamine receptor." Proc. Nat. Acad. Sci. USA 69:2145-2149 (11/83/LS) [18/84/LS].

Kobilka, B. 1992. Adrenergic receptors as models for G protein-coupled receptors. Ann. Rev. Neurosci. 15:87-114.

Putney, J.W. 1992. Inositol Phosphates and Calcium Signalling. Advances in Second Messenger and Phosphoprotein Research. Vol. 26. Raven Press, New York.

Robison G.A., Butcher R.W., and Sutherland E.W. 1967. Adenyl cyclase as an adrenergic receptor. Ann. NY Acad. Sci. 139:703-723.

Sutherland E.W., Rall T.W. & Menon T. 1962. Adenyl cyclase. I. Distribution, preparation, and properties. J. Biol. Chem. 237:1220-1227.

Sutherland E.W. and Rall T.W. 1960. Relation of adenosine-3', 5'-phosphate and phosphorylase to the actions of catecholamines and other hormones. Pharmacol. Rev. 12:265-299.

Sutherland E.W. and Robison G.A. Metabolic effects of catecholamines. A. 1966. The role of cyclic-3', 5'-AMP in responses to catecholamines and other hormones. Pharmacol. Rev. 18:145-161.

Thompson W.J. and Appleman M.M. 1971. Multiple cyclic nucleotide phosphodiesterase activities from rat brain. Biochemistry-USA 10:311-316. (50/82/LS)

Cell Biology: Study Questions

1. List and explain each of the principles in this category.
2. For each of the principles, provide an example *that is not mentioned in this text.*
3. Why does the author say that neurons are physiologically polarized?
4. Which ions are most in need of pumping across a membrane after an action potential? Why?
5. How does the resting asymmetry of ions between inside and outside of a cell relate to the action potential?
6. Why does sodium influx during an action potential stop at around 60 mV, inside positive?
7. How does potassium channel opening affect the action potential?
8. How does myelination speed action potential conduction?
9. Explain the apparent paradox that ion channels and action potentials are "all or none" phenomena, while synaptic potentials are graded?
10. How do electrotonic potentials interact with voltage-gated ion channels?
11. What are the physiological consequences of demyelination?
12. Summarize the events occurring in a synapse during neurochemical transmission.
13. Explain the difference between temporal and spatial summation and give an example of how both could be operative at the same time.
14. Why do we say that algebraic processing occurs in a postsynaptic cell.
15. Distinguish synaptic modulation from synaptic transmission.
16. Why is energy needed for ion pumps to work?
17. What is the significance of the existence of voltage-sensitive calcium channels?
18. Why does calcium rush into the cell when calcium channels open?
19. Why is neurochemical transmission preferable to electrical transmission across synapses?
20. Why does the nervous system have so many neurotransmitters and neuromodulators?
21. What are some of the ways that certain neurochemicals can act as neuromodulators?
22. Why is a membrane receptor called a transducer molecule?
23. What are the needs for axonal transport?
24. How do we know there is axonal transport?
25. Summarize the mechanism for ending neurotransmission.
26. What is the relationship of membrane receptors to ion channels?
27. What is the relationship of membrane receptors to second messengers?
28. How can certain drugs act via membrane receptors?
29. Why do certain neurotransmitters have a family of receptor subtypes?
30. What features of a molecule determine its binding to receptors?
31. Why are receptors made of protein instead of some other kind of molecule?

32. Why does receptor binding affect neurotransmission?
33. What is the relationship of allosterism to neuromodulation?
34. Why do some receptors, such as those for GABA and NMDA, have multiple receptor sites?
35. Explain how equilibrium potential of an ion regulates the flow of ions through an open ion channel.
36. Why are second messenger systems useful and necessary?
37. What are the relationships among postsynaptic receptor, ion channel, and second messenger?

CHAPTER 3

Senses

And when the senses are strong, the thoughts are precise and their conclusions upright. When, on the other hand, the senses become weak, thoughts become unbalanced and their conclusions confused.

—Avicenna, an 11th century Persian.

Each type of physical stimulus to which an animal can respond is registered in the nervous system as a specific **Sensory Modality and Channel.** That is, special detectors and processing pathways are devoted to specific stimuli, such as sound, light, smell, touch, etc. This organization produces **Sensory Selectivity** in which a given detector cell or anatomical pathway registers and processes only a restricted portion of the stimuli in the external world or internal to the body (Figure 3-1).

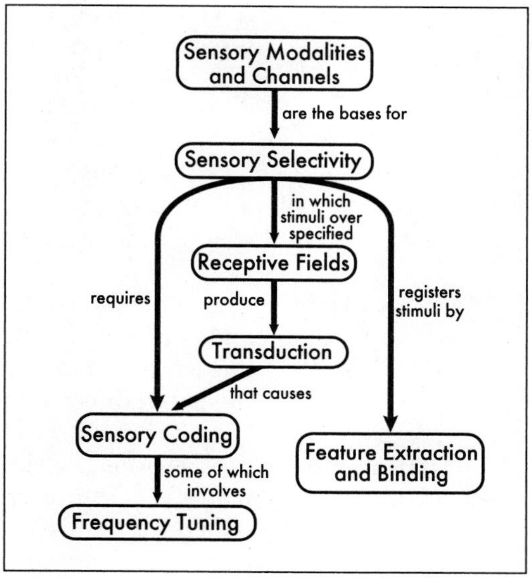

Fig. 3-1 Concept map for the principles involved in sensory function.

Also, a given neuron or place along a sensory pathway can register only a small fraction of the total stimuli in the environment. Thus, the environment is parcelled out by the nervous system into multiple, often overlapping, **Receptive Fields.**

The initial detection of a stimulus requires **Transduction,** which is the conversion from the domain of physics (electromagnetism, pressure waves, etc.) to the domain and language of the nervous system. The language of the nervous system is a schema of **Sensory Coding** wherein the registered stimulus is coded as to its kind and its intensity. For those kinds of stimuli that are rhythmic (light, sound), the coding involves **Frequency Tuning.**

Sensation begins and ends with the processes of **Feature Extraction and Binding.** That is, the total stimulus set is initially extracted and abstracted from the environment into multiple fragments. A visual scene, for example, is deconstructed into millions of lines, edges, and shape abstractions. Then in the final processing and recognition phases, these multiple fragments are bound together in brain circuitry to recreate an abstracted mental representation.

List of Principles

Sensory Modalities and Channels Each type of sensation to which an animal responds is called a modality of sensation (sight, sound, touch, temperature, etc.). For each modality, there are usually specialized sensory cells, or receptors, that are sensitive to that modality and that transduce the stimulus into nerve impulses.
A given sensor system projects only to a part of the processing network; i.e., there is some segregation of function. This segregation can be thought of as separate "channels" that independently code specific features of the stimulus.

Sensory Selectivity Input neurons are specialized, being selectively responsive to a limited part of the physical world. Input neurons have a spatially defined receptive field.

Receptive Field The nervous system monitors its physical world, both within and without, by parcelling it into small regions of space, known as receptive fields. Each sensory system has a range of spatial dimensions that is specific for that sensory system.

Transduction The various physical or chemical stimuli to which sensory neurons respond cause changes in membrane polarization, which in turn trigger one or more electrical "pulses" that propagate as a signal to the spinal cord or brainstem.

	Within the spinal cord and brain, there are successive transductions between chemical and electrical forms of sensory information.
Sensory Coding	Sensory information is coded both quantitatively and quantitatively by the membrane potential responses in sensory neurons. The output of sensory neurons likewise contains quantitative and qualitative codes, in the form of frequency of action potential (impulse) discharge, the intervals between and among impulses, and the pattern of impulse discharge.
Frequency Tuning	The coding of external and internal environment by many sensory receptor cells is frequency specific. That is, those environmental stimuli that fluctuate periodically may be preferentially detected at certain frequencies.
Feature Extraction and Binding	A given neuron in a sensory pathway typically carries only a part of the information in its sensory world. That is, only certain features are extracted and they must be re-integrated in the brain by binding the various features that have been extracted by other sensory neurons into a reconstruction of the original stimulus.

Sensory Modalities and Channels

Each type of sensation to which an animal responds is called a modality of sensation (sight, sound, touch, temperature, etc.). For each modality, there are usually specialized sensory cells, or receptors, that are sensitive to that modality and that transduce the stimulus into nerve impulses.

A given sensor system projects directly only to a part of the processing network; i.e., there is some segregation of function. This segregation can be thought of as separate "channels" that independently code specific features of the stimulus.

Explanation

Sensory selectivity can be thought of as specific "channels" of information flow. A channel can be a specific anatomical pathway that conducts the input of one specific sense modality (pain, touch, temperature, etc.). However, the definition can be extended to include functional aspects that may extend beyond precise anatomical pathways.

Sensory channels have especially well-defined pathways for the early stages of their processing actions. The primary sensory pathway typically contains a chain of three neurons. The first neuron conveys impulses from the periphery into the spinal cord or cranial nerve nuclei. The second neuron projects to cells in lateral and posterior areas of the thalamus; some second neurons terminate on motor neurons in the spinal cord and thus form a basis for simple spinal reflexes. The thalamic third neuron projects to sensory portions of the cerebral cortex. Excitability of neurons in this chain is partially regulated by the sensory cortex, which can either excite or inhibit input from ascending paths.

Because all sensory modalities get converted into nerve impulses, we have to ask: How does the nervous system know what the original sense modality was? A major part of the answer is that each sense modality tends to have its own pathways and circuits in the spinal cord and brain. For instance, if you pinch a dog's toe, the sense modalities of pressure and pain are correctly "interpreted" in the spinal cord, because there is a genetically determined reflex pathway that couples the incoming impulses to spinal motor neurons, which in turn deliver impulses to the correct muscles to produce a withdrawal response. At the same time, there are collateral ascending pathways that allow the sensory impulses to be sent to the brain, so that the brain can be consciously aware of the stimulus – i.e., perceive it.

For a given modality, the ascending pathways have their own topography in the brain. For example, a dog does not "see" odor nor "smell" sights because each kind of stimulus is detected and processed in its own anatomical pathway. Moreover, stimulation of a given receptor, by whatever means, elicits the same sensation; for example, electrical stimulation of the optic nerve elicits sensations of light.

Examples

One well-known example of sensory channels involves hearing, where separate parts of the inner ear mediate the input of specific sound frequencies. The impulses are then routed along defined pathways in the brainstem, thalamus, and temporal part of the cerebral cortex. Likewise, with visual stimuli, separate parts of the visual pathways mediate specific spatial frequencies of light patterns. Such segregation even exists in much simpler sensory systems, such as taste. For example, in rats that are conditioned to avoid the taste of saccharin, only the taste-responsive cells in the brainstem that are sensitive to sweets are affected. Unlike these "sweet" cells, there are central "sodium" cells that respond selectively to the application of sodium applied to the tongue. When a sodium blocker is spread over the tongue, only the central "sodium" and "sweet" cells are affected, while cells responsive to salts, acids, and bitters are unaffected.

Another interesting example of a sensory channel is the frequency-selective response in the visual cortex to counter-phased checkerboard stimuli. That is, if you stimulate the eye with a black and white checkerboard pattern, where the black and white checks reverse positions at specified frequencies, the brain's electrical responses will vary greatly with the frequency of counterphasing.

TERMS

Channel	An anatomical or physiological "pathway" that conducts sensory information of a specific kind.

Related Principles

Conscious Awareness (States of Consciousness)
Frequency Tuning
From Input to Output (Information Processing)
Sensory selectivity
Topographical Mapping (Overview)

References

Campbell, F.W. and Maffei, L. 1974. Contrast and spatial frequency. Sci. Amer. Nov., p. 106-112.

Hasan, Z. and Stuart, D.G. 1988. Animal solutions to problems of movement control:the role of proprioceptors. Ann. Rev. Neurosci. 11:199-223.

Lancet, D. 1986. Vertebrate olfactory reception. Ann. Rev. Neurosci. 9:329-55.

Maffei, L. and Fiorentini, A. 1973. The visual cortex as a spatial frequency analyzer. Vision Res. 13:1255-1267.

Meyer, G.E. and Maguire, W. M. 1977. Spatial frequency and the mediation of short-term visual storage. Science 198:524-525.

Citation Classics

Albe-Fessard D. and Rougeul A. 1958. Activites d'origine somesthesique evoquees sur le corex non-specifique du chat anesthesie au chloralose:role du centre median du thalamus. Electroencephalogr. Clin. Neuro. 10:131-52.

Campbell, F. W., Cooper, G.F., and Enroth-Cugell, C. 1969. The spatial selectivity of the visual cells of the cat. J. Physiol. Lond. 203:223-235.

D'amour, F.E. and Smith, D.L. 1941. A method for determining loss of pain sensation. J. Pharmacol. Exp. Therap. 72:74-79.

Domino E.F., Chodoff P., and Corssen G. 1965. Pharmacologic effects of CI-581, a new dissociative anesthetic, in man. Clin. Pharmacol. Ther. 6:279-91.

Fields, H.L. and Basbaum, A.I. 1978. Brainstem control of spinal pain transmission neurons. Annual Rev. Physiol. 40:217-248.

Garey L.J., Jones E.G., and Powell T.P.S. 1968. Interrelationships of striate and extrastriate cortex with the primary relay sites of the visual pathway. J. Neurol. Neurosurg. Psychiat. 31:135-57.

Janssen P.A.J., Niemegeers C.J.E., Schellekens K.H.L., and Lenaerts F.M. 1967. Is it possible to predict the clinical effects of neuroleptic drugs (major tranquilizers) from animal data? Part IV: an improved experimental design for measuring the inhibitory effects of neuroleptic drugs on amphetamine- or apomorphine-induced "chewing" and "agitation" in rats. Arzneim.-Forsch.-Drug Res. 17:841-45.

Mayer, D.J., Wolfle, T.L., Akil, H. et al. 1971. Analgesia from electrical stimulation in the brainstem of the rat. Science. 174:1351-1354.

Melzack, R. and Wall, P.D. 1965. Pain mechanisms:a new theory. Science. 150:971-979.

Schneider G.E. 1969. Two visual systems. Science 163:895-902.
Tsou, K. and Jang, C.S. 1964. Studies on the site of analgesic action of morphine by intracerebral micro-injection. Sci. Sinica. 13:1099-1109.
Yaksh, T.L. and Rudy, T.A. 1976. Analgesia mediated by a direct spinal action of narcotics. Science. 192:13257-1358.

Sensory Selectivity

Input neurons are specialized, being selectively responsive to a limited part of the physical world. Input neurons have a spatially defined receptive field.

Explanation

Sensory receptors, which are specialized nerve endings or organs, belong to one of two main physiological groups: 1) exteroceptors, that detect stimuli that arise external to the body, and 2) interoceptors, that detect stimuli originating within the body. Exteroceptors include the special senses for vision and hearing, as well as such senses as the vomeronasal sense, olfaction and taste. Exteroceptor functions that are discussed in this chapter are those that affect the outer surface of the body, such as touch, pressure, warmth, and cold.

Interoceptors include sensory units in the viscera, equilibrium sensors in the inner ear, and a special class of receptors devoted to detecting position of limbs and muscle tone, so-called "proprioceptors."

Thus, neurons are selective; some respond only to changes in touch, pressure, or stretch; others to gravity; others to light; others to temperature changes; others to specific chemical senses; others to sound vibrations; and still others to trauma. While many aspects of the physical world are not sensed (e.g., much of the electromagnetic spectrum), that which is detected is achieved with amazing precision and selectivity.

A given input neuron receives input from only certain parts of the external world. The receptive field is related to the topography of the sensory system.

Examples

Certain sensations affecting the skin such as touch, pressure, and temperature can be received from all over the body. But some parts of the body have a much more dense distribution of sensors and thus are more sensitive. Other sensors inside the body can have much more restricted input. For example, the input neurons that monitor blood pressure have their sensors located only in a few blood vessels.

Even in a sensory system such as vision, where the outside world is detected in three dimensions a given primary neuron in the eye only responds to a small fraction of the light signals in the outside world. Efficacy at the system level thus depends on integrating the inputs from many hundreds to thousands of input neurons, each of which monitors only a small fraction of the environment. In the visual areas of cerebral cortex, researchers have shown some of the necessary and sufficient components of an image for activating specific neurons. In the primary visual cortex of monkeys, for example, neurons are specifically sensitive to orientation, size, color, and texture of image components. But in another part of cortex (the inferior temporal area), object selectivity is quite different. Here, if an image of a tiger head, for example, is successively abstracted and the abstracted components presented as stimuli, most of the cells that responded to the complete head will stop responding to extreme simplifications of the image. That is, cells must be presented rather complex images, rather than just simple lines or edges as in other parts of cerebral cortex.

An object feature presented to this inferior temporal area is not recognized by just one cell, but by many cells within a cortical column (see Cortical Columns principle). Cells in a column have overlapping and slightly differing selectivity. This capability enables the brain to register rapidly changing or incomplete object features. To register all components of an image, cells in several columns must respond more or less at the same time (see Feature Extraction & Binding and Rhythmicity & Synchronicity principles).

For any given kind of sensory stimulus, selectivity is also manifest in especially low thresholds for activating a particular kind of stimulus. Thermal receptors are particularly sensitive to temperature changes; chemoreceptors are especially sensitive to chemical changes, etc. Most receptors can respond to stimuli for which they are not specialized, but the thresholds are much higher. For example, the retina of the eye has a very low threshold for responding to light, but it can respond to pressure at a much higher threshold.

TERMS

EEG (Electroencephalogram) The summed voltages that are detected by electrodes that are large enough to detect voltages over a span of tissue that involves many neurons. Typically, this refers to electrical signals detected from electrodes placed on the scalp; these detect the volume-conducted voltages from the underlying cerebral cortex. A more loose definition would include the signals recorded from electrodes implanted into the brain itself, as long as the electrically exposed electrode surface is large enough to detect voltages from many neurons. Extremely small "microelectrodes" detect voltages only from one neuron (when they are close enough to it), and thus that signal would not be called an electroencephalogram. "Unit activity" is the term usually applied to microelectrode recordings.

Receptive Field That part of the environment, either external or internal, to which a given sensor is responsive. (see Receptive Field principle).

Related Principles

Cortical Columns (Information Processing)
Frequency Tuning
Feature Extraction and Binding
Receptive Fields
Rhythmicity and Synchronicity (Information Processing)
Sensory Modalities and Channels
Topographical Mapping (Overview)

References

Barlow, H.B. and Mollon, J.D. (eds.) 1982. The Senses. Cambridge U. Press, New York.

Burgess, P.R., Wei, J.Y., Clark, F.J., and Simon, J. 1982. Signaling of kinesthetic information by peripheral sensory receptors. Ann. Rev. Neurosci. 5:171-87

Hasan, Z. and Stuart, D.G. 1988. Animal solutions to problems of movement control: the role of proprioceptors. Ann. Rev. Neurosci. 11:199-223.

Matthews, P.B.C. 1982. Where does Sherrington's "muscular sense" originate? Muscles, joints, corollary discharges? Ann. Rev. Neurosci. 5:189-218.

Tanaka, K. 1993. Neuronal mechanisms of object recognition. Science. 262:685-688.

Citation Classics

Ciganek, L. 1961. The EEG response (evoked potential) to light stimulus in man. Electroenceph. Clin. Neurophysiol. 13:165-172.

Stevens S.S. 1957. On the psychophysical law. Psychol. Review 64:153-81.

Receptive Field

The nervous system monitors its physical world, both within and without, by parcelling it into small regions of space, known as receptive fields. Each sensory system has a range of spatial dimensions that is specific for that sensory system.

Explanation

The environmental space or body area that can be monitored by sensory systems is called a "receptive field." The size of the field varies with how much innervation it has. Commonly, there is some degree of overlap of receptive fields. Receptive fields can be defined in terms of single neurons: each sensory neuron detects stimuli from a space that is prescribed by the extent of its terminals. Neurons associated with a given receptive field often converge their output onto a fewer number of target cells (Figure 3-2).

In spinal nerves, each dorsal root receives input from a defined region of the body, called a dermatome. Adjacent dermatomes have overlapping

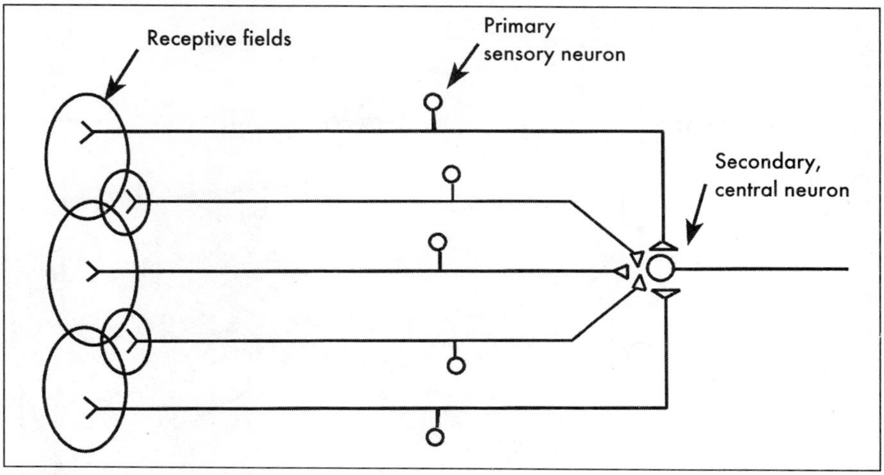

Fig. 3-2 Diagram of the concept of receptive fields, which allows three-dimensional monitoring of the external world or of body parts. The anatomy of the detector arborizations of a neuron dictate the size of its receptive field. Neuronal terminals may overlap certain areas, and thus some overlap of receptive fields often exists. These primary sensory neurons often converge onto a fewer number of second-order neurons located in the spinal cord or brainstem. For these, further projections occur to other neurons in the central nervous system.

receptive fields. Likewise, the sensory parts of cranial nerves have receptive fields from well-defined parts of the head, and one cranial nerve (vagus) has its sensory field defined as the viscera.

The size of a receptive field determines the precision with which stimuli there are abstracted and represented in the nervous system. Large fields are more poorly represented than small ones.

Examples

The retina provides a good example of receptive fields. Experimentally, this can be demonstrated in an immobilized or anesthetized animal in which small beams of light are focussed onto the retina. Recording of action potential discharge in single fibers in the optic nerve reveals where on the retina the receptive field is and also shows that the eye has two kinds of receptive fields that take the form of two concentric zones. In one kind of field, light striking the inner circular zone causes retinal neurons to signal an output; in the outer circular zone, light suppresses output. The other kind of field is just the opposite; activation of the field's periphery triggers an output, while activation of the inner zone suppresses output.

Overlap of sensory information from sensory fields is very conspicuous in the visual system. Even a small dot of light (0.1 mm) covers the receptive fields of many retinal output neurons (ganglion cells), some of which are excited and others inhibited by the small dot of light. Convergence is illustrated by the retina having on the order of 100 million

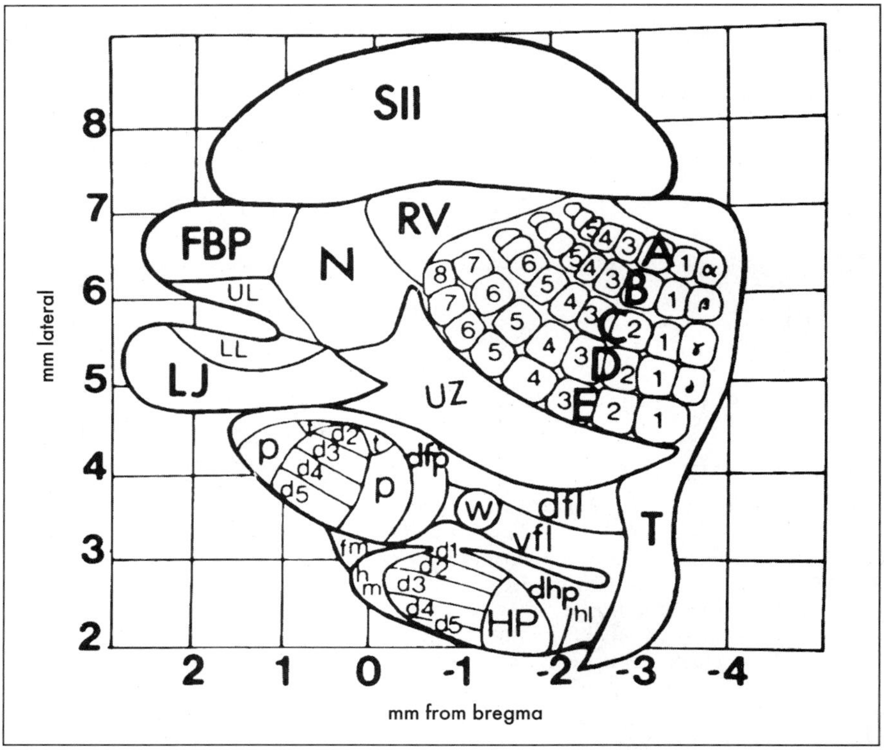

Fig. 3-3 Composite cortical map of the receptive fields for touch, vibration, and electrical stimulation of skin in the anesthetized rat. Abbreviations from caudal to rostral body areas: T = trunk; A-E, 1 to 8 = nystacial vibrissae on snout; hl = hind limb; dfl & vfl = dorsal and ventral forelimb; HP, dl to d5 = digits of hindpaw; UZ = unresponsive zone; RV = rostral small vibrissae; N = nose; P, d2 to d5 = digits of forepaw; LJ = lower jaw. Note how much of the cortex is used to represent the nose and its vibrasse and the paws and how little is used to represent the trunk and limbs. (From Chapin and Lin, 1984).

receptor cells; the optic nerve contains the output of only about 1 million ganglion cells.

The relationship of receptive field size to sensory precision can be illustrated in several ways. In the case of the retina, one part (the fovea) has very small receptive fields and thus permits precise vision, such as the reading of small print. Another example can be found with a dog's ears, which are quite sensitive to pulling and pressure, because there are many nerve fibers there; the receptive fields there are quite small, in that stimulation of a very small point of skin is likely to elicit a response. Conversely, receptive fields over the back are relatively larger, requiring more surface of the back to be stimulated to evoke a response. The nose of a horse or a pig is much more sensitive than most parts of the body because many sensory neurons have terminal arborizations with small receptive fields in the nose (Figure 3-3).

TERMS

Retina	The light-sensitive part of the eye; it has cells (modified neurons) that convert light energy into electrical activity.
Terminal Arborization	The branching of axon terminals; i.e., a given axon terminates in many branches, each of which can make synaptic contact with target cells (neurons, muscles, or glands).

Related Principles

Action Potentials (Cell Biology)
Modularity (Overview)
Sensory Modalities and Channels
Sensory Selectivity
Topographical Mapping (Overview)

References

Allman, J., Miezin, F., and McGuinness, E. 1985. Stimulus specific responses from beyond the classical receptive field: neurophysiological mechanisms for local-global comparisons in visual neurons. Ann. Rev. Neurosci. 8:407-430.

Chapin, J.K., and Lin, C. -S. 1984. Mapping the body representation in the SI cortex of anesthetized and awake rats J. Comp. Neurol. 229:199-213.

Shapley, R., and Lennie, P. 1985. Spatial frequency analysis in the visual system. Ann. Rev. Neurosci. 8:547-583.

Citation Classics

Brown K.T. 1968. The electroretinogram: its components and their origins. Vision Res. 8:633-77.

Gilbert, C.D. 1977. Laminar differences in receptive field properties of cells in cat primary visual cortex. Nature: 268:391-421.

Hartline, H.K. 1940. The receptive fields of optic nerve fibers. Am. J. Physiol. 130:690-699.

Hubel, D.H. and Wiesel, T.N. 1952. Receptive fields, binocular interaction and functional architecture in the cat's visual cortex. J. Physiol. (London) 160:106-54.

Hubel, D.H. and Wiesel, T.N. 1959. Receptive fields of single neurones in the cat's striate cortex. J. Physiol. 148:574-591.

Hubel, D.H. and Wiesel, T.N. 1965. Receptive fields and functional architecture in two nonstriate visual areas (18 and 19) of the cat. J. Neurophysiol. 28:229-89.

Kaneko, A. and Tachibana, M. 1983. Double color-opponent receptive fields of carp bipolar cells. Vision Res. 23:381-388.

Yau, K. -W. 1976. Receptive fields, geometry and conduction block of sensory neurones in the central nervous system of the leech. J. Physiol. 263:513-538.

Transduction

The various physical or chemical stimuli to which sensory neurons respond cause changes in membrane polarization, which in turn trigger one or more electrical "pulses" that propagate as a signal to the spinal cord or brainstem. Within the spinal cord and brain, there are successive transductions between chemical and electrical forms of sensory information.

Explanation

Sensory receptor cells have unstable membrane potentials that can be altered when the appropriate stimulus interacts with certain areas of the membrane. In many cases, the reactive patches of membrane are sodium-channel proteins that mediate depolarization. As with typical neurons, depolarization triggers action potentials that propagate away from the receptor cell. This kind of change occurs irrespective of the physical nature of the stimulus, whether it be sound, light, temperature change, or whatever. Although a given sensory cell is preferentially responsive to one kind of physical stimulus, the transduction process is always the same: a conversion of a physical change into an electrical form.

A succession of transductions occurs as sensory information is routed into the spinal cord and brain and to a motor output. The action potentials in sensory neurons cause the release of chemical "neurotransmitter" in the synapses. If the chemical mediates excitation, the postsynaptic cell will develop graded membrane depolarizations called excitatory postsynaptic potentials (EPSPs). These in turn, if large enough, will mimic the graded receptor potentials of sensory cells and generate action potentials. Other neurochemical transmitters are inhibitory, causing hyperpolarization, known also as an inhibitory postsynaptic potential (IPSP). At the output of a neural circuit, the propagation of action potentials to membrane potential changes in the target cells, either inhibiting them or causing them to transduce the chemical message into a physical action, such as release of secretion for contraction (Figure 3-4).

Examples

The most obvious examples of transduction involve the sensory organs. These respond to physical stimuli in the environment, such as sound, light, heat, etc. This response involves a conversion from the physical energy state that the outside world presents into a representation in the detector neurons. The kinds of physical stimuli can be classified as radiant (sound, light, thermal), mechanical (touch, pressure, stretch, vibration, gravity) and chemical (odors and tastants). The transduction involves a change in the resting membrane potential of the detector neuron. Although varying with type of detector neuron, the typical change is a depolarization, which, if sufficiently great, will trigger the generation of

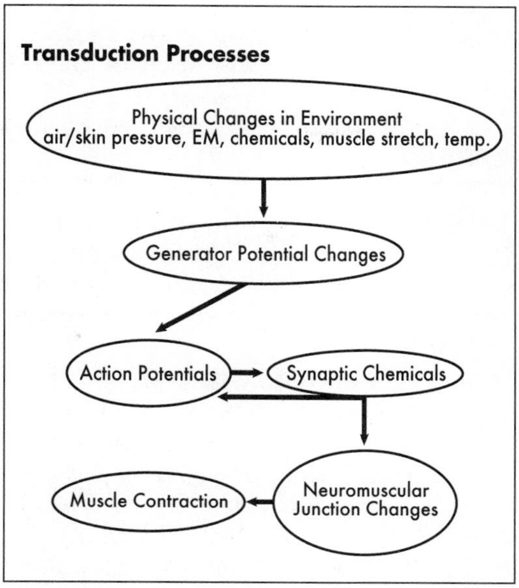

Fig. 3-4 Diagram of the various kinds of transduction processes that occur in the nervous system and at its interfaces with the external world.

action potentials in the fibers of the detector neuron. Detector cells can also be classified according to the environment in which they act: exteroceptors detect changes outside the body (light, sound, etc.), proprioceptors detect changes in body position and muscle and tendon states, and interoceptors detect changes in viscera and blood vessels.

The generation-of-impulse transduction stage is a representation of the magnitude of polarization change in the detector cell. Depending on detector cell type, the pattern of impulse discharge may be characteristic. For example, some pressure-sensitive cells in the skin only discharge impulses when first stimulated, and then they stop discharge. That is why you are more likely to be aware of your clothes when you make a movement. Other detector cells, such as those in muscle, discharge impulses continuously as long as the muscle is stretched. Detector cells release neurochemical transmitters at the terminal ends of their axons. This transduction to chemical form involves many different chemicals, and a given detector cell may have only one or a few of a hundred or so chemical transmitters known to exist in the nervous system. You can even think of the second-messenger chemicals that mediate neurochemical transmission as transducers; one of the G proteins in the light-sensitive pigment, rhodopsin, is actually called "transducin." Ultimately, the neurochemical transmitters may induce yet another transduction process in the target neurons onto which they are released. This transduction process affects whether or not the generation-of-impulse transduction stage ensues, either in processing neurons within the brain or within neurons that send fibers to muscle and glands. The impulses in neuronal fibers going to muscles and glands also trigger a neurochemical transmitter transduction process. But here, the next stage of transduction occurring at muscle or gland is a conversion to the mechanical energy of contraction or secretion.

TERM

Polarization	The voltage difference between inside and outside of a cell. Depolarization refers to decreasing this difference and hyperpolarization refers to increasing it.

Related Principles

Action Potentials (Cell Biology)
From Input to Output (Information Processing)
Information Carriers (Information Processing)
Neurochemical Transmission (Cell Biology)
Nodal Point (Cell Biology)
Reflex Action (Information Processing)
Second Messengers (Cell Biology)
Sensory Selectivity

References

Baylor, D.A., Lamb, T.D., and Uau, K. -W. 1979, The membrane current of single rod outer segments. J. Physiol. 288:589-611.

Brown, H.M., Ottoson, D., and Rydqvist, B. 1978. Crayfish stretch receptor: an investigation with voltage-clamp and ion-sensitive electrodes. J. Physiol. 284:155-179.

Breer, H. Boefkhoff, I., and Tarelius, E. 1990. Rapid kinetics of second messenger formation in olfactory transduction. Nature 345:65-68.

Cory, D.P. and Hudspeth, A.J. 1979. Ionic basis of the receptor potential in a vertebrate hair cell. Nature 281:675-677.

Crawford, A.C. and Fettiplace, R. 1981. An electrical tuning mechanism in turtle cochlear hair cells. J. Physiol. 312:377-412.

Edwards, C. and Ottoson, D. 1958. The site of impulse initiation in a nerve cell of a crustacean stretch receptor. J. Physiol. 143:138-148.

Eyzaguirre, C. and Kuffler, S.W. 1955. Processes of excitation in the dendrites and in the soma of single isolated sensory nerve cells of the lobster and crayfish. J. Gen. Physiol. 39:87-119.

Hunt, C. C., Wilkinson, R.S., and Fukami, Y. 1978. Ionic basis of the receptor potential in primary endings of mammalian muscle spindles. J. Gen. Physiol. 77:683-698.

Kuffler, S.W. 1954. Mechanisms of activation and motor control of stretch receptors in lobster and crayfish. J. Neurophysiol. 17:558-574.

Nakamura, S. and Onodera, K. 1969. Membrane properties of the stretch receptor neurones of crayfish with particular reference to mechanisms of sensory adaptation. J. Physiol. 200:161-185.

Nakamura, S. and Onodera, K. 1969. Adaptation of the generator potential in the crayfish stretch receptors under constant length and constant tension. J. Physiol. 200:187-204.

Nakamura, T. and Gold, G. H. 1987. A cyclic nucleotide-gated conductance in olfactory receptor cilia. Nature 325:442-444.

Trotier, D. and MacLeod, P. 1983. Intracellular recordings from salamander olfactory receptor cells. Brain Res. 268:225-237.
Ye, Q., Heck, G.L., and DeSimone, J.A. 1993. Voltage dependence of the rat chorda tympani response to Na^+ salts: implications for the functional organization of taste receptor cells. J. Neurophysiology. 70:167-178.

Citation Classic

Granit, R. 1955. Receptors and Sensory Perception. Yale University Press, New Haven.
Hirata, F. and Axelrod. J. 1980. Phospholipid methylation and biological signal transmission. Science 209:1082-1090.

Sensory Coding

Sensory information is coded both quantitatively and qualitatively by the membrane potential responses in sensory neurons. The output of sensory neurons likewise contains quantitative and qualitative codes, in the form of frequency of action potential (impulse) discharge, the intervals between and among impulses, and the pattern of impulse discharge.

Explanation

A sense organ generally codes the quantity of information that it receives in terms of the magnitude of the change in its membrane potential (generator potential); this change in potential is in turn associated with a corresponding change in impulse discharge rate. With continued stimulation, most sense organs tend to adapt or habituate, that is, their stimulus-induced change in impulse rate tends to decrease with time.

Certain neurons discharge impulses continually, and qualitative information is carried in some incompletely understood way by changes in the rate of discharge and/or the distribution of patterns of intervals between and among impulses. Other neurons, particularly if they are sensory or motor, may have some rather specific response patterns to an input stimulus: (1) phasic-tonic, in which an initial high rate decreases as accommodation develops to continued stimulation, (2) on, in which the initial response stops abruptly in response to a simultaneously induced inhibitory processes, (3) off, in which the stimulus presence first causes inhibition, followed by a delayed or rebound excitation, and (4) serial on, in which recurrent inhibition can develop oscillatory discharge (Figure 3-5).

The kind or quality of sensation is identified by the specialized sensory receptor, and the coding for that is, as mentioned, maintained throughout large sections of the neural circuitry because the anatomical pathways are segregated. Thus, we can know what kind of information is present often by just knowing where in the nervous system it is. Visual information, for example, is what is present in the retina and in the visual cortex to which it projects.

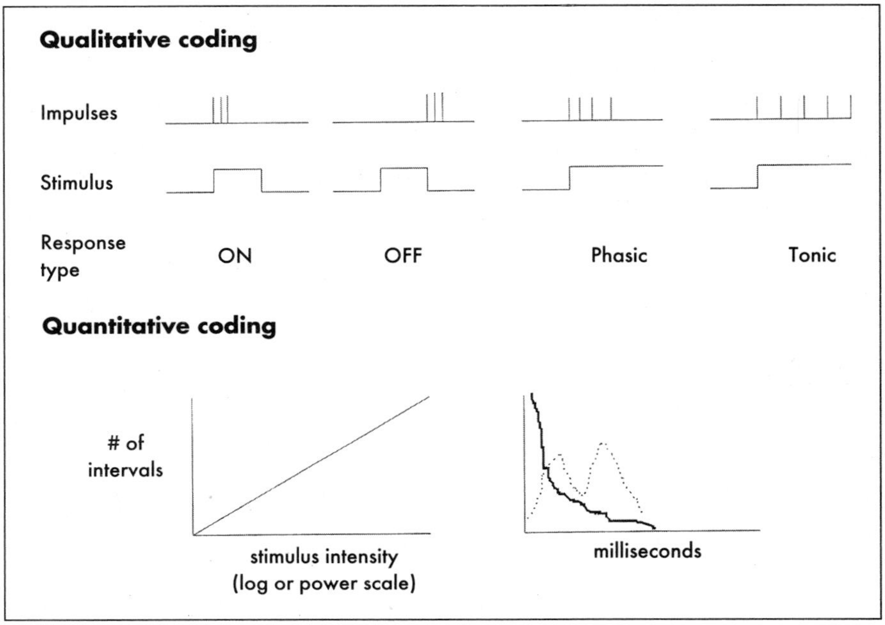

Fig. 3-5 Schematic of sensory qualitative and quantitative coding. ABOVE – A given neuron produces action potentials (impulses) all of the same size. Coding can take place only in the temporal distribution of impulses. Some basic types of qualitative coding by sensory receptor cells are shown at the top. Some receptors fire only when a stimulus comes on; others when stimulus ceases. During continuous stimulation, some cells fire at first and then adapt; others sustain discharge as long as stimulus is on. BELOW – Quantitative coding is most clearly evident in the frequency of firing, which tends to be a linear function of stimulus intensity, with stimulus magnitude expressed either as a logarithmic or power function. The distribution of intervals can also reflect a code. The histogram at lower right can display different types of interval distributions. However, such distributions are calculated without regard to the sequential ordering between and among intervals that often occurs in trains of impulses.

Another coding mechanism appears to exist in the form of the interspike interval patterns among a train of action potentials from a neuron. Such interval coding may be more prominent at higher levels in the nervous system than in the output discharge of a sensory neuron. It is not clear whether interval coding reflects quantitative or qualitative information, or both.

Typically, a sense organ initially codes the quantity of information that it receives in terms of the magnitude of the change in its membrane potential. The membrane potential change tends to be proportional to stimulus intensity. When this change is depolarizing, it promotes impulse discharge, and is therefore often called a generator potential. In many sensory cells, the frequency of impulses generated is proportional to the size of the generator potential. Because the generator potential is proportional to stimulus intensity, this kind of coding is obviously a quantitative indicator of the magnitude of stimulus (Figure 3-6).

In most sensory systems, the more intense the stimulus the higher the rate of impulse discharge along the respective pathways. The dynamic range is so large that it has to be expressed on a logarithmic or power scale. Also, the more intense the stimulus, the more likely it is to activate a large ensemble to receptors.

Examples

Coding for the quality of a stimulus also depends largely on how neurons are organized into networks of the pathways through which nerve impulses are propagated. Some of these paths are "hard-wired," built-in under genetic control during embryonic development. The simplest network is the spinal reflex arc. Information arising from sensory receptors in the skin flows directly back to contract muscles from that area. In a flexion reflex, for example, strong stimulation causes withdrawal of the stimulated limb, and information is usually interpreted as pain at that

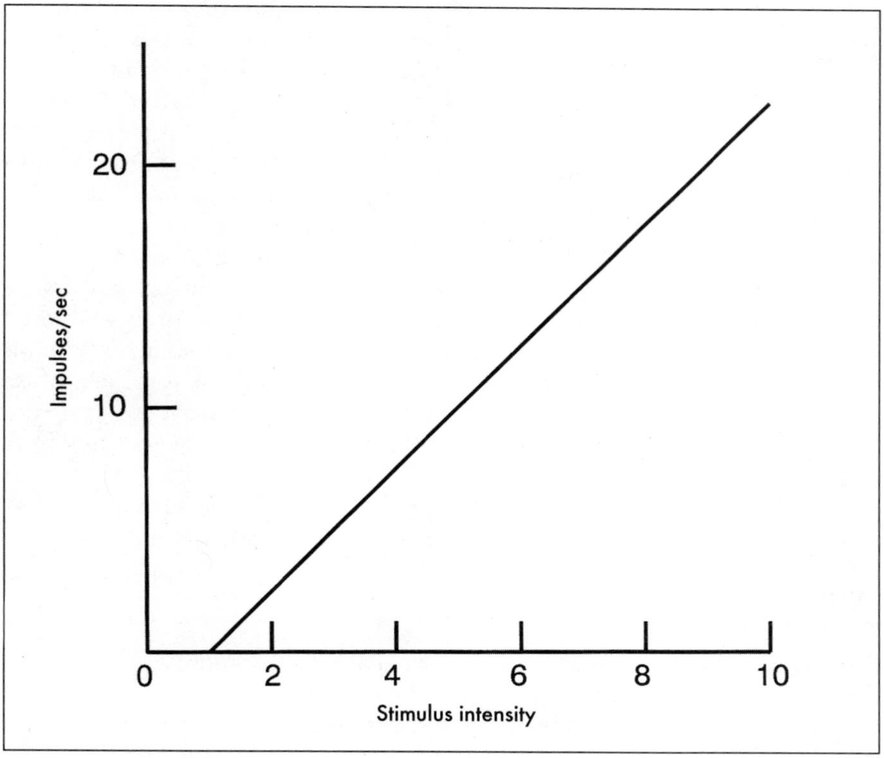

Fig. 3-6 Schematic plot to illustrate the proportional relationship between intensity of a sensory stimulus and the magnitude of receptor neuron response, expressed as number of action potentials ("impulses") per second.

limb because of the hard-wired circuitry. Similarly, animals "see" because axons from the retina propagate impulses to a specific zone of neurons in the cortex at the back of the brain. Sound is not "seen," for example, because auditory receptors do not send impulses to the visual receiving areas of the brain.

Coding is manifest in the graded "generator" potentials that we have mentioned. Nerve impulses, although they occur as brief spikes and thus appear to have the on/off characteristics of digital information, are not a digital code, strictly speaking. Although nerve impulses from a given neuron are typically all of the same size, and can thus be thought of as a digital pulse, some neurons generate clusters of impulses that have smaller impulses toward the end of the burst. Also, the size of the impulse can vary among neurons.

Information is carried in the generation of output pulses (a reversed membrane voltage lasting about 1 msec). The importance of the reversed membrane voltage pulse lies in its being propagated.

In many sensory receptors, the number of output pulses produced is a logarithmic or power function of stimulus magnitude. In people, some sensations are detected over a range of 100,000 to 1. There is, of course, an upper limit (usually less than several hundred per second) on the number of pulses per second that a neuron can generate. This limit derives from the refractory period of a neuronal membrane. While a given neuron is generating an impulse there is a certain time thereafter wherein a new impulse cannot be triggered. This occurs because the impulse is created by a flux of sodium and potassium ions, and these have to re-equilibrate before a new impulse can occur.

A variety of statistical techniques have established that the intervals between and among neurons also are information carriers, especially in central interneurons. The incidence of certain interval patterns can be much higher than chance level, suggesting that some deterministic process is coding information in the pattern of intervals. One line of research suggests that some information is carried in small "bytes" of serially dependent, adjacent interval clusters. Both quantitative and qualitative information can be carried in the discharge pattern of a sensory neuron. Such features include latency to discharge, duration of impulse discharge, and temporal relation of the discharge burst to the stimulus onset or offset.

TERM

Coding	An abstraction of a real-world stimulus that represents the stimulus in another way.

Related Principles

Action Potentials (Cell Biology)
Information Carriers (Information Processing)

Sensory Selectivity
Sensory Modalities and Channels
Transduction

References

Iggo, A. and Klemm, W.R. 1993. Somesthetic sensory mechanisms, p. 787-802. In Dukes' Physiology of Domestic Animals. 11th ed. M. J. Swenson and W. O. Reece, (eds.) Cornell U. Press. Ithaca, New York.

Davis, H. 1961. Some principles of sensory receptor action. Physiol. Rev. 41:391-416.

Dowling, J.E. 1987. The Retina: An Approachable Part of the Brain. Belknap Press, Cambridge.

Kuffler, S.W. 1953. Discharge patterns and functional organization of the mammalian retina. J. Neurophysiol. 16:37-68.

Citation Classics

Adrian, E.D. 1946. The Physical Background of Perception. Clarendon Press, Oxford.

Adrian, E.D. 1949. Sensory Integration. 1st Sherrington lecture. University Press, Liverpool.

Brazier, M.A. b. 1968. The Electrical Activity of the Nervous System. 3rd ed. Williams & Wilkins. Baltimore.

Granit R. 1955. Receptors and Sensory Perception: a Discussion of Aims, Means, and Results of Electrophysiological Research into the Process of Reception. Yale University Press, New Haven, Connecticut. Yale University Press.

Hodgkin, A.L. 1948. Repetitive action in nereve. J. Physiol. (Lond.) 107:165-181.

Katz, B. 1950. Depolarisation of sensory terminals and the initiation of impulses in the muscle spindle. J. Physiol. 111:261-282.

Matthews, B.H.C. 1931. The response of a muscle spindle during active contraction of muscle. J. Physiol. (Lond.). 72:153-174.

Stevens S.S. 1957. On the psychophysical law. Psychol. Review 64:153-81.

Terzuolo, C.A. and Washizu, Y. 1962. Relation between stimulus strength, generator poential, and impulse frequency in stretch receptor of crustacea. J. Neurophysiol. 25:56-66.

Frequency Tuning

The coding of external and internal environment by many sensory receptor cells is frequency specific. That is, environmental stimuli that fluctuate periodically may be preferentially detected at certain frequencies.

Explanation

Some environmental stimuli are periodic. That is, they have various rates, or frequencies. Certain sensory cells have evolved to be especially sensitive to certain portions of the frequency band. They will detect certain frequencies better than others, and some frequencies may not be detected at all. Plotting a curve of response of the sensory cell as a function of stimulus frequency produces a tuning curve that displays the preferred frequency for a given frequency. Advantages of such a sensory scheme include sensitivity. The organism can be especially sensitive to the frequency of a stimulus that has the most biologically adaptive value. In other words, a sensory system does not have to detect everything—just that part of the environment that is biologically most important. Such frequency tuning may not only exist at the peripheral sensory cell level, but may also extend to the higher level function of ensembles of neurons in the cerebral cortex.

Examples

The most obvious example is hearing. We can selectively discriminate different sound frequencies, and there are sound frequencies that we humans cannot hear. Other animals may have different frequency bands that they can hear. Bats, for example, can hear higher frequencies, which they use as a sonar to help navigate in dark caves while chasing flying insect targets.

A similar, but less obvious, case of frequency tuning can be demonstrated in the visual system. Certain cortical pathways ("channels") seem to be selective for certain spatial frequencies. Humans see only a small fraction of the electromagnetic spectrum. However, that fraction is one that adapts us for moving around in our environment and interacting with it effectively.

Certain neurons, such as those organized in columns perpendicular to the surface of auditory and visual neocortex are "tuned" to respond best at certain stimulus frequencies, temporal frequencies of sound stimuli and spatial frequencies of visual stimuli. The receptive fields are "mapped" topographically in the cortex. For certain features, the cortical maps are "computed" in other topographically located cortical regions, and the response projects into another new locus. In bats, for example, cells in one part of the auditory cortex code specifically for such calculated parameters of sound as the target size and distance, and others are mapped for varying velocity.

TERMS

Counterphase Stimulation — Alternation in the stimulus pattern, such as high pitch/low pitch for sound or light/dark for visual stimuli.

Electroencephalogram — Electrical recording of the extracellular voltages that are summated across a large population of actively discharging neurons.

Spatial Frequency — An imprecise but useful definition for visual information is the width of the object. A wide bar stimulus, for example, has low spatial frequency; a narrow line has high spatial frequency. This way of regarding the size of objects is useful because it can be quantified in the frequency domain by the mathematical tools of Fourier analysis.

Related Principles
Feature Extraction and Binding
Receptive Field
Selectivity
Sensory Modalities and Channels

Reference

Shapely, R. and Lennie, P. 1985. Spatial frequency analysis in the visual system. Ann. Rev. Neurosci. 8:547-83.

Feature Extraction and Binding

A given neuron in a sensory pathway typically only carries a part of the information in its sensory world. That is, only certain features are extracted and they must be re-integrated in the brain by binding the various features that have been extracted by other sensory neurons into a reconstruction of the original stimulus.

Explanation

Many stimuli have a variety of features, of which only one or a few can be extracted by a sensory receptor cell and propagated along a sensory pathway into higher centers of the brain. Thus a sensory stimulus is broken down into components so that single neurons can extract and convey certain features of the stimulus. This process is necessary because no one neuron has the "carrying capacity" for all the bits of information associated with most stimuli. Different features of a stimulus activate different populations of neurons. A corollary of this point is that any given neuron may participate at different times and circumstances in carrying more than one kind of information, for it is clear that neurons in the brain are not dedicated to one exclusive function. They are recruited into different ensembles under differing conditions.

The computational problem for the brain thus becomes one of reconstructing an abstraction of the original physical stimulus by binding together all the extracted features being carried by large numbers of neurons. These neurons are functionally linked, but the major unresolved problem is just how they are linked. Two schools of thought prevail currently. One is that at higher levels of the brain, certain neurons are coactivated by different features of the same composite stimulus. Another is that flow of information is routed or gated so that neurons carrying disparate features of the same stimulus converge into common pathways where the information is integrated in common target neuronal ensembles. The two views are not necessarily mutually exclusive.

Examples

Depending on the type of sensory receptor cell, only certain features of the environment are extracted. Examples include touch, pressure, cold, heat, light, sound. There are also internal receptor cells in the body that extract information about the degree that muscles and tendons are stretched.

Many kinds of sensation are chemical. Examples include odors and taste. The bouquet of a fine wine, for example, is a centrally reconstructed representation of the inputs from the various neurons that responded to components of both taste and smell of the wine. There is another chemical sense organ of vertebrate animals, called the vomeronasal organ (VNO), located inside the nasal septum and roof of the mouth. The VNO is sort of a combination taste and smell organ, whose neurons extract features of complex mixtures of chemicals that are in solution. Snakes sample environmental chemicals from other snakes and prey by flicking their tongue into the sense organ. The tip separation of the tongue forks enables them to extract information about spatial gradients in chemical concentrations. How this spatial information or the chemical component information is reconstructed in the brain is not known, except to say that it is apparently not reconstructed in the conscious mind. That is, animals—even mammals—seem not to be consciously aware of the effect of VNO stimuli upon them.

The best studied examples of feature extraction involve the visual system, because there the elements of a visual scene are readily appreciated and amenable to experimental manipulation. For example, one can record from single neurons in the visual cortex while presenting light stimuli of specific geometric pattern and orientation. Bar patterns seem particularly effective because the visual system is very sensitive to edge and contrast features. If a microelectrode is driven into many areas of the visual cortex while the eyes are stimulated with a light-bar stimulus, one finds that a given neuron usually only responds to the bar when it is in a particular orientation in the visual field. Paradoxically, at early stages in the visual pathway, visual information is not extracted and represented as linear segments and boundaries. In the retina and in the relay neurons in the thalamus, the neurons have concentric receptive fields. Also, other areas of cortex respond to complex images.

To reconstruct an image from an ensemble in which each neuronal member is carrying only a specific feature (such as motion, color, brightness, edge orientation, contrast of edges, etc.), neurons cannot obviously change their anatomy with each stimulus. Thus, they must somehow bind their activities together in a dynamic self-organizing process. Recently, it has been shown that certain neurons in the visual cortex, including neurons that may be widely separated, can exhibit coherent firing, especially when a certain visual stimulus has been presented. This is often found in neurons that show either spontaneous or driven oscillations of burst patterns of impulses, with bursts occurring in the general range of 40-60/sec. This has led to the hypothesis that this coherent temporal activity of individual neurons is the mechanism by which the disparate extracted features of a stimulus are bound together to permit reconstruction of the original stimulus.

Visual cortex cells, for example, can be shown to respond to two different components of the same visual scene in much the same way, in terms of their firing rates. However, if the two extracted features are presented at the same time, the cells may display conspicuous synchrony of their firing. That is, they become functionally linked and the assumption is that this is the mechanism by which the two extracted features become linked. By extension, the coherence of many neurons responding to many extracted features of a visual scene serves to reconstruct an abstraction of the original image.

There are some problems with this perspective. Not all neurons in the cortex oscillate, those that do don't always oscillate reliably, and the strength of stimulus features does not necessarily correlate with the degree of oscillation. Likewise, coherent activity is not universal. Depending on the experiment, the number of coherent neurons in the visual cortex during visual stimulation may be quite small. In short, firing oscillation and coherence do not, adequately explain how binding of extracted features can occur.

Another process may contribute to the binding. This is the process of gating, or routing of information flow through certain neuronal ensembles. It may be that different neurons carrying different extracted features of a stimulus have their outputs routed by "control" neurons into selected ensembles where the convergence of inputs into common neuronal pools enables integration and reconstruction of the image. Of course, this may include coherent oscillations of some neurons in those ensembles, and thus the two concepts are not mutually exclusive.

Related Principles

Emergent Properties (Information Processing)
Receptive Fields
Rhythmicity and Synchronicity (Information Processing)
Sensory Selectivity

References

Engel, A.K., Konig, P., Krieter, A.K., and Singer, W. 1991. Interhemispheric synchronization of oscillatory neuronal responses in cat visual cortex. Science 252:1177-1178.

Hubel, D.H. and Wiesel, T.N. 1979. Brain mechanisms of vision. Sci. Amer. 241:150-162.

Löwel, S. and Singer, W. 1992. Selection of intrinsic horizontal connections in the visual cortex by correlated neuronal activity. Science 255:209-212.

Schwenk, K. 1994. Why snakes have forked tongues. Science. 262:1573-1577.

Van Essen, D.C., Anderson, C. H, and Felleman, D.J. 1992. Information processing in the primate visual system: an integrated systems perspective. Science 255:419-423.

Citation Classics

Hubel, D.H. and Wiesel, T.N. 1959. Receptive fields of single neurones in the cat's striate cortex. J. Physiol. (Lond.) 148:574-591.
Hubel, D.H., Wiesel, T.N., and Stryker, M.P. 1978. Anatomical demonstration of orientation columns in macaque monkey. J. Comp. Neurol. 177:361-379.

Senses: Study Questions

1. List and explain each of the principles in this category.
2. For each of the principles, provide an example *that is not mentioned in this text.*
3. What does it mean when we say that the nervous system parcels out the physical world into an assortment of sensory modalities?
4. Why is such parcellation biologically useful?
5. Why do we say that the nervous system has sensory selectivity?
6. What are some of the ways by which the nervous system achieves sensory selectivity?
7. What is the relationship between sensory selectivity and stimulation threshold?
8. What determines the size of a neuron's receptive field?
9. What is the significance of the fact that so many sensory fields overlap?
10. What is the significance of the fact that sensory input converges as it progresses into the nervous system?
11. What is the inevitable result of a sensory transduction process in the nervous system?
12. Summarize the kinds of transduction processes that occur within the spinal cord and brain.
13. How does the nervous system code the quality of a stimulus?
14. How does the nervous system code the quantity or magnitude of a stimulus?
15. Why do you see sights and hear sounds? Why are these and other classes of stimuli not confused?
16. For rhythmic stimuli, what would a sensory tuning curve look like? That is, plot a hypothetical graph of stimulus frequency versus magnitude of response.
17. What would different shapes of such tuning curves say about sensory ability?
18. Since the nervous system breaks down inputs from its external world into fragments and abstracted representations, how does it "know" what the original stimulus was?
19. Why must the nervous system abstract and deconstruct complex stimuli and then reintegrate the components into a representation?
20. Explain the theories for how deconstructed sensory fragments are bound together to reconstruct a stimulus representation.

CHAPTER 4

Information Processing

People's expectations, their scientific preconceptions, influence their judgements. All scientists work from some kind of theoretical framework and interpret evidence in its light. Weak evidence can often be made to fit such a framework, whatever its form.

As always in science, the absence of evidence cannot be taken as evidence of absence.

— Richard Leakey and Roger Lewin

"Information" has multiple meanings, ranging from the colloquial use of the term to the precise equation definitions of Information Theory. Here we use "information" in its colloquial sense, not in the more formal sense of sets of signals in an information channel that involve precise point-to-point, uniquely specific, connections. Certainly, the brain is not a computer nor a Turing machine. As other sections of the book make clear, information in the nervous system takes many forms, ranging from molecular recognition phenomena, to fluctuations in postsynaptic voltage, to patterns of spike trains, to complex dynamic interactions of large ensembles of neuronal populations. What we attempt to focus on in this section are some of the mechanisms by which information processing is accomplished by the nervous system.

It is useful to remember the relationship of information in the nervous system as it gets extracted from the real world. Diagrammatically, we can think of an information pyramid that represents the huge bulk of information in the real-world that gets progressively smaller as information is extracted by a single organism's nervous system (Figure 4-1). The interface between the real-world information and the nervous system are the sensory receptor systems. As we have seen in the chapter on Senses, only certain features of the environment are detected by sensory systems. Some processing occurs in sensory receptor systems themselves, such as the retina of the eye, and this further reduces the amount of information that

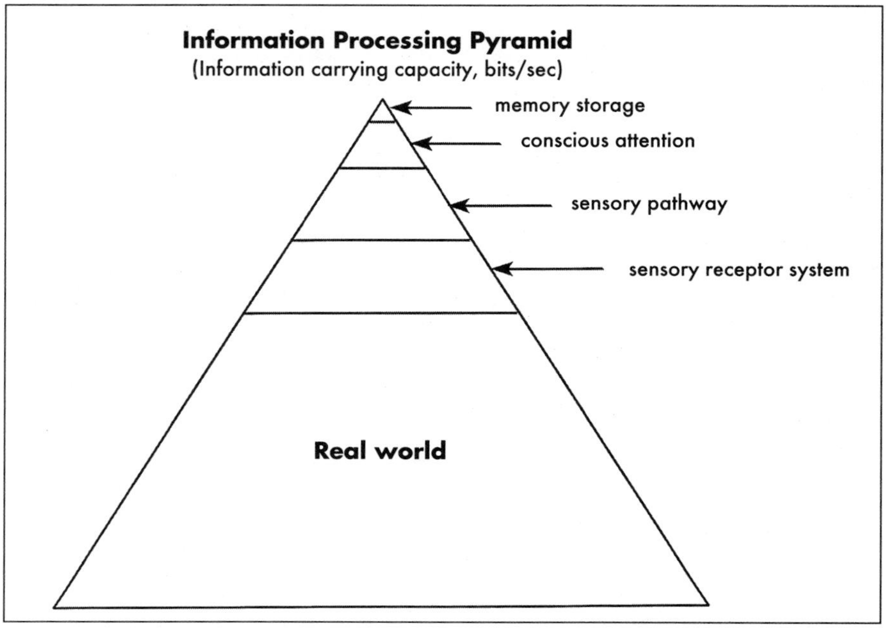

Fig. 4-1 Diagram to represent the relative abundance of "information" present in the real world and its reduction as information is extracted and abstracted at various levels of processing in the nervous system.

will be "considered" by the brain. Much of the information that does get passed on to the brain is processed unconsciously—in the background, so to speak. The human mind is consciously aware of only a small portion of the information that its brain is processing. Finally, the amount of information that gets stored in the form of accessible memories is a still smaller fraction of that which is consciously perceived.

In a biological system, information processing is the process by which an organism operates on information in the environment to produce an appropriate behavior. One common-language idea of information in the context of an organism's behavior takes many forms. For example, physical forces of nature (light, gravity, energy, mass) all have properties that describe the nature of the forces. An organism's task is to function successfully in an environment of such forces, whose properties are continuously changing. Thus, the organism has to process the information about these forces to generate appropriate (i.e., successful) behavior.

As we have seen in the earlier modules of principles, neurons are adapted to receive input from other neurons, and after some modifications within the receiving neuron, relay that information to other neurons. We have seen that the organism has an array of sensors for monitoring the information content in the environment. The focus in this section is on how neurons act on various kinds of information before transmitting it on to other neurons.

This rather simplistic notion is not universally held. One problem for traditional information processing views is that they invite thinking about nervous system function as a simple linear sequence of

stimulus → processing → output

Indeed, linearity is not the fundamental property of nervous system function. More and more, neuroscientists are coming to view nervous processes as nonlinear, deterministic processes that can be described with fractal geometry and chaos theory. Nonetheless, there is no intrinsic requirement that information processing has to be linear.

In the 1930s Peter Anokhin introduced the view that behavior is not simply the response to stimuli but rather a realization of many functional systems. According to this "functional systems theory," there are no sensory or motor mechanisms or special central or peripheral processes. Rather, there are functional systems whose elements interact to produce the behavior. Organism and environment are viewed as a functional unity and thus stimulus-response relations are regarded as inappropriate units of analysis.

A major basis for the Anokhin perspective is that even at the simple level of receptive fields, the behavioral response to activation of a given receptive field is variable. That is, the response to stimulation depends on the behavioral context in which a stimulus occurs. Such variability has been amply demonstrated in many kinds of receptive fields. However, how such observations invalidate the idea of stimulus and response is not clear to me—nor to most neuroscientists. Stimulus-response relations in the nervous system and functional systems do not have to be mutually exclusive. Another basis for the functional systems view is that the nervous system has ample ability to adjust both central and peripheral neurons to the demands of the behavioral context. Thus, all neural activity is said to be active—as opposed to reactive. However, this distinction seems contrived. Organisms are both active and reactive. Reactivity is a fundamental building block of nervous system function, as has been amply demonstrated in many experiments in which portions of the nervous system have been isolated, as in spinal cord transection. Neither active nor reactive mode negates the existence of "information" in the physical world nor the importance to an organism of being able to operate on that information as the dynamical mechanism for generating appropriate and successful behavior.

In many ways, these holistic, systems-level perspectives provide just another way to say that the whole is greater than the sum of its parts. We should have no problem with that. Nonetheless, the whole is not independent of its parts, and understanding of its parts may prove to be prerequisite for understanding the whole. The traditional, reductionistic view of information processing has enormous heuristic value, which should be self-evident from what has been presented herein about cell functions, sensory receptors, and transduction of sensory stimuli, and what is to be presented about how the nervous system operates on the

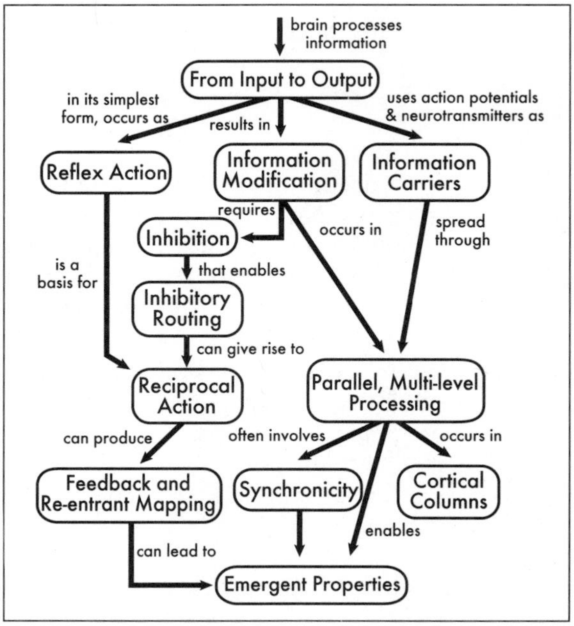

Fig. 4-2 Concept map for the information processing reactions of the nervous system.

"information" it receives in the ongoing genesis of behavior. Even if the idea of information processing is ultimately supplanted by a more holistic theoretical framework, for the moment we can justify our indulgence in the traditional perspective, because that approach has (1) led us to our present understanding of the nervous system, and (2) is a parsimonious and efficient way to teach what we know about how brain processes **From Input to Output** (Figure 4-2).

At the neuronal level, information processing consists of the specific chemical and physical transformations that occur when the nervous system detects, analyzes, and responds to changes in environment. That analysis may involve filtering or averaging of information, so that some is lost. Information may be transformed or erased, especially when contrasted and compared with the stored information that we call memory. Ultimately, information may be added to existing memory or may emerge from the analysis process in the form of output for appropriate operation of glands and muscles.

Processing reactions take place at all levels of the nervous system, from the coding in sensory receptors, to synaptic reactions and associated patterns of impulse generation, to the routing through neuronal networks, to profuse interactions of widespread neuronal subsystems. At each level, there can be substantial **Information Modification.** The most obvious **Information Carriers** are neuronal action potentials and the associated molecular interactions involved with neurotransmitters. The simplest form of processing at a circuit level is **Reflex Action,** where sensory input more or less induces an automated response, often with only a minimal amount of processing. When a reflex circuit is arranged so that the output goes to another reciprocally connected circuit, the basis is laid for **Reciprocal Action,** wherein the two circuits can interact, often in mutually regulating ways.

All processing manifests itself as **Information Modification,** and such modification typically requires the activity of certain strategically placed neurons that can cause **Inhibition.** Inhibitory mechanisms create the opportunity for **Inhibitory Routing.** In reciprocally connected circuits that have some inhibitory output, each reciprocal member can cyclically pace the activity of the other. Inhibitory routing also helps direct information flow selectively through certain circuits and not others, helping to ensure that there is **Parallel, Multi-level Processing.**

When reciprocal action is occurring in large ensembles of linked neuronal circuits, there is a functional basis for **Feedback and Re-entrant Mapping.** Feedback occurs when part of a neuron's output is led back into that same neuron. Re-entrant mapping is a similar idea at the population level: some of the output of one population of neurons projects through mapped pathways into a second ensemble, which after a certain processing delay, sends a re-entrant input into the first ensemble.

Parallel, multiple-level processing is most profound in the cell-dense cerebral cortex, where processing often involves segregation of sensory information into **Cortical Columns.**

Activity in multiple pathways often exhibits **Rhythmicity and Synchronicity** because the widespread circuits are common synchronizing inputs and because the distributed circuits are connected so that they can influence each other. All of the foregoing processing phenomena, and particularly re-entrant mapping and parallel distributed processing, give rise to the many **Emergent Properties** of the nervous system.

List of Principles

From Input to Output

Nervous system processes can be summarized as follows:

Input → Processing → Output

↓

Memory

As input leads to output, processing actions cause information within the neurons to be facilitated or disfacilitated, inhibited or disinhibited. Further, the information can summate, temporally or spatially. Such processing addresses the neuronal changes that were produced by previous experience and thus produces appropriate motor output and behavior.

Information Carriers

In the nervous system, "information" occurs in many neurochemical and microanatomical forms. Sensory inputs are converted from their original analog state to a quasi-digital state of nerve impulses. Information propagates as a pattern of electrical activity. The anatomical pathways (circuits) that are engaged in this pattern are selected and recruited by the nature of the input. Subsequent processing involves a succession of conversions, along pathways that may be quite specific for certain kinds of information. The conversions between action potential pulses and graded postsynaptic potentials are typically mediated by specific neurochemical secretions and receptor interactions. Ultimately, the output of the nervous system occurs once again in analog form, in the form of muscle contractions or glandular secretions.

Reflex Action

Reflex Action at its most basic level, is a relatively simple involuntary and stereotyped response to specific stimulation. Reflex action results from relatively invariant connectivity between the sensory input neurons and the motor output neurons.

Many "higher" nervous system functions are based on the integrated interactions of several or more primitive reflexes, and they are also involuntary and stereotyped. These compounded reflexes may be summed or even occur as a series of linked or chained reflexes, which many people prefer to describe as "reactions" or "responses."

Reciprocal Action

Some neuronal ensembles interact reciprocally, often with mutually exclusive effects. Commonly, this is achieved by action in one ensemble suppressing activity in the reciprocally coupled ensemble. These coupled, mutually regulating neuronal populations often are the basis for rhythmic, periodic functions and behaviors.

Information Modification

Information can be modified in various ways. Some of these mechanisms include active inhibition, filtering, summation, occlusion, and gating.

Inhibition

Inhibition is a major feature of the central nervous system. It serves to organize and control function. At the cell level, the basic mechanisms can be postsynaptic and presynaptic. At the circuit level, inhibition may feed forward, feed back, and may spread laterally. With appropriate coupling to targets, inhibitory phenomena may underlie oscillatory activity in neuronal ensembles.

Inhibitory Routing Inhibition is a primary mechanism by which the flow of action potentials is selectively routed into certain pathways. The inhibition may be of the feed-forward type or feedback type.

Parallel, Multi-level Processing Many of the brain's functions are conducted in widely distributed, parallel pathways, where several-to-many assemblies are simultaneously interacting, Neural function proceeds at many levels from genetic expression to molecular communication among neurons, to consciousness and behavior. Information from one level often affects functions at other levels.

Feedback and Re-entrant Mapping Many neural circuits are organized so that some of the output is fed back to regulate the input. Additionally, large ensembles of certain neurons may send a mapped output into a second ensemble, which after some processing delay sends a re-entrant input into the first ensemble.

Cortical Columns The outer mantle of tissue (cortex) over both the cerebellum and the cerebrum has highly complex organization. The cells are functionally arranged in distinct ways that include a vertical, columnar organization in which the circuitry includes inhibitory neurons and negative feedback that help to regulate population activity within the column.

Rhythmicity and Synchronicity Many neurons discharge impulses in a synchronously coupled way that produces functional rhythms. This coherent activity can magnify or suppress the output, depending on the phase relationships. Additionally, synchronous firing in large populations of neurons, especially if they involve widely separated neurons that have the capacity to participate in diverse functions, can lead to emergent properties.

Emergent Properties Cooperativity among ensembles of neurons creates emergent properties, at least in more advanced brains, levels, stages, and states. The properties of a whole ensemble do not exist in and cannot be simply summed from the properties of the parts. Emergent properties of neuronal systems are fundamental to how these systems work.

References

Alexandrov, Y. and Jarvilehto, T. 1993. Activity versus reactivity in psychology and neurophysiology. Ecological Psychology. 5: 85-103.

Edelman, G.M. 1987. Neural Darwinism: The Theory of Neuronal Group Selection. Basic Books, New York.

Shannon, C.E. 1948. A mathematical theory of communication. Bell System, Technical Journal. 26: 379-623.

From Input to Output

Nervous system processes can be summarized as follows:

Input → Processing → Output
↓
Memory

As input leads to output, processing actions cause information within the neurons to be facilitated or disfacilitated, inhibited or disinhibited. Further, the information can summate, temporally or spatially. Such processing addresses the neuronal changes that were produced by previous experience and thus produces appropriate motor output and behavior.

Explanation

Input from the external environment as well as within the body is brought in via sensory systems, whereupon that information is processed both serially and in parallel. The processing integrates a given sensory input with other concurrent stimuli, with memories of past experiences, and with the existing situational context (which includes the role of emotion).

Changes in physical environment are transduced (converted to another form) by sensory receptors, and this transduction must be coded. The initial coding is achieved by a relatively steady change in the voltage across the membrane of the receptor; the relation between stimulus intensity and magnitude of change in receptor voltage is often logarithmic.

For most sensory receptors the next stage of coding occurs as the steady receptor voltage change triggers the discharge of impulses—brief (about 1 msec) transmembrane voltage changes—that move from the receptor via the attached neuronal process into the central nervous system. Impulses code information in various ways: by latency between stimulus and impulse discharge, by the number of impulses in a burst, by the interval between and among impulses in a burst, or by rate of change in impulse frequency.

The next stage of information transfer, or processing, occurs when the impulses from receptors reach other neurons. The message transfer begins with the release of chemical secretions ("neurotransmitters") that diffuse across the intercellular gap (the synapse) to react with the membrane of the neuron that contacts the receptor. The neurotransmitter then causes a relatively slow change in transmembrane voltage, somewhat analogous to the change that stimuli induce in sensory receptor organs. Finally, this slow voltage change, if of the right polarity, may trigger the neuron to discharge impulses to the next neurons in line or to the effector organs (glands and muscles). The slow voltage change must reach a certain magnitude, or threshold, before impulses can be triggered. This scheme allows a great deal of processing to occur at this level. Because a given neuron receives information from as many as a thousand other neurons, the slow voltage changes contributed by each can interact, adding or subtracting to determine whether or not the threshold for impulse discharge is reached.

Input information may lead almost directly to output instructions; such simple function is usually confined to simple reflexes in the spinal cord, such as the knee jerk, in which the neuron carrying sensory input connects directly to the output (or motor) neuron. Even at this level, however, processing still occurs in the sensory receptor's membrane voltage, in the chemical release in the synapse, in the membrane voltage of the motoneuron, and even at the neuromuscular junction.

More commonly, many nerve cells must process input signals before they generate output signals. For example, the routing of sensory impulses is regulated by inherited functional pathways for such signals and their complex interconnections. Also, the brain compares incoming information with stored information of previous experiences (memory) before processing and generating output signals; often, this process occurs consciously.

Neurons can only excite or inhibit other neurons (or bias the excitability of each other through modulatory transmitters). Given these two basic actions, a chain of neurons can produce only a limited response. A pathway can be facilitated by excitatory input; removal of such input constitutes disfacilitation. A pathway may also be inhibited by inhibitory input. Removal of such input constitutes disinhibition, which, if other sources of excitation are present in the inhibitory input, can augment excitation.

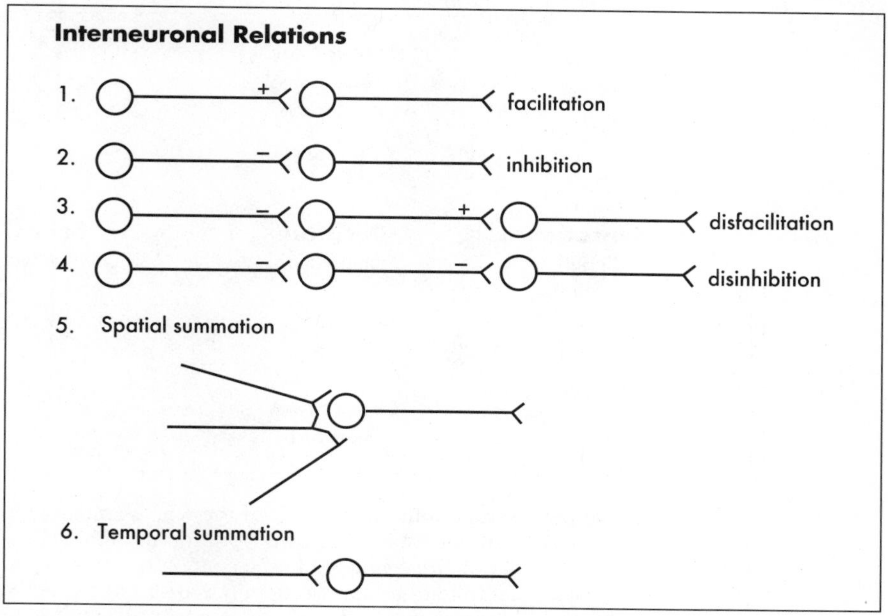

Fig. 4-3 Diagram of the basic ways that neurons can interact with each other in the process of converting input to output.

When a given target neuron receives inputs from multiple sources, those inputs can be spatially summated if the inputs arrive closely enough in time before the influence of each has decayed. If a target neuron receives input from a single axon terminal and that input occurs repeatedly at short intervals, the inputs will summate temporally (Figure 4-3).

Examples

The simple spinal reflex is the simplest example of processing of sensory input into motor output. The role of processing and memory is not very large in such a simple reflex, but processing is a factor. For example, when spinal motor neuron discharges impulses in response to sensory input and activates a motor neuron, collaterals from the axon feedback onto the motor neuron to suppress further output activity. This is a form of negative feedback processing.

In other circuits, the feedback may be positive. If such output is interrupted periodically, the net result is oscillation. In the human brain, the most obvious circuit arrangement of this kind involves linkages between the thalamus and occipital cortex that control the rhythmic electroencephalographic activity known as alpha rhythm.

At higher levels of organization, processing and memory become dominant forces in determining output. At this level, sensory inputs, which may be registered as conscious sensations, may be perceived in the consciousness

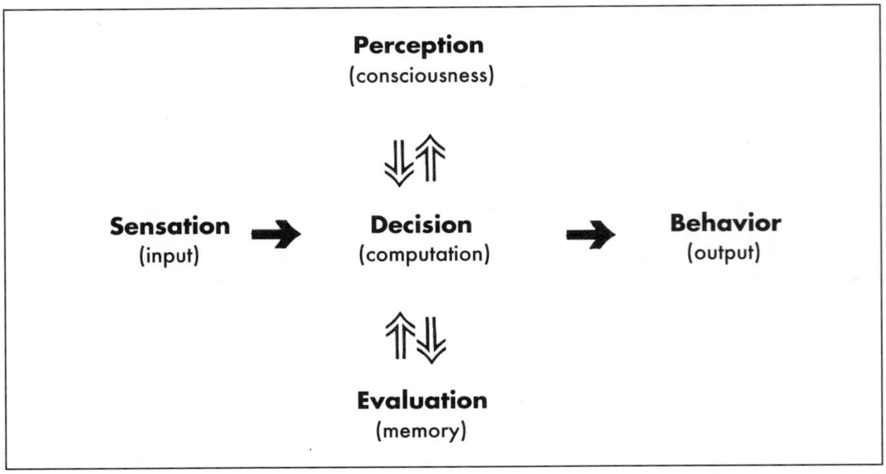

Fig. 4-4 Diagram of the input/output relations at the whole-animal level. Sensory inputs require decisions about the kind of behavioral output that should occur. Perceptions about the sensation are evaluated in the context of prior experience, and the underlying neural computations constitute the decisions about behavioral output. (From Melvin J. Swenson and William O. Reece, editors: Dukes' Physiology of Domestic Animals, 11th ed. Copyright 1993 by Cornell University. Used by permission of the publisher, Cornell University Press.)

and evaluated, either consciously or unconsciously, in the context of past experience. This evaluation, whether conscious or not, is accomplished by complex spatio-temporal synaptic computations that lead to decisions that determine the output, which is behavior (Figure 4-4).

TERMS

Feedback A portion of the output of a circuit that is "fed back" as part of the input to that circuit. Negative feedback tends to suppress the influence of input, while positive feedback magnifies the influence of the input. Commonly, negative feedback helps to achieve normal function, while positive feedback is often a sign of disease or deterioration.

Related Principles

Conscious Awareness (States of Consciousness)
Homeostatic Regulation (Overview)
Feedback and Re-entrant Mapping
Inhibition
Inhibitory Routing
Memory Processes (Learning and Memory)
Nature of Information
Reflex Action
Rhythmicity (Overview)
Synchronicity

References

Georgopoulos, A.P. 1986. On reaching. Ann. Rev. Neurosci. 9:147-70

Hasan, Z., and Stuart, D.G. 1988. Animal solutions to problems of movement control: The role of proprioceptors. Ann. Rev. Neurosci. 11:199-223.

Knudsen. E.I., du Lac, S., and Esterly, S.D. 1987. Computational maps in the brain. Ann. Rev. Neurosci. 10:41-65.

Posner, M.I. and Peterson, S.E. 1990. The attention system of the human brain. Ann. Rev. Neurosci. 13:25-42.

Shapley, R. and Lennie, P. 1985. Spatial frequency analysis in the visual system. Ann. Rev. Neurosci. 8:547-83.

Soechting, J.F. and Flanders, M. 1992. Moving in three-dimensional space: Frames of reference, vectors, and coordinate systems. Ann. Rev. Neurosci. 15:167-91.

Information Modification

Information can be modified in various ways. Some of these mechanisms include active inhibition, filtering, summation, occlusion, and gating.

Explanation:

Processing reactions take place at all levels of the nervous system, from the coding in sensory receptors, to synaptic reactions and associated patterns of impulse generation, to the routing through neuronal networks, to profuse interactions of widespread neuronal subsystems. Modification can occur at all points.

While there are many elegant and mathematical ways of explaining information processing, at the neuronal level it consists of the specific chemical and physical transformations that occur when the nervous system detects, analyzes, and responds to changes in environment. At all stages, information can, and usually is modified. The processing may involve filtering or averaging of information so that some is lost. Information may be transformed or erased, especially when contrasted and compared with the stored information that we call memory. Ultimately, information may be added to existing memory or may emerge from the analysis process in the form of output instructions for appropriate operation of glands and muscles.

Many interesting analogies can be drawn between certain principles of computer and engineering technology and information modification phenomena in the nervous system. Such properties as triggering, gating, switching, and synchronizing occur in the nervous system. Also observable are amplitude discrimination, filtering, amplification, linear-to-logarithmic signal transformation, waveform generation, and frequency modulation. Basic arithmetical operations occur, such as counting, averaging, integration, differentiation, sign inversion, correlation, coincidence detection, delay, phase shift, and interval measurement.

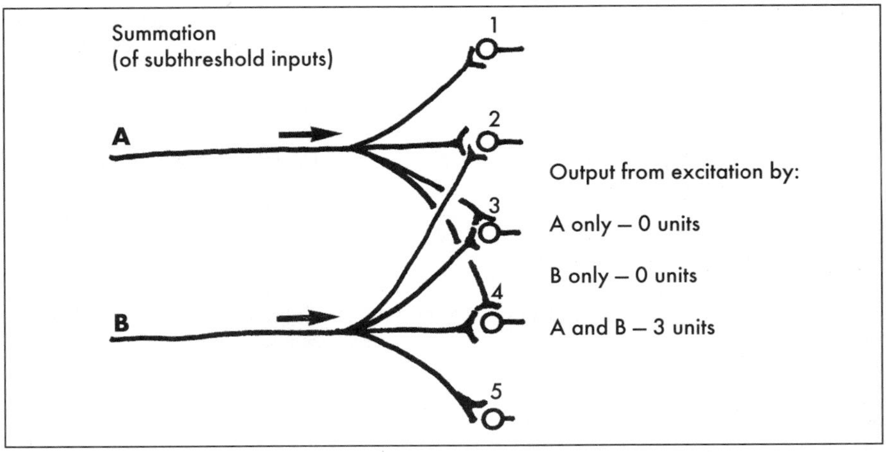

Fig. 4-5 Example of the spatial summation that can occur when a target neuron receives input from more that one source. In the example, assume that a target neuron must receive two excitatory inputs to reach the threshold for becoming activated. When input neurons A and B are both active, they can summate spatially to activate three output neurons, whereas no output is generated when only A or B is active.

Examples:

One of the most obvious ways that information is modified is by summation of inputs. If inputs have overlapping projections to common targets, the effect on one input pathway can be augmented by simultaneous input from another input pathway. This is called spatial summation (Figure 4-5). On the other hand, if inputs with overlapping projections to common targets are supramaximally active at the same time, the output can be LESS than might have been expected if the activity in the two pathways were staggered in time. Rather than summation, there is what is called occlusion (Figure 4-6).

One example of occlusion occurs in the abdominal ganglion of the mollusc, *Aplysia*. More than 90% of the neurons are active during reflex withdrawal of the gill. These same neurons are also active during respiratory pumping and during small spontaneous gill contractions. Selective behavior can occur because the temporal pattern of activity is different for each of the three behaviors. The advantage of shared circuitry is obvious: fewer neurons are needed to produce a range of behaviors. Conversely, given a fixed number of neurons, more diverse processing and behaviors become possible.

An example of information filtering is in a complex visual scene, which can be filtered into its various spatial frequency components, allowing us to see the gross outline of an object (its low spatial frequencies), as well as its fine detail (its high spatial frequencies). Neurons along the visual pathway are selectively responsive to certain spatial frequencies.

An example of gating is where an excitatory input could trigger inhi-

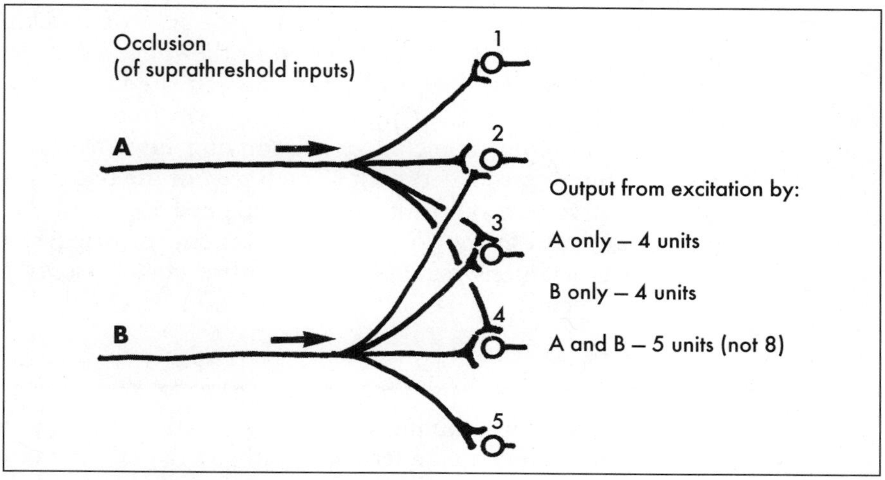

Fig. 4-6 Example of the occlusion that can occur when input pathways are supramaximally active at shared target neurons. In this example, assume that activity at one synapse is sufficient to generate an output. Thus, input A activates four target units, as does input B. However, little is gained when both inputs are active because they share three of the same target neurons.

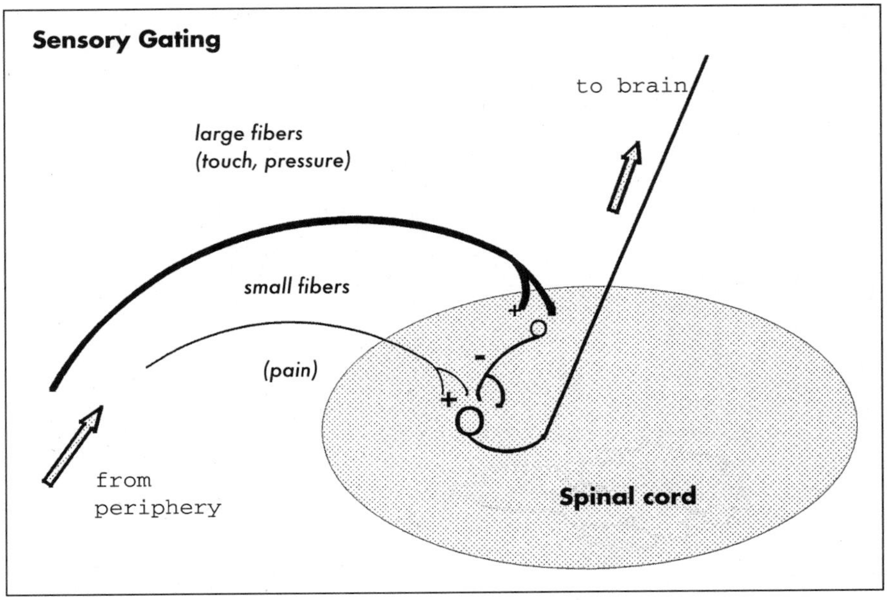

Fig. 4-7 Gating theory for mechanism of pain regulation in the spinal cord. Small diameter sensory nerves mediate pain sensations by exciting certain spinal cord neurons that relay painful inputs to the parts of the brain that perceive pain. If large diameter fibers, which mediate coarse pressure and vibration sensations, are activated, they may shut off the pain pathway by their excitation of inhibitory neurons.

bition along another parallel pathway so that one kind of input is preferentially transmitted (Figure 4-7). An example of blocking is where pain-producing stimuli can be blocked in the spinal cord by descending active inhibitory influences or by reflexly activated inhibitory pathways in the spinal cord (see the Pain Perception principle).

An example of enhancing or magnifying information can be seen when the stimuli associated with a small part of the body is not only sent to its respective portion of the sensory mapped region of the cortex, but parallel pathways in the core of the brainstem can trigger an activating system that heightens activity of the entire cortex (see the Readiness Response principle).

TERMS

Gating A selective routing of information flow in a neuronal circuit through certain alternative paths at the expense of others. Usually determined by strategic location of inhibitory neurons.

Occlusion A functional response that is less than what might be expected. In this case, we refer to inputs that have overlapping projections to shared target neurons. If the inputs are supramaximally active at the same time, the total output can be less than what would be expected if input pathways were sequentially activated.

Summation A functional response that is more than what might be expected. In this case, inputs that have overlapping projections to shared target neurons might not affect all targets if the input were sequential. However, if the input in two or more such inputs is synchronous, there may be summation of synaptic responses so that more of the output target neurons are likely to be affected.

Related Principles
From Input to Output
Information Carriers
Inhibition
Inhibitory Routing
Pain Perception (States of Consciousness)
Transduction (Senses)
Two Basic Actions (Cell Biology)

References

Barlow, H.B. 1953. Summation and inhibition in the frog retina. J. Physiol. 119:69-88.

Carr, C.E. 1993. Processing of temporal information in the brain. Ann. Rev. Neurosci. 16:223-43.

Moshvon, J.A., Thompson, I.D., and Tolhurst, D.J. 1978. Spatial summation in the receptive fields of simple cells in the cat striate cortex. J. Physiol. 283:53-77.

Zeki, S. and Shipp, S. 1988. The functional logic of cortical connections. Nature 335:311-317.

Information Carriers

In the nervous system, "information" occurs in many neuro-chemical and microanatomical forms. Sensory inputs are converted from their original analog state to a quasi-digital state of nerve impulses. Information propagates as a pattern of electrical activity. The anatomical pathways (circuits) that are engaged in this pattern are selected and recruited by the nature of the input. Subsequent processing involves a succession of conversions, along pathways that may be quite specific for certain kinds of information. The conversions between action potential pulses and graded postsynaptic potentials are typically mediated by specific neurochemical secretions and receptor interactions. Ultimately, the output of the nervous system occurs once again in analog form, in the form of muscle contractions or glandular secretions.

Explanation

Input to the nervous system comes in the form of physical phenomena, such as light, sound, temperature, pressure, etc. Many features of these stimuli are analog (i.e., continuously varying). The specialized neurons that detect changes in the environment (including internal environment of the body as well) undergo a change in membrane potential that corresponds to the kind and magnitude of the sensory stimulus. This membrane voltage change, if of sufficient amount and polarity, can trigger the generation of pulsatile nerve action potentials. These impulses are propagated into the nervous system by the cells that generate them.

But then what? A central issue of the neuron doctrine, that each neuron is a distinct functional unit, is the question of how neurons "communicate" with each other. What is the information carrier? There are only two possibilities: electrical or chemical communication. For many years the prevailing view was that of electrical ("electrotonic" or "ephaptic") communication, wherein an impulse in one neuron, for example, acted as an electrical stimulus to trigger the genesis of an action potential in a target neuron. In some tissues, electrical conduction, via diffusing ions, is clearly the mechanism of communication. We know this because the transfer of activity from one cell to an adjacent cell occurs almost instantly, far too quickly to have been accomplished by chemical means.

More commonly, the information carrier inside the nervous system is a chemical neurotransmitter. At the axon terminal of such sensory cells a gap (synapse) exists between the sensory cell and its proximate target neuron. At this point, sensory information is transduced into neurochemical forms, where specific neurochemicals are released into the synapses. The neurochemicals alter the resting membrane voltage of the postsynaptic cell. If the voltage change is of the right polarity and of sufficient magnitude, nerve impulses are triggered and propagated to other target neurons.

Akin to the relationship between potential and kinetic energy, we can think of nervous system information as existing in static and propagated form. The static (storage) form of information is contained in number and kind of synaptic structures, both presynaptic and postsynaptic. The biochemical "machinery" responsible for making, storing, releasing, recognizing, responding to, and terminating neurotransmitters all are part of the information store of the nervous system.

Propagation of information within a neuron takes the form of "explosive" electric pulses, called action potentials. These are in fact short reversals of resting membrane voltage, and the flux of ionic current that creates this voltage reversal is actually an electrical pulse. Such pulses occur in focal regions of membrane, where they act as a stimulus to trigger adjacent membrane regions into the same kind of voltage reversal. Thus, an action potential spreads, point-to-point, along the surface of neuronal membrane.

The post-synaptic response to neurochemical transmitters is electrical. The postsynaptic resting membrane voltage at the point where neurotransmitter acts may become more or less polarized as a consequence. If

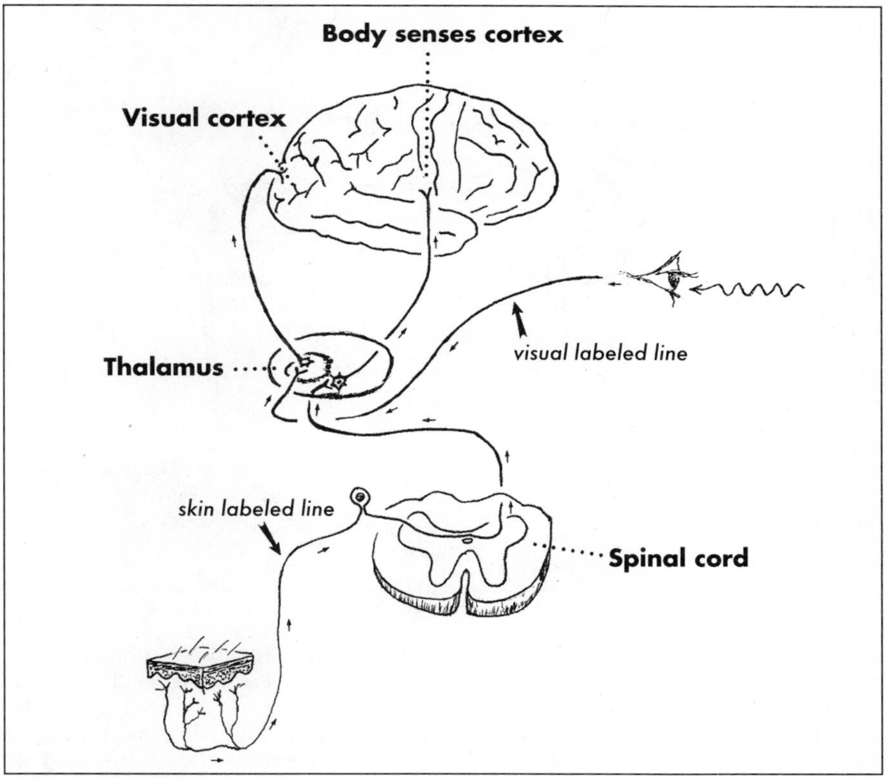

Fig. 4-8 Diagram illustrates, how the circuitry can constitute a carrier for specific kinds of information. For example, certain pathways carry sensory information from the skin, while others carry visual information. Thus, one pathway is a carrier of skin sensation, while the other is a carrier for visual input. In short, anatomy helps to define the nature of the information.

the response is in the depolarization direction (i.e., an excitatory postsynaptic potential, EPSP), an action potential will be generated if the appropriate depolarization threshold is reached. This process is antagonized by inhibitory postsynaptic potentials (IPSPs).

"Information" is carried not only by the sequence of neurochemical and electrical changes along neuronal changes. The circuitry that is accessed by the sensory input is also an information carrier (Figure 4-8). That is, the sensory world is coded in large part by the identity of neurons, as defined by their central pathways and connections. Ensembles of widely distributed neurons that respond to sensory input constitute recognition arrays. This is the basis for the so-called "labeled line" theory. To the brain, neurons have labels, and their pathways are labeled lines that conduct messages equivalent to "something is going on in line so and so" (Figure 4-9).

Functional circuits may be predetermined by the genetic controls that specify the nerves of the body and their connections to structures within the central nervous system. However, within the central nervous system,

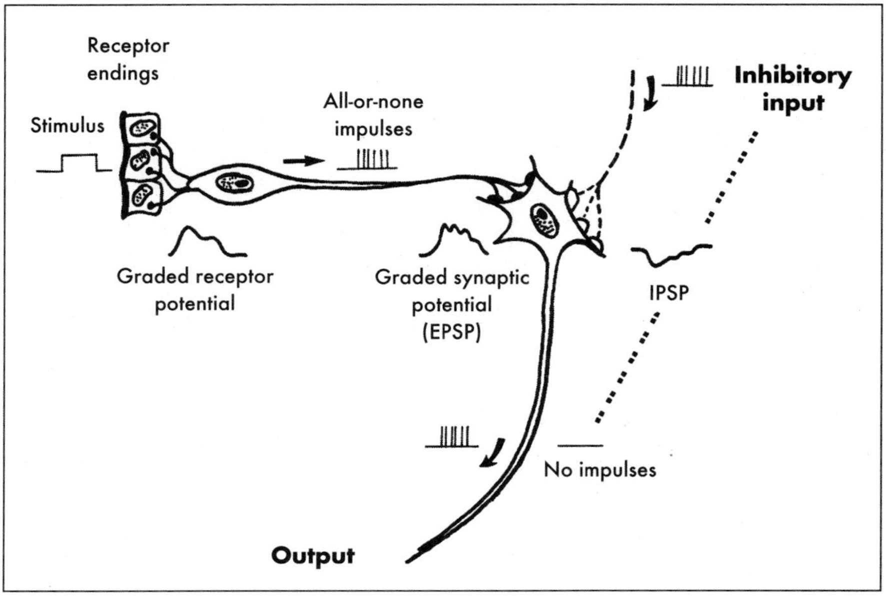

Fig. 4-9 Diagram of some of the transduction processes that occur as sensory information from the physical world is converted into a succession of electrical, chemical, and mechanical changes within the body. Physical stimuli are transduced in sensory cells into graded changes in membrane potential. These, if large enough, generate action potentials that are propagated into the spinal cord or brainstem. At the postsynaptic target neurons, the action potentials are transduced into chemicals that induce graded changes in membrane polarization. If the change is excitatory (EPSP), action potentials may be generated, and these in turn will release chemicals onto other target cells. Ultimately, at gland or muscle cell targets, the action potentials will again be transduced into chemical signals, and these in turn are transduced into mechanical actions, such as secretion or muscle contraction. (From Melvin J. Swenson and William O. Reece, editors: Dukes' Physiology of Domestic Animals, 11th ed. Copyright 1993 by Cornell University.

many circuits are unspecified (not predetermined). Rather, they are selected by sensory experience.

Thus, the newborn nervous system does not necessarily have a pre-set mechanism to "read" a sensory information code so much as it adaptively responds as the nature of the sensory input selects which neural circuits will be engaged to represent and act on the information. Sensation can be said to "sculpt" its own circuitry. Finally, there are also other ways in which information is carried and modified. Some significant amounts of neurotransmitter and especially hormones act at distances that extend beyond a single synaptic cleft with geometry of axonal and dendritic terminal arborization. Another final influence on information is the net electrical field. In some areas of the brain, the summed voltages generated in the process of carrying information are capable of modifying activity within the neurons that generated those extracellular fields.

At some point, this sequence of transduction mechanisms engages neurons that lead to output (i.e., glandular secretions or muscle contraction). Thus, we see that nervous system output involves transductions from the nerve impulses in motor neurons to a release of chemicals to a depolarization of the resting membrane potential of muscle to mechanical contraction and force.

In short, transduction in the nervous system involves multiple conversions along prescribed pathways of information from physical forces in the environment to secretions of glands and contraction of muscle. While all these transformations may seem unduly complex and unnecessary, each stage of transduction affords opportunity for modulation, and that is the proximate basis for the astonishing flexibility and computational power of the nervous system.

Examples

When a cat hears a nearby dog barking, a whole constellation of transduction processes are activated. The sound waves in air are transduced into mechanical movements within the inner ear. These in turn generate changes in resting membrane voltage of specialized cells that respond to sound waves. The resting membrane voltages, which are analog signals, will trigger nerve impulses, which are quasi digital. These nerve impulses propagate along defined pathways into the brainstem and thalamus. At each relay point, where there is a synapse between successive neurons in the input pathway, impulses become converted and coded in the form of neurochemical secretions. These secretions diffuse across the synapse, bind stereospecifically to complementary receptor molecules, trigger a change in resting membrane voltage, and lead to generation of new impulses in the postsynaptic neurons. This successive change from membrane depolarization to nerve impulses to neurochemical secretions continues through many synapses. Ultimately, an output pathway conducts similar transduction processes toward muscle and gland output. For this example, the output nerve impulses in other pathways lead to neurochemical release on gland cells, depolarizing them, promoting release of secretions, which allows a frightened cat to spit and release adrenalin into the bloodstream. Similarly, output nerve impulses depolarize smooth muscle around hair follicles, stimulating muscle contraction and allowing the hair to stand up on the back. Similarly, output nerve impulses in yet other pathways depolarize many skeletal muscle fibers, causing them to contract, which is the basis for a tense postural tone of a frightened cat.

An example of how neuronal activity carries information that influences its own activity can be seen with a single neuron. The propagation of action potentials down an axon occurs because a small zone of depolarization creates an electrical field that is strong enough to act as an electrical stimulus to adjacent parts of the cell membrane. Thus, action potentials move down the axon in a way that is analogous to a burning fuse. At a population level, the extracellular currents generated by large ensembles of neurons may create an extracellular electrical field that is large enough to bias the firing sensitivity of the cells in those ensembles. Experiments in the hippocampus of animals has shown that this part of the brain generates large oscillating fields in the range of 4-10/sec. Whether or not these cells respond to inputs varies with the phase of their oscillating extracellular voltage field. A similar phenomenon may also occur with the so-called alpha rhythm in the human electroencephalogram, but this has not been verified conclusively. Alpha rhythm correlates inversely with stimulation. Alpha rhythm is most prominent when the eyes are shut and visual stimulation is thus reduced.

TERMS

Analog — A continuously varying parameter, as opposed to a digital para-meter. Examples include temperature, light intensity, sound intensity, and frequency.

Digital — A parameter that has a finite value that can be characterized as on-off or yes-no or all-or-none. Nerve impulses are not exactly like digital values in a digital computer, but they are all-or-none, and they normally do not vary in size within a given neuron.

Transduction — The conversion of one energy form to another: such as light waves to nerve impulses, sound waves to analog changes in membrane voltages, the molecular signature of odorants to nerve impulses, etc.

Related Principles

Action Potentials (Cell Biology)
Electrotonus (Cell Biology)
Epigenetics (Development)
Information Modification (Information Processing)
Neurochemical transmission (Cell Biology)
Plasticity (Development)
Second Messengers (Cell Biology)
Transduction (Senses)

References

Black, I.R. 1991. Information in the Brain: A Molecular Perspective. MIT Press. Cambridge, Massachusetts.
Carr, C.E. 1993. Processing of temporal information in the brain. Ann. Rev. Neurosci. 16:223-43.
McGeer, P.L., Eccles, J., and McGeer, (eds.) G. 1987. Molecular Neurobiology of the Mammalian Brain. 2nd ed. Plenum Press, New York.

Kandel, E.R., Schwartz, J.H., and Jessell, T.M. 1991. Principles of Neural Science. 3rd ed. Elsevier, New York.
Shapley, R. and Lennie, P. 1985. Spatial frequency analysis in the visual system. Ann. Rev. Neurosci. 8:547-83.

Citation Classics

Adrian, E.D. 1914. The all-or-none principle in nerve. J. Physiol. 47:460-474.
Brazier, M.A.B. 1968. The Electrical Activity of the Nervous System. Williams & Wilkins, Baltimore.
Erlanger, J. and Gasser, H.S. 1937. Electrical Signs of Nervous Activity. U. Pennsylvania Press, Philadelphia.
Fulton, J.F. 1949. Physiology of the Nervous System. 3rd ed. Oxford Univ. Press, London.
Hodgkin, A.L. 1964. The Conduction of the Nervous Impulse. Thomas, Springfield.
Lorente de Nó, R. 1947. A Study of Nerve Physiology. Rockefeller Institute, New York.
Ruch, T.C. and Patton, H.D. 1965. Physiology and Biophysics, 19th ed. of Fulton and Howell's Textbook of Physiology. Saunders, Philadelphia.
Sherrington, C.S. 1947. The Integrative Action of the Nervous System, 2nd ed. Cambridge University Press, Cambridge.

Reflex Action

Reflex Action, at its most basic level, is a relatively simple involuntary and stereotyped response to specific stimulation. A reflex action results from relatively invariant connectivity between the sensory input neurons and the motor output neurons.

Many "higher" nervous system functions are based on the integrated interaction of several or more basic reflexes, and they are also involuntary and stereotyped. These compounded reflexes may be summed or even occur as a series of linked or chained reflexes, which many people prefer to describe as "reactions" or "responses."

Explanation

Nerve cells can have only two main actions on other nerve cells: excitatory or inhibitory. Because neurons are organized into networks, excitatory and inhibitory effects can be orchestrated into complex patterns of information flow. Spinal reflexes have the simplest organization. All reflexes are more or less automated networks of neurons that link sensory receptor cells to central neurons in the brain or spinal cord to motor nerves. Most reflexes are genetically determined by the anatomical organization of neurons that mediate the reflex (Figure 4-10).

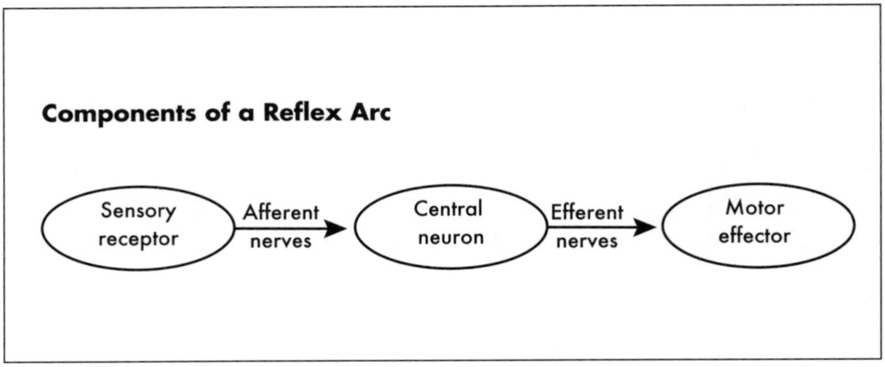

Fig. 4-10 Diagram of the basic components of a reflex system.

The neural circuitry for reflexes is commonly referred to as a "reflex arc," primarily because the reflex circuits in the spinal cord have an arc-like anatomy; note, however, that there are also cranial reflexes. Five neural elements commonly make up a reflex arc: (1) a receptor organ, (2) an afferent neuron, (3) interneurons in the cord or brain, (4) efferent neurons, and (5) an effector organ, which may be a skeletal muscle, smooth muscle, or gland. In some of the simpler (monosynaptic) reflexes, an interneuron may not be involved.

Reflexes are, of course, subject to modification from neural systems outside the immediate circuitry that mediates the reflex. Reflex action may be enhanced or suppressed from outside circuits that have shared connections.

Examples

Several specific reflexes are routinely used in clinical neurological examinations. Names given to reflexes often indicate their basic function. For example, in the withdrawal reflex in which a dog withdraws its foot in response to toe pinch, each activated sensory neuron sends input into the spinal cord that automatically relays excitation back into nerves that excite flexor muscles of that leg. Some reflexes may also have inhibitory components. In the flexion reflex, for example, sensory input excites some neurons in the cord that inhibit the antagonists (extensors) of the flexor muscles of that limb. This organization of neurons is called reciprocal innervation, and occurs in many kinds of reflexes (Figure 4-11).

There are primitive locomotor reflexes, often called central pattern generators, that generate crude automatic and stereotypical movement patterns. Such reflex systems occur in the brainstem and spinal cord. Intentional motor acts are generated elsewhere in the brain, and intentional movement control is often mediated by way of these primitive locomotor reflexes.

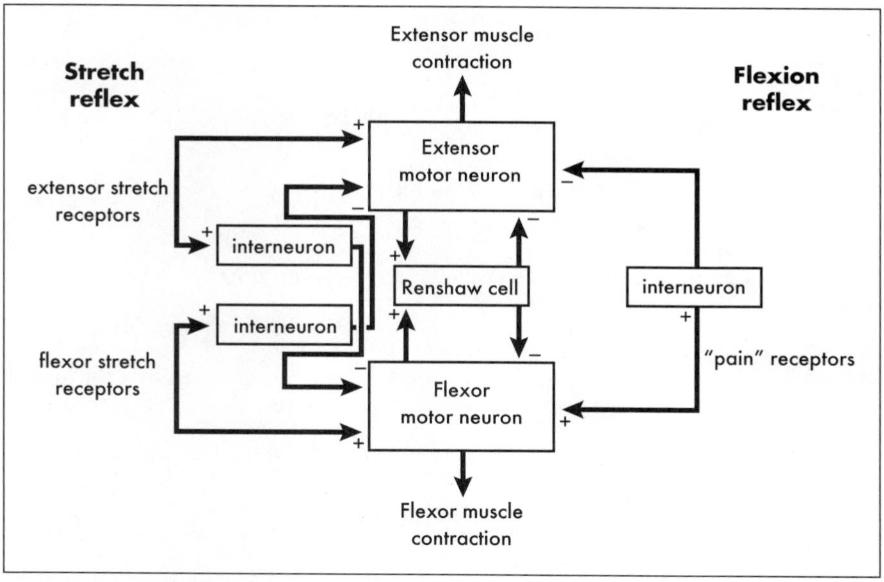

Fig. 4-11 Example of mutually regulating reflexes in the spinal cord. When stretch receptors in an extensor muscle are excited by stretch, the appropriate extensor motor neuron causes contraction of that extensor muscle. Then, via inhibitory interneurons, the antagonistic flexor muscle is inhibited so that a limb, for example, achieves extension. Also, concurrently, an interneuron pool of Renshaw cells is excited, which in turn inhibits and shuts off the activity of the motor neuron, so that extension is not excessive. Comparable phenomena occur when stretch activates stretch receptors in flexor muscles. In flexion reflex phenomena, painful stimuli cause a withdrawal of the affected body region. In limbs, for example, flexors are directly excited to contract, while at the same time, via an inhibitory interneuronal pool, the extensors are inhibited. (From Melvin J. Swenson and William O. Reece, editors: Dukes' Physiology of Domestic Animals, 11th ed. Copyright 1993 by Cornell University. Used by permission of the publisher, Cornell University Press.)

TERMS

Afferent
A term for direction of impulse flow. With reference to a given neuron, inputs are considered afferent.

Efferent
A term for direction of impulse flow. Outputs of a given neuron are considered as efferents.

Central Pattern Generator
A pool of neurons that generates a stereotyped pattern of muscle contractions and movement.

Reciprocal Inhibition
Mutual inhibitory action between two neurons or two pools of neurons.

Related Principles

Fixed Action Patterns (Motor Activity and Control)
From Input to Output
Homeostatic Regulation (Overview)
Reciprocal Action
Sensory Modalities and Channels (Senses)

References

Eaton, R.C. (ed.) 1984. Neural Mechanisms of Startle Behavior. Plenum, New York.

Gewecke, M. and Wendler, G. (eds.) 1985. Insect Locomotion. Parey, Berlin.

Grillner, S. and Wallen, P. 1985. Central pattern generators for locomotion, with special reference to vertebrates. Ann. Rev. Neurosci. 8:233-261.

Hasan, Z. and Stuart, D.G. 1988. Animal solutions to problems of movement control: the role of proprioceptors. Ann. Rev. Neurosci. 11:199-223.

Lundberg, A. 1979. Multisensory control of spinal reflex pathways. Prog. Brain Res. 50:11-28.

Pearson, K.G. 1993. Common principles of motor control in vertebrates and invertebrates. Ann. Rev. Neurosci. 16:265-97.

Citation Classics

Katz, B. 1949. The efferent regulation of the muscle spindle in the frog. J. Exp. Biol. 26:201-217.

Matthews, B.C.H. 1931. The response of a muscle spindle during active contraction of a muscle. J. Physiol. 153-174.

Renshaw, B. 1940. Activity in the simplest spinal reflex pathways. J. Neurophysiol. 3:373-387.

Sherrington, C.S. 1898. Decerebrate rigidity and reflex coordination of movements. J. Physiol. 22:319-332.

Reciprocal Action

Some neuronal ensembles interact reciprocally, often with mutually exclusive effects. Commonly, this is achieved by action in one ensemble suppressing activity in the reciprocally coupled ensemble. These coupled, mutually regulating neuronal populations often are the basis for rhythmic, periodic functions and behaviors.

Explanation

Some neural ensembles antagonize each other. Reciprocal action in the nervous system occurs when one neuronal population that is active has part of its output directed to suppress activity in a connected population that in turn can suppress activity in the first population.

Many neurons receive feedback from the very neurons to which they project. Because a certain delay is imposed through the synapses that link

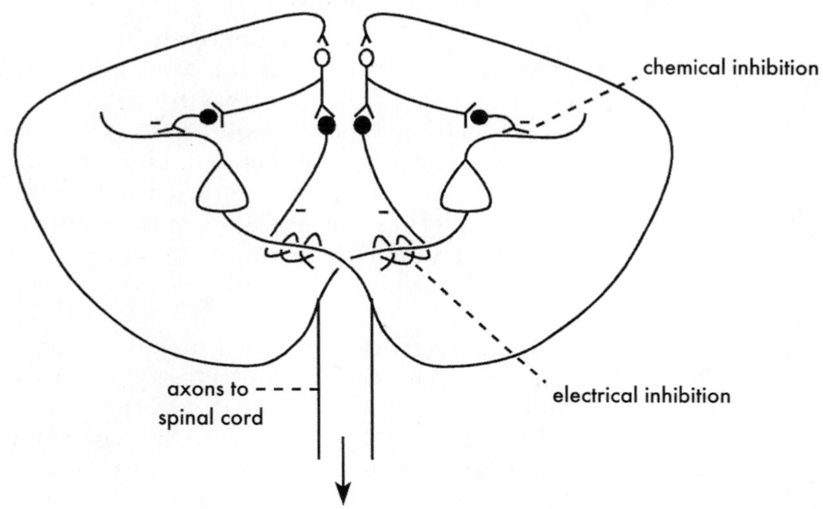

Fig. 4-12 Schematic diagram of reciprocal inhibition, as occurs in the motor control system for movements of the tail of the fish. Inhibitory neurons (indicated as black circles) are seen to presynaptically inhibit motor neurons of the contralateral side. When one Mauthner motor neuron activates tail muscles that supply the opposite side, collaterals from that neuron activate other brainstem neurons that temporarily inhibit the Mauthner cell controlling muscle on the other side of the tail. In total, the reciprocal actions create the automatic movement of tail fin from side to side.

reciprocally coupled ensembles, the output from such systems tends to be rhythmic. As one area consumes a certain amount of time in activating another, its target population takes another finite amount of time in inhibiting the population that excited it. Then another finite amount of time occurs during recovery from the inhibition. The net consequence can be a very clear rhythm.

This property also tends to produce a check on runaway function, so that a given physiological or behavioral function does not become excessively dominant (see also the Homeostasis principle).

Examples

The easiest way to identify neuronal ensembles that function reciprocally is to look for body functions and behaviors that are mutually exclusive or rhythmic. One of the simplest rhythmic behaviors that has a simple reciprocal mechanism is the tail beating motion of swimming fish. For forward propulsion to occur, the tail must swing first to one side and then the right. This is accomplished from two "command neurons," called Mauthner cells, in the brainstem that alternate their firing patterns that activate the muscles on the contralateral side of the tail. The controlling circuitry is based on the fact that collateral axons from one Mauthner cell cross over to excite inhibitory neurons that affect the Mauthner cell that controls the muscles on the other side of the tail. Thus, when one Mauthner cell fires, it makes contralateral tail muscles contract, while ipsilateral tail muscles do not contract because their controlling Mauthner cell is inhibited. One special feature of this circuitry is that there are two kinds of inhibitory neurons, one acting synaptically on Mauthner cell dendrites and the other kind acting through electrical fields around the axon (Figure 4-12).

For higher levels of functioning, this organizing principle provides checks and balances that regulate functions within reasonable limits. Some of the functions and behaviors that are regulated in this way are respiration, blood pressure, heart rate, various visceral secretions and movements, temperature control, eating, drinking, sleep, and approach and avoidance drives. Paired control centers are often reciprocally inhibitory, but in some cases, such as sleep and approach-avoidance centers, the mutual regulation may be more complex.

Obviously, peripheral stimuli associated with taste, smell, stomach contractions, metabolic rate, and body temperature are important in regulating appetite and thirst. In the brain, a group of neurons in the lateral hypothalamus seems to promote the drive for eating, while a group of neurons in the ventromedial hypothalamus promotes satiety; both areas are under many influences from other brain areas. Both noradrenergic and cholinergic transmitter systems help to regulate appetite and thirst (Figure 4-13).

Other good examples of reciprocal control include breathing and heart rate. More complex examples of mutual regulation include control of body temperature, appetite, and thirst.

Each of these functions is controlled, at least to some extent, by reciprocally acting populations of neurons. In the case of respiration and heart rate, the neuronal populations are located in the brainstem (medulla oblongata). Body temperature, appetite, and thirst are controlled by reciprocally acting populations of neurons in the hypothalamus.

Rhythmic pattern generation in some neural circuits is a property of the circuitry, not of individual pacemaker neurons. Other pattern generating circuitry is driven by pacemaker neurons. What paces the pacemakers? Often it is reciprocal interaction, wherein a neuron is captured into pacemaker mode by positive feedback from its target neurons: A fires B, B fires A, A fires B, etc.

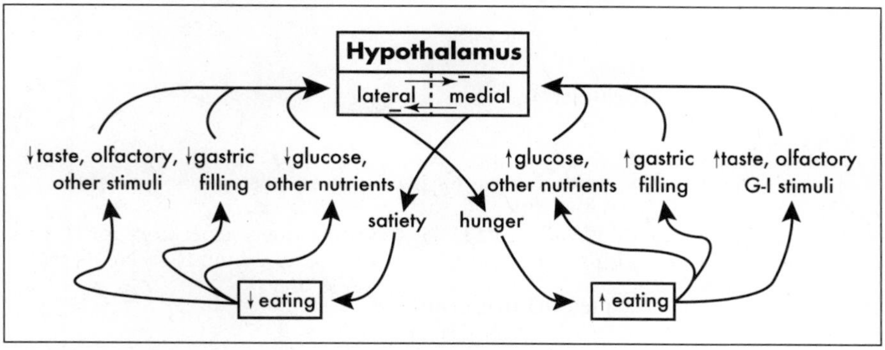

Fig. 4-13 Diagram of the interrelations of various influences in the regulation of appetite. Lateral and medial parts of the hypothalamus inhibit each other, and together they form a central control for appetite regulation. Active neurons in the lateral hypothalamus initiate hunger, eating, increased blood glucose, etc. (right-hand side of drawing), which in turn activate medial hypothalamic neurons to inhibit the lateral neurons. Opposite kinds of regulatory action appear on the left, where active medial hypothalamic neurons depress appetite and reduce taste stimuli, etc., which in turn promote activity in lateral hypothalamic neurons to promote appetite. (From Melvin J. Swenson and William O. Reece, editors: Dukes' Physiology of Domestic Animals, 11th ed. Copyright 1993 by Cornell University. Used by permission of the publisher, Cornell University Press.)

TERMS

Contralateral Refers to the opposite side of the body.

Ipsilateral Refers to the same side of the body.

Pacemaker A neuron or group of neurons that fires rhythmically to drive or pace the activity of its targets, which could be another group of neurons or a set of muscles.

Reciprocal Action The idea that two neurons or pools of neurons are mutually influencing, with mutually exclusive consequences. Commonly, this means that when one group of neurons is active, part of their output inhibits another group of neurons that otherwise would produce antagonistic functions of behaviors.

Related Principles

Feedback and Re-entrant Mapping
Homeostatic Regulation (Overview)
Inhibition
Inhibitory Routing
Modularity (Overview)
Reflex Action

Rhythmicity (Overview)
Synchronicity
System Modulation

Reference

Delcomyn, F. 1980. Neural basis of rhythmic behaviors in animals. Science. 210:492-498.
Grillner, S. and Wallen, P. 1985. Central pattern generators for locomotion, with special reference to vertebrates. Ann. Rev. Neurosci. 8:233-262.

Inhibition

Inhibition is a major feature of the central nervous system. It serves to organize and control function. At the cell level, the basic mechanisms can be presynaptic or postsynaptic. At the circuit level, inhibition may feed forward, feed back, and may spread laterally. With appropriate coupling to targets, inhibitory phenomena may underlie oscillatory activity in neuronal ensembles.

Explanation

Imagine what brain function and behavior would be like without inhibition. The nervous system would run amuck if all neuronal action were excitatory. Control and modulation of function would be impossible. Inhibited targets may be other neurons, glands, or smooth muscle (but never a direct inhibition of skeletal muscle).

Inhibition at the cell level occurs only in one of two ways: presynaptic or postsynaptic. In presynaptic inhibition, presynaptic terminals are persistent depolarized, with the consequent effect that they cannot continue to discharge or release transmitter while being maintained in a depolarized state. Postsynaptic inhibition occurs in the synapses, which by virtue of having appropriate membrane receptors exposed to certain neurotransmitters, respond to input with *hyper*polarization instead of depolarization. Hyperpolarization is a state of the neuron being more electronegative, relative to extracellular fluid, than in the normal resting state. Thus, such neurons are less likely to discharge impulses and more-than-usual excitatory input must be received to reach the depolarization threshold for generating impulses.

At the circuit level, inhibitory influences may feed forward from the neurons of origin to act on targets or may feed back to inhibit the neurons that activated the inhibition in the first place. Feed-forward inhibition is often expressed laterally in a circuit, so that flow of information is enabled in part of the circuit while being suppressed in adjacent, lateral parts of the circuit. Feed-back inhibition, on the other hand, acts on the neurons that activated inhibitory cells in the first place (Figure 4-14).

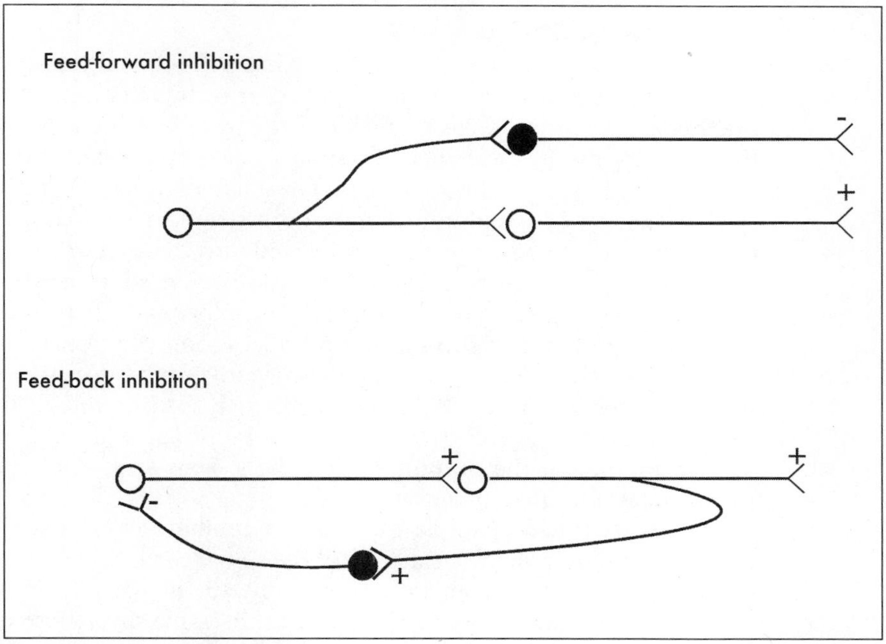

Fig. 4-14 Diagram of the simplest circuits for feed-forward inhibition and feed-back inhibition. Black circles represent cell bodies of inhibitory neurons.

Examples

Single Cell Mechanisms

Presynaptic Inhibition. The presynaptic terminal of axons converts electrical signal into chemical signals when that terminal is depolarized sufficiently to release neurotransmitter. Although depolarization is essential for the release of large quanta of transmitter, transmitter release does not continue if the presynaptic terminals are clamped in a persistently depolarized state. Such clamping can occur if a presynaptic axon terminal receives sustained excitatory input from another axon.

Postsynaptic Inhibition. One well-studied example of presynaptic inhibition is the Mauthner cell of fish (see Reciprocal Action principle). Each Mauthner cell receives input from the eighth cranial nerve on both sides—and from the Mauthner cell on the opposite side. Presynaptic chemical inhibition occurs in the large axons on the distal part of the lateral dendrites of the Mauthner cells. Stimulation of the eighth cranial nerve causes excitation on the Mauthner cell on the same side, but presynaptically inhibits the contralateral Mauthner cell. This, along with electrical inhibition of axon collaterals of each Mauthner cell, prevents simultaneous contraction of tail muscles on both sides. In this way, a fish flicks its tail alternately in each direction. Clearly, this is how propulsion is achieved.

Circuit-level Mechanisms

Feed-forward or Lateral Inhibition. One well-studied example of lateral inhibition is the enhancement of visual contrast by lateral inhibition occurring within the retina. Cells that are excited by light inhibit their neighbors, thus magnifying the difference in responsiveness between cells that are light activated and their neighboring cells. Thus, it should be evident that significant processing of visual information occurs right in the retina, before the representation of a visual scene is projected into central circuits.

Feed-back or Recurrent Inhibition. Feed-back inhibition often occurs as part of a servo-system based on negative feedback. That is, the output of the nervous system or a sub-set of it produces some physiological or behavioral consequence that in turn supplies negative (inhibitory) feedback that suppresses the excitatory forces that generated the physiological or behavioral effect. In the process, over-activity in input parts of the circuit is suppressed by part of the output, so that the system as a whole is regulated within a certain range of activity.

The best-known example of recurrent inhibition is the so-called Renshaw cell system in the ventral horn of the spinal cord. When a spinal motor reflex is activated by incoming sensory input, the motor neurons in the cord activate appropriate muscles. However, at the same time those motor neurons have collateral axonal branches that feedback an inhibitory influence on that same motor neuron. The consequence of such an effect is that muscle contraction is time limited. It does its job and then quits.

If inhibitory neurons are coupled in the proper network arrangement, physiological and behavioral rhythms may emerge. One kind of network oscillator emerges when neurons are reciprocally connected to each other with inhibitory neurons. Oscillation occurs when there is a constant maintained excitation of the reciprocally inhibitory neurons and there is simultaneously a build-in process that allows the activity of single neurons to decrease after prior activity. Arrangements such as this underlie the ability of central pattern generator circuits to produce rhythmical patterns of function and behavior.

TERM

Renshaw Cells Small interneurons located in the ventral grey matter of the spinal cord. These cells are excited by collaterals of active motor neurons and send inhibitory influences back onto the motor neurons that activated them.

Related Principles

Information Modification
Inhibitory Routing
Nodal Point (Cell Biology)
Reciprocal Action
Two Basic Actions (Cell Biology)

References

Eccles, J.C. 1982. The synapse: from electrical to chemical transmission. Ann. Rev. Neurosci. 5:325-39.

Furshpan, E.J. and Furukawa, T.Y. 1962. Intracellular and extracellular responses of several regions of the Mauthner cell of the goldfish. J. Neurophysiol. 25:732-771.

Furukawa, T.Y. and Furshpan, E.J. 1963. Two inhibitory mechanisms in the Mauthner neurons of goldfish. J. Physiol. 26:140-176.

Stent, G.S., Kristan W.B. Friesen, W.O., Ort, C.A., Poon, M., and Calabrese, R.L. 1978. Neuronal generation of the leech swimming movement. Science 200:1348-1357.

Stillito, A.M. 1979. Inhibitory mechanisms influencing complex cell orientation selectivity and their modification at high resting discharge levels. J. Physiol. 289:33-53.

Suga, N., O'Neill, W.E., and Manabe, T. 1978. Cortical neurons sensitive to combinations of information-bearing elements of bisonar signals in the mustache bat. Science. 200:778-781.

Citation Classics

Brock, L.G., Coombs, J.S., and Eccles, J.C. 1952. The nature of the monosynaptic excitatory and inhibitory processes in the spinal cord. Proc. R. Soc. B 140:170-176.

Coombs, J.S., Eccles, J.C., and Fatt, P. 1953. The action of the inhibitory synaptic transmitter. Aust. J. Sci. 16:1-5.

Coombs, J.S., Eccles, J.C., and Fatt, P. 1955. The specific ionic conductances and the ionic movements across the motoneuronal membrane that produce the inhibitory postsynaptic potential. J. Physiol. 130:326-373.

Renshaw, B. 1941. Influences of discharge of motoneurons upon excitation of neighboring motoneurons. J. Neurophysiol. 4:167-183.

Inhibitory Routing

Inhibition is a primary mechanism by which the flow of action potentials is selectively routed into certain pathways.

Explanation

Coordinated responses to sensory input can occur because action potentials arising from a stimulus are selectively routed into certain pathways and not others. Likewise, complex motor commands require selective routing. Strategically placed inhibitory neurons provide a basis for routing and switching the flow of information into specific circuits.

When inhibition occurs at specific points in an interactive circuit of neurons, it can act as a gate to route information flow (Figure 4-15). As discussed in the Inhibition module, the various kinds of inhibitory circuits include feedback and feed-forward inhibition. Both kinds can serve to route information.

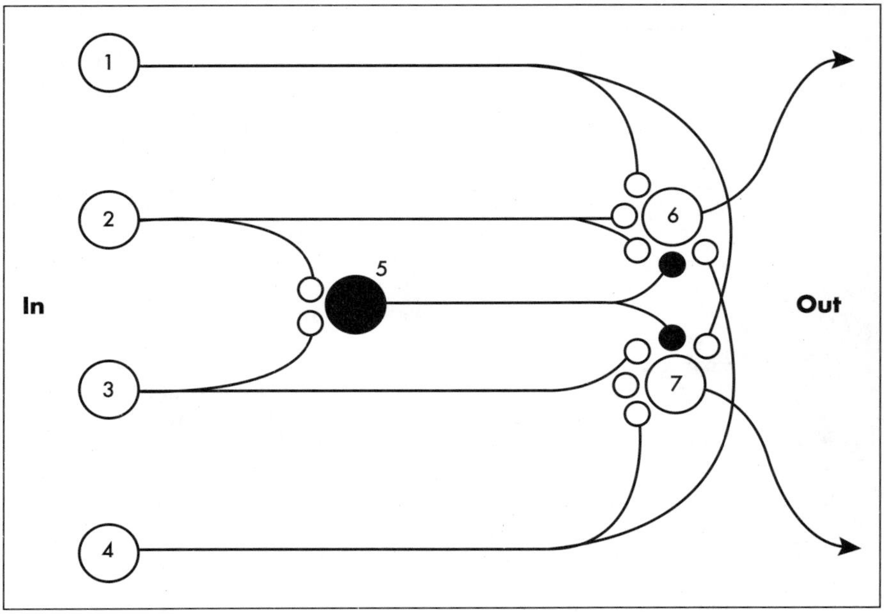

Fig. 4-15 Diagram of a pool of excitatory neurons and inhibitory neuron (black circles) to illustrate the effect of circuit design and threshold for excitation on routing of information. Assume that impulses begin in the four input neurons on the left and that a net of at least two excitatory terminals must be active to release enough neurotransmitter to reach the threshold for exciting subsequent neurons. Input from neuron 2, for example, would excite neuron 6 because at least two axon terminals are active. Input from 2 and 3, however, would not have an output because the inhibitory neuron would also be activated, and its inhibitory transmitter would cancel part of the effect of the two active excitatory terminals at 6 and 7. Still further complexity could be achieved, as it is in real brain circuits, by drawing in branches that feed back excitatory or inhibitory influences on other neurons in the circuit. (From Klemm, W.R. 1972. Science, the Brain, and our Future. Bobbs-Merrill, New York.)

In circuits where the inhibition is fed back into the "input" neurons, a synchronization of the activity of many neurons can occur; this process is believed to underlie the production of large, rhythmic electroencephalographic (EEG) waves such as alpha-waves (8-10/sec) from the visual cortex of humans and theta-waves (4-7/sec) from the hippocampus of mammals.

Examples

A good illustration of selective routing can be seen in the simple spinal withdrawal reflex. When a painful stimulus is delivered to the foot, action potentials are delivered to the flexor muscles of the leg to cause withdrawal of the foot. Excitation does not spread into the nerves that activate the extensors of that leg. Indeed, if such were the case, both extensors and flexors would contract and withdrawal from the pain could not occur. Extensors do not get excited because they are actually inhibited by a feed-forward mechanism. Thus, action potentials are routed in some paths and not others.

As circuitry becomes more complex, inhibitory neurons can provide an array of alternative pathways, depending on the anatomical connections and the thresholds needed to activate a given neuron.

Other examples are shown in the modules on Pain Perception and Inhibition.

Related Principles

Inhibition
Pain Perception (States of Consciousness)
Synchronicity
Two Basic Actions (Cell Biology)

References

Burrows, M. 1980. The control of sets of motoneurones by local interneurones in the locust. J. Physiol. 298:213-233.

Carkowska, J., Jankowska, E., and Sybirska, E. 1981. Common interneurones in reflex pathways from group 1a and 1b afferents of knee flexors and extensors in the cat. J. Physiol. 310:367-380.

Deliagina, T.B., Feldman, A.B., Gelfand, I.M., and Orlowsky, G.N. 1975. On the role of central programs and afferent inflow in the control of scratching movements in the cat. Brain Research. 100:297-313.

Getting, P.A. 1989. Emerging principles governing the operation of neural networks. Ann. Rev. Neurosci. 12:185-204.

Parallel, Multi-level Processing

Many of the brain's functions are conducted in widely distributed, parallel pathways, where several-to-many assemblies are simultaneously interacting. Moreover, many of the neurons in these assemblies are shared by different circuits. Many of these assemblies are topographically mapped. Neural functions proceed at many levels, from genetic expression to molecular communication among neurons, to consciousness and behavior. Information from one level often affects functions at other levels.

Explanation

Many of the targets of parallel inputs connect to each other via other pathways. Thus, these target ensembles are interacting simultaneously with their parallel input. Recall from the Circuit Design module that many circuits diverge. This produces a series of parallel, distributed pathways whereby information gets distributed more or less simultaneously to multiple targets. In many cases, these target ensembles are reciprocally connected to each other. Thus, the processing that is occurring *within* each distributed target is subject to further modification by interactions with the processing that is occurring *among* the various targets.

Some neural computations occur in maps—arrays of neurons in which the tuning of neighboring neurons for a particular parameter value varies predictably. Each such array is tuned slightly differently, and they operate in parallel on inputs. The information represented in such maps is transformed almost instantly into a place-coded ("topographic") probability distribution. Of the map systems that have been discovered are those for visual line orientation and direction of motion, auditory maps of amplitude and time interval, and motor maps of movements.

The tuning of individual neurons for mapped parameter values is broad, relative to the range of the map. Thus, any given stimulus of the right type tends to activate neurons throughout the map, but the locations of peak activity code the precise information about the mapped parameter.

Examples

As a simple example of parallel, distributed processing, consider the neuronal populations in the abdominal ganglia of the mollusc, *Aplysia*. The movements in the gill withdrawal reflex seem to be mediated by a single, large distributed network, rather than by small dedicated circuits. Very few neurons discharge during only one or two of the various behavioral contexts in which gill withdrawal occurs. As many as 90% of the abdominal ganglion cells become activated during any kind of gill movement, either reflex gill withdrawal, small spontaneous contractions of the gill, or respiratory pumping. However, the total numbers of action potentials and their firing patterns are specific for each of the three behaviors. If two or more behaviors were made to occur at the same time, new types of firing patterns appear, indicating an interaction in which there is competition for shared elements in the distributed network. In a distributed system, temporary functional circuits emerge that are specific to certain stimuli or motor commands. The advantage of this property is that the same neurons and their connections can be used in various ways. Studies in lobsters suggest that given neurons can be reassigned dynamically to circuits that mediate different behaviors.

In higher animals, the projections of neurons in the locus coeruleus provide another example (Figure 4-16). These neurons, whose cell bodies are clustered together in the brainstem, send norepinephrine-containing fibers to many sites. These target ensembles have other connections with each other.

Vision provides one of the best-studied examples of distributed processing. Visual stimuli are processed more or less simultaneously in about 25 neocortical areas in the monkey. An additional seven areas are visual-association areas. These 32 areas have 305 known connections among themselves. Most of these areas are reciprocally connected, providing an

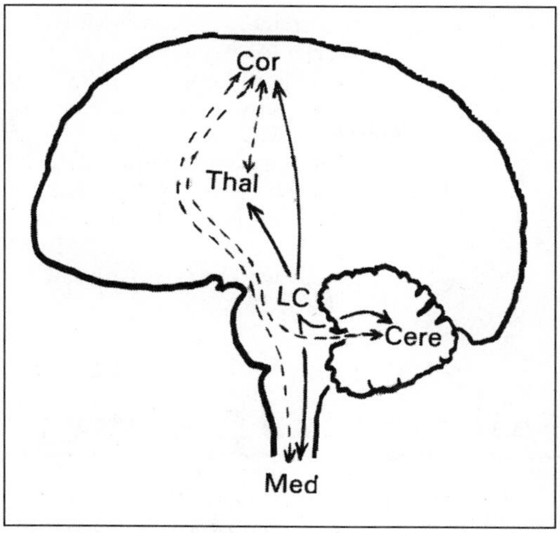

Fig. 4-16 Illustration of parallel, distributed processing emerging from a group of brainstem neurons known as the locus coeruleus (LC). LC neurons send divergent output (solid lines) to multiple structures: cortex (Cor), thalamus (Thal), cerebellum (Cere), and medulla (Med). Many of the target ensembles connect also to each other *(dashed lines)*.

anatomical basis for re-entrant mapping of visual cues. There are also extensive connections with nonvisual areas of the cortex. Sixty pathways link visual processing to 13 areas in the somatosensory/motor cortex.

In the hippocampus, a brain area that seemingly has nothing to do with vision, there are "place cells" that fire only when an animal is located in a specific region of three-dimensional space. In the temporal cortex of monkeys, certain cells respond selectively to specific features in a visual scene. For example, some cells fire when the monkey sees another monkey's hand, others fire when a face is seen, etc.

Another, much simpler example of parallel, distributed processing can be seen with the system for homeostatic regulation of viscera (see the Visceral Control principle).

Related Principles

Circuit Design (Overview)
Emergent Properties
Feedback and Re-entrant Mapping
Lateralization
Reciprocal Action
Visceral Control (Motor Activity and Control)

References

Aersten, A. 1993. Spatio-temporal aspects of brain function. Elsevier, Amsterdam.

Alexander, G.E., DeLong, M.R., and Strick, P.L. 1986. Parallel organization of functionally segregated circuits linking basal ganglia and cortex. Ann. Rev. Neurosci. 9:357-381.

Felleman, D.J. and Van Essen, D.C. 1991. Distributed hierarchical processing in the primate cerebral cortex. Cerebral Cortex. 1:1-47.

Getting, P.A. 1989. Emerging principles governing the operation of neural networks. Ann. Rev. Neurosci. 12:185-204.

Harris-Warrick, R.M. and Marder, E. 1991. Modulation of neural networks for behavior. Ann. Rev. Neurosci. 14:39-57.
Shapley, R. and Lennie, P. 1985. Spatial frequency analysis in the visual system. Ann. Rev. Neurosci. 8:547-583.
Simmers, J. Meyrand, P., and Moulins, M. 1995. Dynamic networks of neurons. Amer. Sci. 83:262-268.
Van Essen, D.C. and Zeki, S.M. 1978. The topographic organization of rhesus monkey prestriate cortex. J. Physiol. 277:193-226.
Wu, J.-Y., Cohen, L.B., and Falk, C.X. 1994. Neuronal activity during different behaviors in Aplysia: a distributed organization? Science. 263:820-823.

Feedback and Re-entrant Mapping

Many neural circuits are organized so that some of the output is fed back to regulate the input. Additionally, large ensembles of certain neurons may send a mapped output into a second ensemble, which after some processing, sends a re-entrant input into the first ensemble.

Explanation

Feedback is an intrinsic property of self-regulating systems, whether biological or machine. The basic requirement is that there is a control system that acts much like a home-heating thermostat. The principle is that there is some setting for the system's function that serves as a reference by which input perturbing signals are compared. The control system adjusts the output based on the disparity between the reference signal and the perturbing signal, taking into account the feedback information about the system's current output.

In most cases of normally operating servo systems, the feedback is negative. That is, it suppresses runaway activity. If the feedback were positive, the controller gain might progressively increase, resulting in a lack of control (Figure 4-17).

At a higher level of operation in a system with multiple, interacting, distributed components, there is a process of re-entry, wherein recursive signalling occurs between separate networks along their ordered anatomical connections. The extensive reciprocal innervation of many brain areas is the underlying basis for re-entrant, recursive signalling. The idea is that input stimuli to one ensemble of neurons is conducted to another ensemble, which in turn sends a re-entrant input back into the first ensemble (Figure 4-18).

Examples

Normal function of nervous systems requires negative feedback. If a nervous system had no way to suppress activity, it could not route information in certain pathways nor could it prevent an excitatory input from

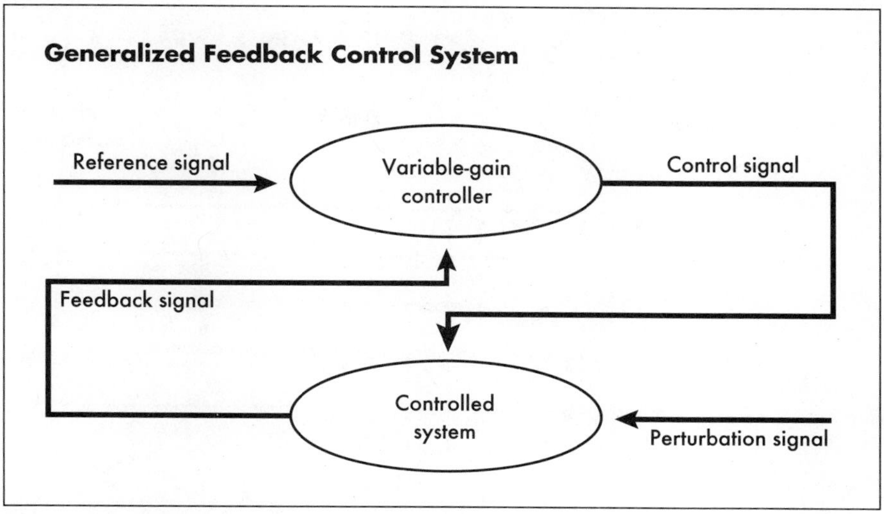

Fig. 4-17 Diagram of a simple self-regulating (servo) system. The "controller" part of the system can vary in gain to adjust the output to the system under control. The controlled system is normally under the influence of external perturbations, the influence of which is signalled by feedback from the controlled system to the controller. The controller compares the feedback signal with the set point reference signal and adjusts the controller's output accordingly to minimize the mismatch.

growing out of control. A good example of what happens where there is no feedback is epilepsy. A small excitatory stimulus produces excitatory responses that continue to build until they become out of control.

Re-entrant mapping in the nervous system is most evident with the topographical maps in the neocortex. The visual system of the monkey, for example, has over 30 different maps, each containing a representation for a different visual feature, such as color, movement, edge orientation, etc.). As groups of neurons are activated in one map, an output of that map then engages another group of neurons in another map. Re-entrant signalling between such maps serves to strengthen and magnify the functional relationship between such linked ensembles. Thus, we see that the unit of selective engagement is not a single neuron, but rather an ensemble of many neurons (Figure 4-19).

TERMS

Recursive Signalling — The idea that signals flow repeatedly between two or more neurons or neuronal ensembles.

Re-entrant Mapping — The idea that reciprocally connected, topographically mapped neuronal ensembles receive a re-entrant input that has been relayed from other ensembles that were activated by an earlier output from the ensemble receiving the re-entrant input.

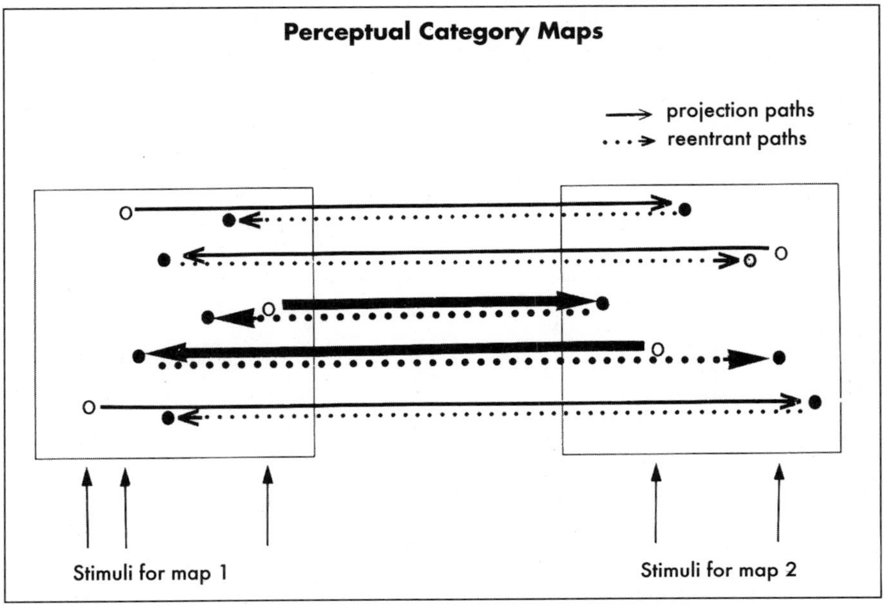

Fig. 4-18 Diagram of the idea of re-entrant mapping, which is a feedback of activity from large arrays of neurons that are topographically mapped into the arrays of neurons that initiated the activity. Here we illustrate the concept that one mapped set of circuits responds to a topographic sensory input and projects an output (solid lines) to another mapped region, which in turn has its own mapped sensory input. Neurons in each mapped region send re-entrant output to the other mapped area (dashed lines). If certain pathways are used repeatedly, they become facilitated (heavy solid and dotted lines) so that they become the representation (memory) for a given set of stimuli. Such re-entrant mapping is thought to be a factor in the conscious perception of categories of stimuli. (Based on the concepts of Edelman, 1989.)

Fig. 4-19 Drawing of interacting cortical areas, where neuronal maps of visual scenes can participate in re-entrant mapping with other kinds of mapped areas in the cortex. There are even re-entrant connections between the visual cortex and the brainstem relay system for vision, the lateral geniculate body (LGN).

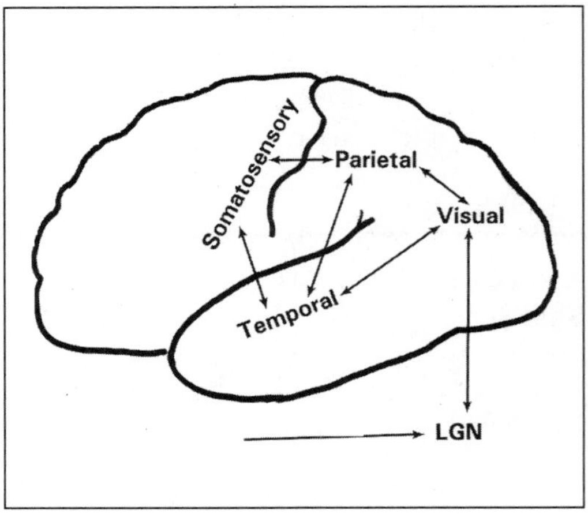

Related Principles
From Input to Output
Reciprocal Action
Topographical Mapping (Overview)

References

Edelman, Gerald M. 1989. The Remembered Present. Basic Books, New York.
Edelman, G.M. 1992. Bright Air, Brilliant Fire. Basic Books, New York.
Knudsen, E.I., duLac, S., and Easterly, S.D. 1987. Computational maps in the brain. Ann. Rev. Neurosci. 10:41-66.
Middlebrooks, J.C., Dykes, R.W., and Merzenich, M.M. 1980. Binaural response-specific bands in primary auditory cortex (AI) of the cat: topographical organization orthogonal to isofrequency contours. Brain Res. 181:31-48.

Cortical Columns

The outer mantle of tissue (cortex) that covers the cerebrum has highly complex organization. The cells are functionally arranged in distinct ways that include a vertical, columnar organization in which the circuitry includes inhibitory neurons and negative feedback that help to regulate population activity within the column.

Explanation

Anatomical studies have identified different types of neurons in the cortex and their interconnections. Physiological studies involving stimulation of one neuron and recording from one of its targets have identified which connections are excitatory and which are inhibitory.

Anatomically, the cerebral cortex has a crude horizontal layering of cells, but there is a distinct vertical orientation of axons and dendrites, both of which can receive inputs from other regions of the cortex, the brainstem, and particularly the great sensory relay system in the thalamus. The cortex has unusually high cellular densities, and the number of interconnections is uncountable. This leads most neuroscientists to believe that the great bulk of information processing in the brain occurs in cortical tissue (Figure 4-20).

Examples

Cerebral cortex cells are of two recognizable types: (1) pyramidal cells, which have large and elongate cell bodies and whose axons constitute the major output of the cortex, and (2) stellate cells, which are smaller and have dense, but short, axonal and dendritic projections.

The outer layers of cerebral cortex are of special interest for two reasons: (1) they are the proximate source of the summated voltages that

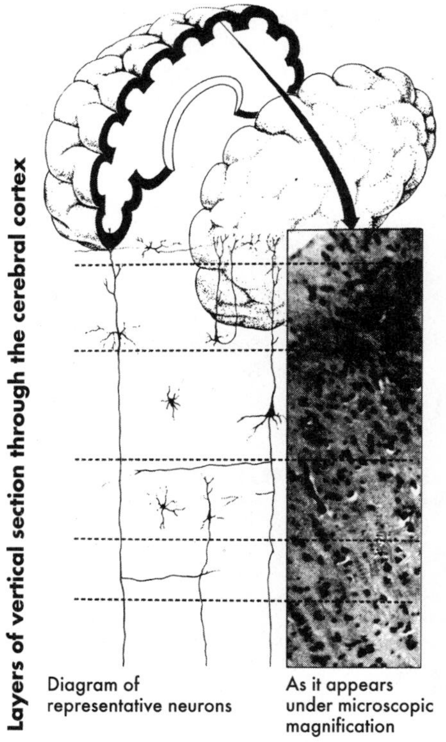

Fig. 4-20 Cerebral cortex. The greatly simplified drawing at the left shows that neurons in the cortex are oriented vertically from the surface, with cell bodies located in layers. At the right is a photograph of human cortex, as seen magnified about 80 times. Black dots are the cell bodies of neurons (the large, elongated black splotches are collapsed blood vessels). Neuronal branches, as seen in the diagram, are not evident because they do not stain well and because there are so many and they are so densely packed. The grey background of the photograph is actually the densely packed matrix of millions of neuronal branches, axons and dendrites. (From Science, the Brain, and our Future, by Klemm W.R., Bobbs-Merrill, New York.)

constitute the electroencephalogram (EEG), and (2) the apical dendrites, which form the outermost part of the cortex and receive massive sensory inputs from the brainstem and the thalamus (Figure 4-21).

Because the neurons in the cortex tend to be oriented vertically, with dendrites projecting into a neuron-free zone on the surface, the electrical fields can summate over space to become large enough to be detectable through the insulating barriers of skull and scalp. In other words, the electroencephalogram, recorded from the scalp surface, is a continuously fluctuating summation of the extracellular voltage fields (Figure 4-22).

It is not possible to discern the nature of microcircuitry from looking at histological slides. This has to be determined by single-cell recordings, where, for example, one stimulates one cell and determines which other cells respond. Painstaking studies of this kind have determined that there are clearly definable microcircuits in which the elements have been characterized as to whether they are excitatory or inhibitory on their targets.

Physiological studies have indicated that discrete vertical arrays have homogenous function and that complex processing and decision tasks are accomplished by simultaneous function in a large set of column arrays,

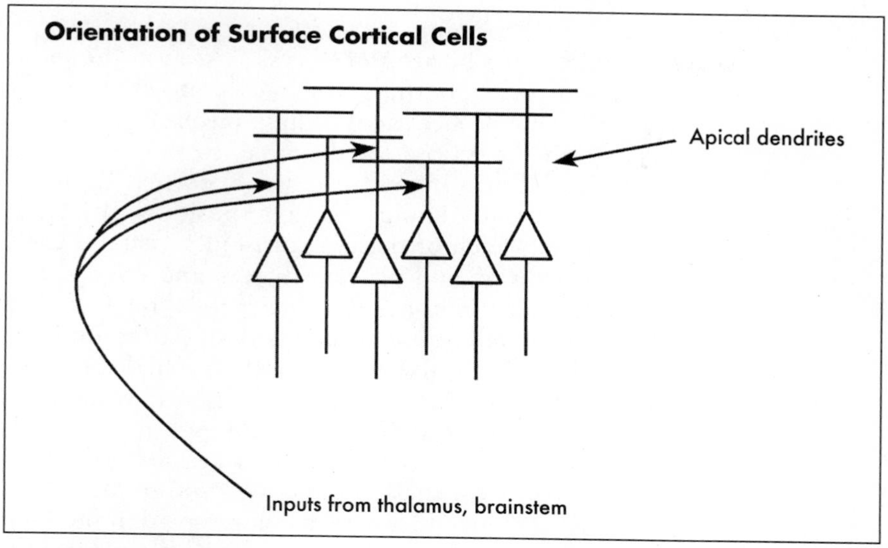

Fig. 4-21 Diagram of the orientation of cortical cells nearest the surface of the brain. The apical dendrites extend vertically to the surface where they branch in many directions parallel to the surface. The apical dendrites receive many axon terminals from neurons in the thalamus and brainstem. Electrical polarization changes in these dendrites create electrical dipoles that contribute to the electroencephalogram.

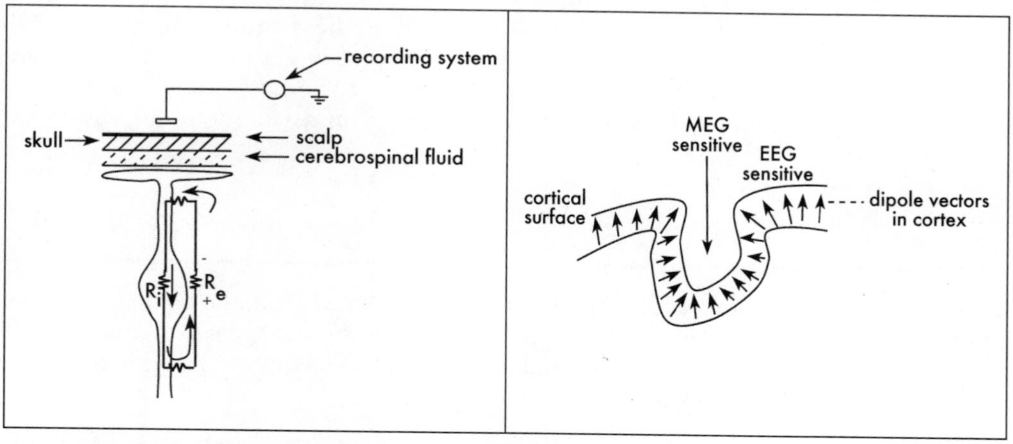

Fig. 4-22 LEFT – Diagram showing how one cortical neuron contributes to the electroencephalogram (EEG). Ionic currents flowing across neuronal membrane as it is depolarized create voltages as they pass through the internal resistance of the cytoplasm (Ri), the transmembrane resistance (not labeled), and the resistance of the extracellular fluid (Re). EEG electrodes are typically placed in the equivalent of extracellular space and thus they monitor the voltage developed from ionic currents flowing through extracellular fluid. RIGHT – Vertical orientation of adjacent neurons creates electrical dipoles that orient perpendicular to the cortical surface. Activity in invaginated areas is not readily detected in the EEG, but is detected in a magnetoencephalogram (MEG).

with each array performing one specific aspect of the task. Columns do intercommunicate via horizontal fibers, as seen in the figures. In addition, in cerebral cortex, outputs from a column on one side of the brain include a projection to a cortical column on the other side of the brain. Projections from the retina, for example, to the visual cortex are not simply represented as a map on the visual cortex. Each retinal area is analyzed for different stimulus features (position, contrast, line orientation, color, etc.). Each of these features in a visual scene is extracted by individual neurons and processed over and over again in column after column, and again in neighboring cortical regions. Within a given vertical column, usually measuring less than 0.5 mm in diameter, most of the cells will extract only one feature of a visual stimulus. For example, if one stimulates a cat's eye with a dark bar while recording from different points in the same functional column, responses will occur in those neurons only when the bar is in a certain orientation. In some poorly understood way, these interactions within and among columns re-integrates the extracted features as represented in individual neurons and "binds" them back together to reconstruct a representation of the original image.

The organization of the cerebellar cortex is slightly different from that of the cerebral cortex. For example, there are fewer layers, and they are more orderly. The surface is dominated by apical dendrites from huge pyramidal cells, which are densely packed into a distinct layer. In addition, beneath this pyramidal cell layer, near the white matter, there is a unique layer of densely packed cell bodies, called the granule-cell layer, whose axons project toward the surface with extensive interconnections with the pyramidal cell dendrites. Some estimates are that a single pyramidal cell receives contacts from as many as 80,000 granule cells.

TERMS

Electro-encephalogram — A recording of the summated voltages from the tissue underlying the electrodes. Typically, an electrode registers voltages over a span of a few millimeters, and the voltage strength is on the order of 10 to 40 microvolts.

Homologous — For our purposes, it means anatomically comparable structures. For example, a region on one side of the brain has a counterpart on the other side.

Thalamus — A large mass of neurons that lies on both sides of the midline, below the cerebral cortex. It has many distinct groups of neurons, and many of these receive topographically mapped inputs for specific kinds of sensations.

Related Principles
Emergent Properties
Inhibition
Inhibitory Routing
Parallel, Multi-level Processing
Rhythmicity and Synchronicity

References
Allman, J., Miezin, F., and McGuinness, E. 1985. Stimulus specific responses from beyond the classical receptive field: neurophysiological mechanisms for local-global comparisons in visual neurons. Ann. Rev. Neurosci. 8:407-430.

Brody, H. 1955. Organization of the cerebral cortex IV. A study of aging of the human cerebral cortex. J. Comp. Neurol. 102:511-556.

Creutzfeld, O.D. 1983. Cortex Cerebri. Springer-Verlag, Berlin.

Gilbert, C.D. 1983. Microcircuitry of visual cortex. Ann. Rev. Neurosci. 6:217-247.

Jones, G. and Peters, A.A. (eds.) 1985. The Cerebral Cortex. Plenum, New York.

Mountcastle, V.B. 1957. Modality and topographic properties of single neurons of cat's somatic sensory cortex. J. Neurophysiol. 20:408-434.

Zeki, S. and Shipp, S. 1988. The functional logic of cortical connections. Nature. 335:311-317.

Citation Classics
Gray, E.G. 1956. Axo-somatic and axo-dendritic synapses of the cerebral cortex: an electron microscope study. J. Anat. 93:420-433.

Grey, C.M. and Singer, W. 1989. Stimulus-specific neuronal oscillations in orientation columns of cat visual cortex. Proc. Natl. Acad. Sci. USA. 86:1698-1702.

Hubel D.H. and Wiesel T.N. 1952. Receptive fields, binocular interaction and functional architecture in the cat's visual cortex. J. Physiol. Lond. 160:106-154.

Hubel D.H. and Wiesel T.N. 1965. Receptive fields and functional architecture in two nonstriate visual areas (18 and 19) of the cat. J. Neurophysiol. 28:229-289.

Leao A.A.P. 1944. Spreading depression of activity in the cerebral cortex. J. Neurophysiol. 7:359-390.

Lorente de No R. 1934. Studies on the structure of the cerebral cortex. II. Continuation of the study of the ammonic system. J. Psychologie and Neurologie 46:113-177.

Rhythmicity and Synchronicity

Many neurons discharge impulses in a synchronously coupled way that produces functional rhythms. This coherent activity can magnify or suppress the output, depending on the phase relationships. Additionally, synchronous firing in large populations of neurons, especially if they involve widely separated neurons that have the capacity to participate in diverse functions, can lead to emergent properties.

Explanation

Rhythms govern such key functions as breathing, the heartbeat, sleep, certain rhythms in the brain electrical activity (alpha and theta rhythms), and various hormonal systems. The rhythms are created and sustained by neuronal rhythmic pattern generator circuits. Several mechanisms operate. One is mutual inhibition, wherein two neurons or pools of neurons are under continuous excitatory input and their outputs inhibit each other (Figure 4-23). For oscillation to develop, there must be additional physiological factors such as adaptation or postinhibitory rebound. Thus, first one and then the other neuron is active.

Sequential disinhibition is another mechanism used by central pattern generators. Here, functionally distinct pools of neurons receive tonic excitatory input, and interactions within a single pool are excitatory. Some of the pools of neurons have inhibitory interactions with neurons in other pools. Oscillation develops when activity in one pool inhibits neurons in a second pool. Neurons in a third pool, which had formerly been inhibited by the second pool of neurons, are thus released from inhibition (i.e., disinhibited).

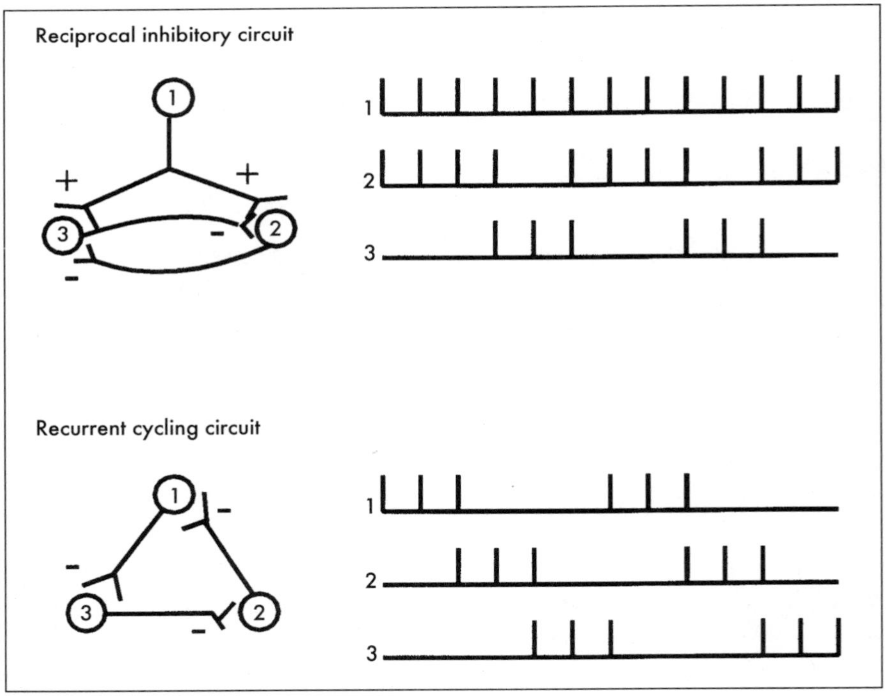

Fig. 4-23 Illustrations of how rhythmic behaviors can be generated by simple neuronal circuits. ABOVE – reciprocal inhibition between neurons 2 and 3 cause the action potential discharge of each one to be periodically interrupted. BELOW – Because all three neurons inhibit each other, they can, if otherwise spontaneously active or driven from other inputs, display a cycle of firing.

Negative feedback can also cause oscillation. Here, an excited target neuron has collateral branches that feed back to the originating excitatory neuron via an inhibitory interneuron. Thus, the output causes a transient inhibition of the circuit's input neuron. Large pools of such circuits are well documented in the spinal cord and in the part of the brain most involved in forming memories, the hippocampus.

Certain neurons seem to have intrinsic oscillating properties, and this commonly recruits a population of neurons into synchronous oscillation. Synchronous activity in neurons leads to summated voltages in the field potentials, as can be recorded in the electroencephalogram (EEG). The EEG is a vector sum of the extracellular voltages generated from millions of neurons. The origin of these potentials are currents from glial cells, action potentials, and postsynaptic potentials (mostly the latter). The frequency range of synchronous activity can range from about 1 to 40 EEG waves per sec. Simultaneous recording of EEG from several sites can disclose underlying synchronicity by spectral coherence, which is a measure of the correlation of specific frequency components in the spectrum of frequencies that occurs in the EEG.

Phase spectra and coherence spectra of the EEG are typically flat or only gently sloping over a range of several octaves; this is incompatible with a model of discrete oscillators. This suggests that the brain has many oscillator systems.

The spatial extent of coherence of synchronous activity is quite variable. At high electrical frequencies, only localized clusters of neurons may be linked into synchronous activity. But at low frequencies, the entire cortex may be involved.

The functional consequence of such oscillation is not known. It may vary with the parts of the brain and the circuitry involved. However, getting neurons to fire together is a potent way to magnify their effect on neurons that are targets of their output. Effects at the targets summate because of the synchronous input. Possible functions include the gating or routing of nervous impulses, facilitating throughout when the extracellular field potential is of one polarity and inhibiting throughout at the opposite polarity. Another possibility is that oscillation binds widely scattered parts of neural circuitry into a cohesive system to represent a given set of stimulus properties, memory, and motor program.

Examples

Oscillatory activity can be observed at both the cellular and population level. Oscillating single cells are evident from a burst-firing pattern, with periodic intervals separating bursts. Oscillating populations of cells are evident from the extracellular field potential, which has recurrent voltage waves that are relatively large and of long duration.

Burst-firing patterns of certain thalamic cells occur with the same periodicity as the large cortical EEG waves known as spindles that occur during sleep and certain forms of sedation. Likewise, the EEG alpha waves from the visual cortex are correlated with certain burst-firing neurons in the thalamus. Similar burst-firing of neurons in the septal nucleus of the brain occurs during the same time and with the same periodicity as the large EEG waves known as theta activity in the hippocampus; these septal cells are pacemakers for driving theta activity.

When a moving bar in a visual scene is presented to an animal's visual system, many neurons located along a column of tissue perpendicular to the surface fire synchronously when the bar has a certain orientation. Different sets of neurons have different orientation tuning preferences. This synchronous firing of groups of cells produces a rhythmic extracellular electrical field, fluctuating at about 40 times/sec. When a stimulus bar is presented and field potentials are recorded in several separate visual maps that have overlapping receptive fields, the groups of neurons in the two maps show a mutual oscillation that is phase locked at about 40 Hz.

Studies of visual pathways in kittens reveal that about ⅓ of the neurons oscillate when the eyes are stimulated with flashed bars or drifting gratings. The corresponding EEG could display oscillations of synchronous activity in the range of 30-80 waves/sec. The frequency of synchronous activity seems to change with maturation.

TERMS

Coherence A time-locked relationship between events. In this case, we refer to cellular electrical activity that has some fixed phase relationship between or among the generators. In the case of the EEG, there are mathematical techniques that permit measurement of the degree of coherence—in specified frequency bands—between neuronal generators.

Septum A cluster of neurons (medial and lateral septal nuclei) that lie along the midline of the brain, just anterior to the thalamus. The medial septum is a major source of input fibers to the hippocampus, and it mediates the driving force for rhythmic, oscillatory EEG activity of the hippocampus, known as "theta rhythm."

Related Principles

Emergent Properties
Ensembles of Dynamic Neural Networks (Learning and Memory)
Feedback and Re-entrant Mapping
Parallel, Multi-level Processing
Reciprocal Action
System Modulation

References

Barlow, J.S. 1993. The Electroencephalogram: its Patterns and Origins. MIT Press, Cambridge, Mass.

Bullock, T.H. 1993. Integrative systems research on the brain: resurgence and new opportunities. Ann. Rev. Neurosci. 16:1-15.

Bullock, T.H. 1993. Brain waves: is synchronization a sign of higher function? Is the EEG basically rhythmic? p. 288-305. In How Do Brains Work? Birkhauser, Berlin.

Carr, C.E. 1993. Processing of temporal information in the brain. Ann. Rev. Neurosci. 16:223-243.

Engel, A.K., Konig, P., and Singer, W. 1991. Direct physiological evidence for scene segmentation by temporal coding. Proc. Nat. Acad. Sci. 88:9136-9140.

Friesen, W.O. and Stene, G.S. 1978. Neural circuits for generating rhythmic movements. Ann. Rev. Biophys. Bioeng. 7:37-61.

Getting, P.A., Linnard, P.R., and Hume, R.I. 1981. Central pattern generator mediated swimming in Tritonia I. Identification and synaptic interaction. J. Neurophysiol. 44:151-165.

Ghose, G.M. and Freeman, R.D. 1992. Oscillatory discharge in the visual system: does it have a functional role? J. Neurophysiol. 68:1558-1579.

Hall, J.C. and Rosbash, M. 1988. Mutations and molecules influencing biological rhythms. Ann. Rev. Neurosci. 11:373-394.

Jacklett, J.W. 1981. Circadian timing by endogenous oscillators in the nervous system: toward a cellular mechanism. Biol. Bull. 160:199-227.

Lopes da Silva, F. 1991. Neural mechanisms underlying brain waves: from neural membranes to networks. Electroenceph. Clin. Neurophysiol. 79:81-93.

Shapley, R. and Lennie, P. 1985. Spatial frequency analysis in the visual system. Ann. Rev. Neurosci. 8:547-583.

Singer, W. 1993. Synchronization of cortical activity and its putative role in information processing and learning. Ann. Rev. Physiol. 55:349-374.

Steriade, M. and Jones, E.G. 1990. Thalamic Oscillations and Signalling. John Wiley & Sons, New York.

Tank, D.W., Celperin. A., and Kleinfeld, D. 1994. Odors, oscillations, and waves: does it all compute? Science 265:1819-1820.

Citation Classics

Bradley P.B. and Elkes J. 1957. The effects of some drugs on the electrical activity of the brain. Brain 80:77-117.

Brown K.T. 1968. The electroretinogram: its components and their origins. Vision Res. 8:633-677.

Ciganek L. 1961. The EEG response (evoked potential) to light stimulus in man. Electroencephalogr. Clin. Neuro. 13:165-172.

Garey L.J. Jones E.G. and Powell T.P.S. 1968. Interrelationships of striate and extrastriate cortex with the primary relay sites of the visual pathway. J. Neurol. Neurosurg. Psychiat. 31:135-157.

Gray C.M. and Singer, W. 1989. Stimulus-specific neuronal oscillations in orientation columns of cat visual cortex. Proc. Natl. Acad. Sci. USA. 86:1698-1702.

Petsche H., Stumpf C. and Gogloak G. 1962. The significance of the rabbit's septum as a relay station between the midbrain and the hippocampus. I. The control of hippocampus arousal activity by the septum cells. Electroencephalogr. Clin. Neurol. 14:202-211.

Emergent Properties

Cooperativity among ensembles of neurons creates emergent properties, at least in more advanced brains, levels, stages, and states. The properties of a whole ensemble do not exist in and cannot be simply summed from the properties of the parts. Emergent properties of neuronal systems are fundamental to how these systems work.

Explanation

Edelman has developed a very interesting concept that the interactions of a variety of topographical maps is the underlying cause of the emergent property we know as consciousness. In this view, neural maps are the main operational system of the nervous system. The best studied such maps deal with auditory, visual, somatosensory and associative areas of the cerebral cortex. Each of these mapped areas is interconnected. Dynamical interactions among maps are coordinated by an integrative process called "re-entry." Re-entry involves a dynamic signalling between maps along ordered connections that are parallel and recursive. Each map is considered to be a selective system that categorizes or classifies the sensory input that it receives. Such categorization requires no a *priori* "knowledge" of what the sensory information is. Computer simulations of neural networks support the basic premise that such mapped systems can become self-organizing and distinguish sensory input.

Mapping begins with the mapping and categorization of sensory input. At the next level, there is perceptual mapping, in which recurrent activity in sensory maps develops preferred pathways that represent the perception of stimulus. Finally, there is "global mapping," which involves recurrent activity of all the mapped domains within the brain, including subcortical motor and limbic systems.

In this schema, the basic operational units of the nervous system are not single neurons but rather local groups of strongly interconnected neurons. Membership in any given local functional grouping is determined by synaptic connection strengths, which are subject to change. As distinct from the large neuronal groups that make up a map, these smaller groups are the units of "selection." That is, specific patterns of sensory input "select"" a subset of neurons and their interconnections to represent the categorization of that input. Maps are subject to a degree of reorganization in response to a dynamic competitive process of neuronal group selection.

These neuronal groups exhibit "degeneracy" (or redundancy) in that different circuits can have similar functions or that a given neuronal group can participate in multiple functions, depending on the current state of intrinsic and extrinsic connections.

Mapping phenomena seem to have a straightforward relationship to sensory and motor processing. But for certain "higher" functions such as emotions and memory, it is not clear what, if any, role is played by mapping. Emotions are clearly controlled by a complex, multiple interconnected set of structures in the so-called limbic system. Point-to-point mapping in many parts of this system is not evident. Another example is with memory. Numerous experiments suggest that memory "is not a thing in a place, but rather a process in a population." The explanation must reside in the fact that so much of the brain circuitry is highly interconnected.

Examples

Some of these ideas have been examined and supported with neural networks in the form of a selective recognition automaton that has a body form and a set of rather interesting behaviors. This automaton is based on principles of neuronal group selection in response to repeated stimulus. Cells within a certain group develop certain strengthened connections, while connections to cells outside the group are progressively weakened. Once neuronal groups have been formed, they demonstrate a number of interesting properties, particularly a large degree of cooperativity among cells within the group and competition with cells in other groups. Network stimulations show that each group tries to expand the size of its receptive field at the expense of neighboring groups.

At the single cell level, it can sometimes be shown that when a cell receives convergent input from several sources, the combined effect can be greater than the sum of the isolated inputs. But emergent properties are much more evident when ensembles of neurons are considered.

Emergent phenomena are difficult to understand but are readily apparent. One obvious example is sleep. Most neurons in the brain are still active in sleep, and it is not at all clear from observing single cell firing patterns why sleep occurs. Another example is visual perception. As explained elsewhere, a given neuron that is responsive to visual information only extracts a very limited part of a visual image, such as an edge, or a color. No one knows how the pooled activity of all such neurons serves to reconstruct the original image in the "mind's eye" (but see the Rhythmicity and Synchronicity principle).

These fragmentary responses in diverse, distributed networks are somehow bound together to emerge as a reconstructed representation of the whole original visual scene. Moreover, the scene can be consciously perceived.

Related Principles

Feedback and Re-entrant Mapping
Rhythmicity and Synchronicity
Topographical Mapping (Overview)

References

Abeles, M. 1991. Corticonics. Cambridge University Press, Cambridge.
Edelman, G.M. 1989. The Remembered Present: A Biological Theory of Consciousness., Basic Books, New York.
Edelman, G.M. 1992. Bright Air, Brilliant Fire. On the Matter of the Mind. Basic Books, New York.
Levin F. 1990. Mapping the Mind: The Intersection of Psychoanalysis and Neuroscience. Lawrence Erlbaum Assoc., Hillsdale, New Jersey.
Reeke, G.N., Jr., Finkel, L.H., and Edelman, G.M. 1990. Selective recognition automata, P. 203-226. In An Introduction to Neural and Electronic Networks, ed. by S.F. Zornetzer, J.L. Davis, and C. Lau. Academic Press, New York.

Information Processing: Study Questions

1. List and explain each of the principles in this category.
2. For each of the principles, provide an example *that is not mentioned in this text.*
3. Explain the difference between facilitation and disinhibition; between inhibition and disfacilitation.
4. What is the relationship between information processing and memory?
5. Explain the difference between temporal and spatial summation.
6. List the ways that information is carried and registered in the nervous system.
7. What is the advantage of having multiple forms of information in the nervous system? What are the disadvantages?
8. Explain the labeled line theory and its implications for information carriers.
9. In a complex sensory system such as vision, which functions are mostly under genetic control and which are more dependent on sensory experience?
10. Summarize the transduction processes that occur as a baseball batter swings at a pitched ball.
11. Under what conditions can electrical fields act as information carriers?
12. What are the general net consequences of two pools of neurons that are mutually antagonistic?
13. How does reciprocal action contribute to homeostasis?
14. What is the difference between presynaptic and postsynaptic inhibition? What are the implications of this difference?
15. What is the difference between feed forward and feed back inhibition?
16. Draw three different neuronal circuit diagrams that show how inhibition routes information.
17. Primitive organisms, such as coelenterates, have very limited ability to route information. How is that reflected in their behavioral capabilities?

18. How do negative and positive feedback differ in their functional consequences?
19. Distinguish feed back and re-entry, with examples.
20. Why do we think that the most complex information processing occurs in cortical tissue?
21. Where does the EEG come from? Is it an epiphenomenon or does the EEG signal have effects of its own? Explain and defend.
22. What are some functional consequences of coherent neural activity? Give examples.
23. Give some examples of synchronous activity at the cell level and at the population level.
24. How does the frequency of coherent electrical activity in neuronal populations affect the spatial extent of the coherent activity?
25. What do we mean by emergent properties of the nervous system?
26. What are some examples of emergent properties?
27. What are the advantages of parallel, distributed processing over having a single central processing locus?
28. What are the hallmarks of reflex action?
29. What are the advantages of reflex action?
30. Explain how information is modified by inhibition, filtering, summation, occlusion, and gating.

CHAPTER 5

States of Consciousness

Mind in nature is a property of particular brains with particular histories; that is, of particular phenotypes with particular brain areas and structures capable of the kind of memory that leads to consciousness

—George Edelman

Another way of looking at the issue of what is the mind, is the view expressed by Allan Hobson. He says that "during waking, the external world provides input that our brain-mind orders via orientational unity, feature of figural continuity, and cause-effect linearity. The waking brain-mind achieves orientational stability via a combination of accurate monitoring of these reliably continuous inputs and by the constant updating of recent memory. In dreaming, not only are the external cues lost, but recent memory mechanisms are impaired. As a result, the normal organization and stability of mental processes is no longer operative, creating the mentalistic instability that is characteristic of dream bizarreness".

At the core of the mechanisms of various states of consciousness is what we might call a **Readiness Response,** which is a global brain and body activation produced by biologically significant stimuli, either exogenous or endogenous. As part of this brain activation, **Conscious Awareness** occurs, at least in higher animals. Awareness can be focussed to produce **Selective Attention.** Conscious, selective attention of our inner and outer worlds is the basis for human **Cognition,** that unique human ability to think about past, present, and future (Figure 5-1).

Conscious awareness is also what enables us to have **Pain Perception,** which is a response to noxious stimuli that not only causes us to avoid such stimuli, but which produces a feeling that can only be registered consciously.

When the tonic excitatory drive that actives the brain diminishes, the brain can enter the state of unconsciousness and body quiescence that we call **Sleep.** At various times during sleep, the brain state can be interrupted to convert into a unique state that enables **Dreaming.**

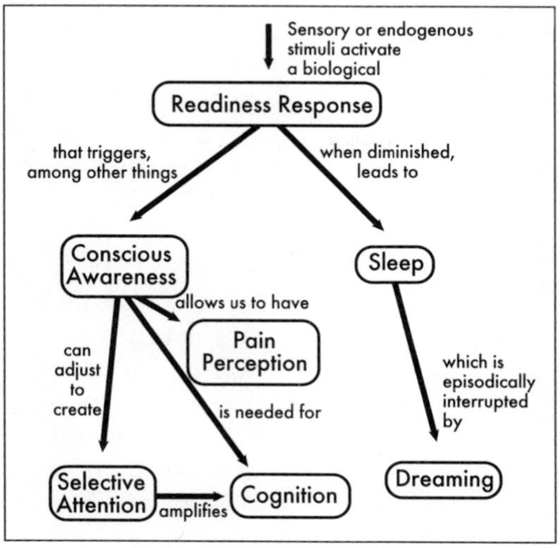

Fig. 5-1 Concept map of the various states of consciousness.

List of Principles

Readiness Response Widespread populations of neurons in the brainstem govern the responsiveness to sensory input, the state of consciousness and alertness, activation of many visceral and emotive systems, the tone of postural muscles, and the orchestration of primitive and locomotor reflexes. The brainstem mobilizes a host of sensory, motor, and visceral responses to biologically significant stimuli to produce a generalized "readiness response."

Conscious Awareness In humans, at least, the nervous system makes us aware that we are aware. That is, it "knows that it knows." This state is created, perhaps as an emergent property, by the interaction of the cerebral cortex and the brainstem reticular formation.

Consciousness and perception are mutually implicated. There is no consciousness without perception and no perception without consciousness. Similar mutual implications exist with consciousness and other mental states, such as beliefs and intentions.

Selective Attention Attentiveness to stimuli or situational context is selective and involves multiple interdependent, activated neuronal populations. Each area may make its own contribution to the stimulus or situation analysis. Because real-world stimuli and situations comprise much more information than the nervous system can accommodate, the brain adapts to this sensory overload by selectively attending at any given moment to only a small subset of the stimulus/situation and then distributing an appropriate abstracted output to appropriate output targets Past learning helps the nervous system to "decide" which features of the stimulus or situation are most salient.

Cognition	Cognition (which we may loosely regard as thinking) occurs in multiple brain areas, wherein each area makes its contribution to the analysis or task, yet each area is interdependent on the others. The cognitive process involves local computation in parallel, distributed sites, followed by re-entrant or delayed inputs back into those local process areas. The process depends on conscious awareness, is amplified by selective attention, is plastic, and involves progressive recruitment of other local areas over time.
Pain Perception	Pain is a sensory *perception,* occurring in the consciousness. Neuronal connectivity ambiguities can alter or distort the perception of pain.
Sleep	Sleep is a behavioral quiescence, generally presumed to produce rest. In higher animals, sleep is more than just quiescence, with special neural functions that generate and sustain it, along with characteristic neurophysiological changes.
Dreaming	Dreaming is a unique stage of sleep in which the brain creates its own inner consciousness that is disconnected from awareness of events in the external world. The brain is activated and produces physiological signs that are quite distinct from ordinary sleep.

Readiness Response

Widespread populations of neurons in the brainstem govern the responsiveness to sensory input, the state of consciousness and alertness, activation of many visceral and emotive systems, the tone of postural muscles, and the orchestration of primitive and locomotor reflexes. The brainstem mobilizes a host of sensory, motor, and visceral responses to biologically significant stimuli to produce a generalized "readiness response."

Explanation

Various populations of neurons in the brainstem are nodal points between sensory input and motor output. These populations have overlapping sensory fields and govern the responsiveness to sensory input, the state of consciousness and alertness, activation of many visceral and emotive systems, the tone of postural muscles, and the orchestration of primitive and locomotor reflexes. Collectively, these effects permit an adaptive response to biologically significant stimuli—in short, a "readiness response." Behavioral readiness is a state of preparedness for making appropriate behavioral responses to environmental contingencies.

Activation of the brainstem, particularly the reticular formation and the neurons surrounding the central canal in the midbrain, engages a constellation of sensory, integrative, and motor responses for adaptive response to novel or intense stimuli.

We can think of a readiness response as including visceral, behavioral, and mental activation. When the animal is aroused by sensory input, all relevant systems are activated by reflex action. The brainstem, particularly its reticular formation, has long been accepted as mediating such components of readiness as arousal and orienting. The brainstem also mediates many of the other components of readiness by generating a global mobilization that includes enhanced capability for selective attention, cognition, affect, learning and memory, defense, flight, attack, pain control, sensory perception, autonomic "fight or flight," neuroendocrine stress responses, visuomotor and vestibular reflexes, muscle and postural tone, and locomotion.

In a readiness response, the cerebral cortex is excited, enhancing consciousness and arousal level (Figure 5-2). Concurrently, muscle tone is enhanced, preparing the body for forthcoming movement instructions. Simultaneously, the limbic system is activated, which gives an emotional "flavor" to the situation and allows new stimuli to be evaluated in the context of memories. Also at the same time, neurons of the hypothalamus and the autonomic nervous system mobilize the heart and other visceral organs for the so-called fight-or-flight situations. This conglomeration of responses makes an animal *ready* to respond rapidly and vigorously, if necessary, to biologically significant stimuli. These multiple reflex-like responses are for the most part very obvious during startle reactions of animals or humans. Less intense stimuli may not evoke a full-blown readiness reflex because the brain can quickly determine whether or not a response of great intensity is appropriate to the stimuli.

Brainstem reticular formation cells receive collateral sensory inputs from all levels of the spinal cord, including such diverse sources as cutaneous receptors of the body and head, Golgi tendon organs, aortic and carotid sinuses, several cranial nerves, olfactory organs, eyes, and ears, in addition to extensive inputs from other brainstem areas, the cerebellum, and the cerebral cortex. Thus the brainstem reticular formation (BSRF) is ideally situated to monitor and respond to a variety of stimuli that can be biologically significant.

When BSRF neurons are stimulated by sensory input of any kind, they relay excitation through numerous reticular synapses to the brainstem and

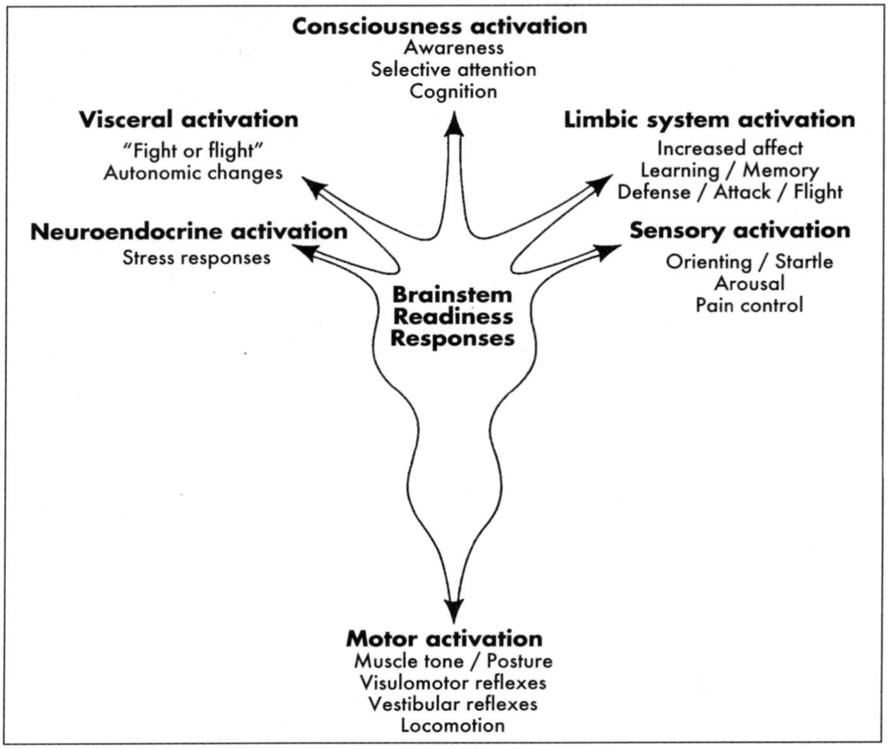

Fig. 5-2 Diagram of the physiological components of the readiness response. (From Klemm, W.R. 1993. Neurophysiology of Consciousness, pp. 105-145. In Dukes' Physiology of Domestic Animals, 11th ed. Edited by Melvin J. Swenson and William O. Reece, Cornell U. Press. Ithaca, New York.)

its extension into the thalamus and finally activate widespread zones of the cerebral cortex. In contrast, sensory information that arrives in the cortex via the main sensory paths passes through relatively few synapses and arrives at specific and very discrete zones of the cortex (sensory cortex).

The BSRF responds in similar ways to any sensory stimulus, whether from the skin, eyes, ears, or whatever, to "awaken" the cerebral cortex so that it can respond to and process stimuli. While the obvious conclusion is that the reticular formation creates arousal by direct excitation of higher centers, one cannot yet rule out the possibility that the excitation is indirect and results from a release from inhibition.

Examples

Some components of the readiness response are more evident than others. Consider orienting. If you hear a sudden, loud noise, most likely you will reflexly turn your head toward the sound and become tense. You may or may not be aware of many visceral changes, such as an immediate rise in pulse rate and blood pressure.

Another good example to which most people can relate is found in a sleeping cat who is suddenly startled into arousal by a dog barking nearby. The cat leaps to its feet, orients to the dog, becomes extremely tense (including arching of the back and extension of the limbs). The hair may rise and the cat may hiss and prepare to lash out its claws toward the dog. Clearly, the cat is mobilized for total body response to the threat.

One reason that the readiness response is a total body response is that the BSRF gets extensive input from various brain regions, particularly the neocortex and limbic system. Such input can be a major influence on behavior. For example, the cortical and limbic system activities that are associated with the distress of a newly weaned puppy probably supply a continuous barrage of impulses to the BSRF, which in turn continually excites the cortex to keep the pup awake and howling all night.

The role of the BSRF in these arousing responses can be demonstrated by direct electrical stimulation at many points within the brainstem reticulum. Such stimulation activates the neocortex (indicated by small, high-frequency waves in the EEG), the limbic system (rhythmic theta activity in the hippocampus), and postural tone (increased electrical activity of muscles). Additionally, many visceral activities are activated via the spread of BSRF excitation into the hypothalamus.

All readiness response components seem to be triggered from the BSRF, the central core of the brainstem. Evidence that the BSRF performs an important function in readiness includes: (1) Humans with lesions in the BSRF area are lethargic or even comatose; (2) Surgical isolation of the forebrain of experimental animals causes the cortex to generate an EEG resembling that seen in sleep; (3) Direct electrical stimulation of the BSRF has unique abilities to awaken sleeping animals and to cause hyperarousal in awake animals; (4) BSRF neurons develop a sustained increase in discharge just *before* behavioral and EEG signs of arousal.

Some recent studies have implicated cholinergic neurons in the pons in the EEG arousal component of the readiness response. These neurons appear to be under tonic inhibitory control of adenosine, a neuromodulator that is released during brain metabolism. This may relate to the mental stimulating properties of caffeine and theophylline, which act by blocking adenosine receptors.

TERMS

Affect — More or less equivalent to emotions. A term that is often reserved for animals, in order to seem less anthropomorphic.

Autonomic Nervous System — That subdivision of the nervous system that regulates (unconsciously) visceral activities. This includes regulation of the cardiovascular, respiratory, and digestive systems.

Brainstem That part of the brain that is interposed between the spinal cord and the forebrain. It is generally considered to include the medulla, pons, midbrain, and some scientists would also include the hypothalamus.

EEG (Electroencephalogram) Electrical potentials recorded from a population of neurons, typically recorded from electrodes placed on the scalp.

Limbic System A collection of neuronal groups in the brain that collectively act to govern emotions. Important structural areas include the hypothalamus, hippocampus, septum, amygdala, and parts of the cerebral cortex (cingulate, piriform, entorhinal).

Reticular Formation A massive collection of neurons in the central core of the brainstem. Many of these neurons are small with only local interactions, while others have long ascending and descending fiber projections.

Thalamus A mass of neuronal clusters along the midline of the brain, lying underneath the cerebral cortex. Many of these neuronal clusters are topographically mapped specific sensory pathways. Here, we are talking about the reticular portion of the thalamus, which is the anterior extension of the brainstem reticular formation. This part of the thalamus projects nonspecifically to widespread parts of the cortex.

Related Principles

Conscious Awareness
Neurotransmission
Neurohormonal Control (Overview)
Pain Perception
Receptive Fields (Senses)
Selective Attention
Reflex Action (Information Processing)

References

Eaton, R.C. (ed.) 1984. Neural Mechanisms of Startle Behavior. Plenum, New York.

Hobson, J.A. and Brazier, M.A.B. (eds.) 1980. The Reticular Formation Revisited. Raven Press, New York.

Klemm, W.R. 1990. The readiness response, pp. 105-145. In Brainstem Mechanisms of Behavior, edited by W.R. Klemm and R.P. Vertes, John Wiley & Sons, New York.

Klemm, W.R. 1992. Are there EEG correlates of mental states in animals? Neuropsychobiology. 26:151-165.

Rainnie, D.G., Grunze, H.C.R., McCarley, R.W., and Green, R.W. 1994. Adenosine inhibition of mesopontine cholinergic neurons: implications for EEG arousal. Science. 263:689-692.

Steriade, M. and McCarley, R.W. 1990. Brainstem Control of Wakefulness and Sleep. Plenum Press, New York.

Citation Classics:

Bradley P.B. and Elkes J. 1957. The effects of some drugs on the electrical activity of the brain. Brain 80:77-117.

Ciganek L. 1961. The EEG response (evoked potential) to light stimulus in man. Electroencephalogr. Clin. Neuro. 13:165-172.

Leao, A.A.P. 1944. Spreading depression of activity in the cerebral cortex. J. Neurophysiol. 7:359-390.

Moruzzi, G. and Magoun, H.W. 1949. Brain stem reticular formation and activation of the EEG. EEG Clin. Neurophysiol. 1:455-473. (Reviewed in *Current Contents*, October 5, 1981. Moruzzi, G., and Magoun, H.W.: Untitled)

Petsche H., Stumpf C., and Gogloak G. 1962. The significance of the rabbit's septum as a relay station between the midbrain and the hippocampus. I. The control of hippocampus arousal activity by the septum cells. Electroencephalogr. Clin. Neuro. 14:202-211.

Starzl, T.E., Taylor, C.W., and Magoun, H. 1951. Collateral afferent excitation of the reticular formation of the brainstem. J. Neurophysiol. 14:479-496.

Conscious Awareness

In humans at least, the nervous system makes us aware that we are aware. That is, it "knows that it knows." This state is created as an emergent property by the interaction of the cerebral cortex and the brainstem reticular formation.

Consciousness and perception are mutually implicated. There is no consciousness without perception and no perception without consciousness. Similar mutual implications exist with consciousness and other mental states, such as beliefs and intentions.

Explanation

Nervous systems exhibit different levels of responsiveness. At the most elemental level of vertebrates there are simple reflexes operating within the spinal cord or cranial nerve nuclei. More complex, compound reflexes involve integrated functioning of several reflexes. Then, there is mental awareness of environmental stimuli. Finally, in humans at least, there is awareness of awareness (i.e., consciousness). The brain operates along a continuum of states ranging from alert wakefulness to sleep to coma.

Consciousness can be defined many ways. Certainly, consciousness includes being aware of the sensory world. But in that sense, the brain of even relatively primitive animals produces behavior that indicates that the organism's nervous system is "aware" of the environment. One distinction is that in consciousness the brain is aware that it is aware. One operational consideration of conscious awareness seems to be that it emerges from a set of neural processes that continues throughout a stimulus condition—and outlasts it! In short, a conscious brain lives in the past, present, and future.

To be conscious is to have sensations, that is, mental representations of something happening outside of the brain. For humans, sensations occur in association with one or more of the five senses (sight, sound, smell, taste, touch/pressure). You cannot have consciousness without sensory awareness, although such awareness may be indirectly generated by the brain as images that simulate physical stimuli. The same idea applies to conscious thoughts, which typically are "heard" as voices in the head or "seen" as mental images. We can speak of the "mind's eye, mind's ear, mind's nose, etc."

An important distinction has to be made between the *receiving* of sensory information ("sensation") in the brain and in *perceiving* it. Seeing and perceiving are distinctly different things, and this point can be extended to sensory phenomena in general. Perception is also context dependent. Perception always has an element of interpretation or emotional response. Conscious perceptions can be distorted by ambiguous stimuli or certain drugs or by the limitations of resolution and sensitivity of the sensory receptor cells.

Consciousness arises as an emergent property from the interaction of the cerebral cortex and the central core of the brainstem, the brainstem reticular formation (BSRF) (see the principle, Readiness

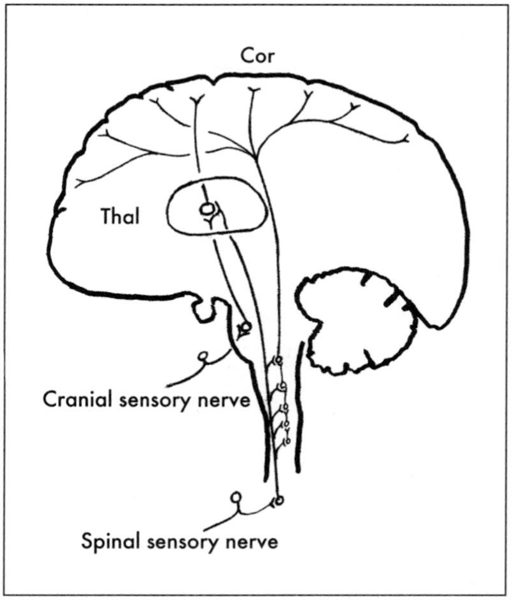

Fig. 5-3 The major pathways in the brain that are crucially involved in the genesis of consciousness: cortex (Cor), thalamus (Thal), and brainstem reticular formation (BSRF). Sensory inputs enter specific thalamic nuclei, whereas nonspecific thalamic areas receive indirect innervation from polysynaptic pathways in the BSRF. The specific thalamic nuclei relay sensory input to specific zones of the cortex ("sensory cortex"). At the same time, collateral branches of the sensory input fibers activate the BSRF, which in turn activates all of the cortex, not just the relatively small sensory portion, to induce a state of consciousness in which the sensory input can be evaluated consciously.

Response). When the BSRF is active, it provides a tonic excitatory drive for the whole cortex. In a sense, the reticular formation can be said to "arouse" cortical cells to be more receptive to sensory information arriving over the primary pathways. That same arousal effect operates on internally generated images, memories, and thoughts. Conversely, depression of reticular activity leads to behavioral sedation. Destruction of the reticular formation causes permanent unconsciousness and coma (Figure 5-3).

There is no universally accepted explanation for conscious perception. However, one appealing working hypothesis is that consciousness involves a categorization of sensory inputs at a perceptual level as a result of sensory inputs into large ensembles of topographically mapped circuitry. Such "locally mapped" responses may be extended by interactivity with other mapped circuits, leading to more "global" mapping that allows concept categorization. Expression of globally mapped concepts yields language, behavior, and consciousness (Figure 5-4).

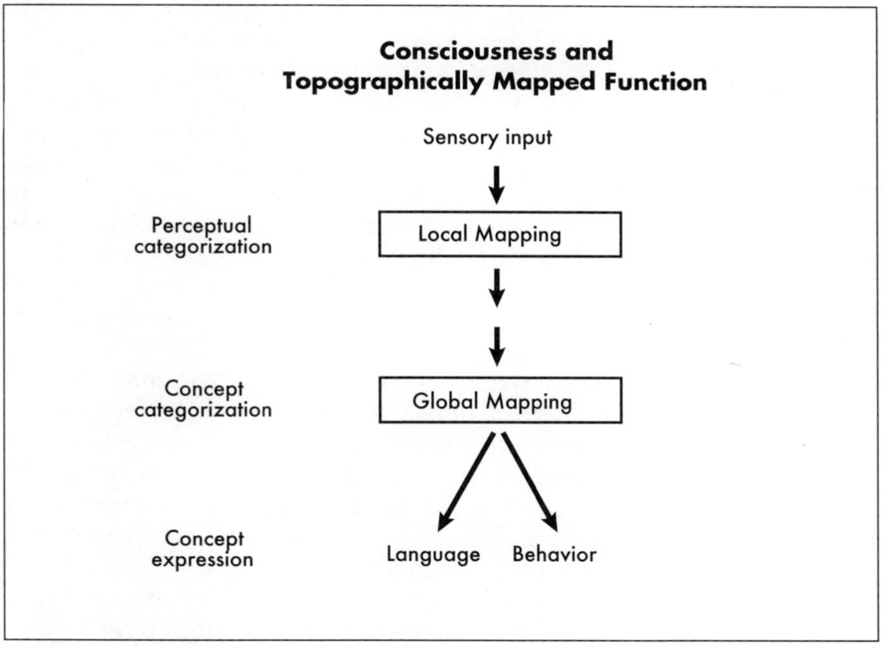

Fig. 5-4 Diagram of Edelman's idea that consciousness arises as an emergent property of local and global maps in the brain. Mapped sensory input is first represented in restricted mapped circuitry that enables recognition of the input. Then, interactions with other mapped regions of the brain produce a global mapping that enables categorizing the input in terms of concepts. Concepts then are expressed as language and/or behavior. (Based on "The Remembered Present," by G. Edelman. 1989. Basic Books, New York.)

Examples

Elemental kinds of responsiveness to stimuli that do not have to involve consciousness include such simple spinal reflexes as the knee jerk response to muscle stretch and flexor responses to painful stimuli. Some important cranial nerve reflexes include the pupillary constriction to light and swallowing and coughing reflexes. Well-known compound reflexes, which also can operate at subconscious levels, include postural reflexes mediated from the spinal cord to the brainstem and cerebellum and then back to the spinal cord. Orienting reflexes often involve body orientation to novel or startling stimuli.

Consciousness and awareness do not necessarily go together. Humans can react to stimuli to which they are consciously unaware (as in scratching a mosquito bite while concentrating on another task). In anesthesia, experiments have shown that electrical stimulation can evoke electrical responses along brain sensory pathways that are as large as or even larger than those seen during consciousness. Disease or malfunction of certain brain areas, for example, can lead to cognitive deficits of which the conscious brain may be unaware. For example, certain lesions in the occipital region of the cortex can cause Anton's syndrome, a state of blindness in which the patient is unaware of the deficit, vigorously denying it in the face of conflicting evidence. Wernicke's aphasia is the classic language-based example of unawareness of cognitive deficits. Lesions in certain temporal cortex regions cause these patients to have incomprehensible speech, yet they are unaware of the problem. Another classic syndrome is Korsakoff's disease, which is a condition associated with medial temporal lobe dysfunction in which there is severe memory deficit of which the subject may be unaware. Closed head injuries may lead to state of unawareness of memory deficit, personality change, or abrasive social behavior.

More recently, we have come to realize that humans can exhibit what is called "implicit learning" in which new knowledge and capabilities are acquired without conscious awareness. In one experimental demonstration of this point, subjects were shown a set of alphabetical character strings that were generated by a rule of an artificial grammar: for example, TSXS, TSSXXVPS, and PVV. After memorizing 20 such strings, subjects were shown a number of new strings and asked to identify those that conformed to the grammar. Performance was much better than chance levels, even though the subjects did not consciously know what the rule of grammar was. In short, while the brain was consciously engaged in memorizing the sets of letters, it was unconsciously determining underlying patterns that revealed the grammatical rule that generated the letter sets. Modern neuroscience is acquiring the capability to discuss unconscious mental life without making any reference to the vague and difficult-to-test theories of Freud. This point is also elaborated in the Learning and Memory section.

TERMS

Brainstem Reticular Formation	The central core of the brainstem, extending from the spinal cord to the thalamus. It is called reticular, because most of its neurons are small, with numerous axonal and dendritic branches.

Related Principles

Cognition
Emergent Properties (Information Processing)
Memory Kinds Reflect Memory Processes (Learning and Memory)
Memory Processes (Learning and Memory)
Readiness Response
Reflex Action (Information Processing)
Selective Attention
Sensory Modalities and Channels (Senses)

References

Bruce, V. and Green, P.R. 1990. Visual Perception. Physiology, Psychology and Ecology. Lawrence Erlbaum, Hillsdale, New Jersey.

Crick, F. 1994. The Scientific Search for the Soul. Scribner, New York.

Eccles, J.C. 1994. How the Self Controls the Brain. Springer-Verlag, Berlin.

Edelman, G.M. 1987. Neural Darwinism: The Theory of Neuronal Group Selection. Basic Books, New York.

Edelman, G.M. 1992. Bright Air, Brilliant Fire. Basic Books, New York.

Eimas, P.D. (ed.) 1990. Neurobiology of Cognition. MIT Press, Cambridge, Massachusetts.

Flanagan, O. 1992. Consciousness Reconsidered. MIT Press, Cambridge, Massachusetts.

Humphrey, N. 1992. A History of the Mind. Evolution and the Birth of Consciousness. Simon & Schuster, New York.

Joseph, R. 1992. The Right Brain and the Unconscious. Discovery of the Stranger Within. Plenum Press, New York.

Milner, A.D. and Rugg, M.D. 1992. The Neuropsychology of Consciousness. Foundations of Neuropsychology Series. Academic Press, New York.

Nagal, T. 1993. Experimental and Theoretical Studies of Consciousness. Ciba Foundation Symposium 174. John Wiley & Sons. New York.

Penrose, R. 1994. Shadows of the Mind: A Search for the Missing Science of Consciousness. Oxford University Press, Oxford, England.

Reber, A.S. 1993. Implicit Learning and Tacit Knowledge. An Essay on the Cognitive Unconscious. Oxford U. Press, New York.

Wagner, M.T. and Cushman, L.A. 1994. Neuroanatomic and neuropsychological predictors of unawareness of cognitive deficit in the vascular population. Arch. Clin. Neuropsychol. 9:57-69.

Citation Classics

Bradley, P.B. and Elkes, J. 1957. The effects of some drugs on the electrical activity of the brain. Brain. 80:77-117.

Leao A.A.P. 1944. Spreading depression of activity in the cerebral cortex. J. Neurophysiol. 7:359-390.
Rogers C.R. 1957. The necessary and sufficient conditions of therapeutic personality change. J. Consult. Clin. Psychol. 21:95-103.
Stevens S.S. 1957. On the psychophysical law. Psychol. Review 64:153-181.
Szasz T.S. The myth of mental illness. Amer. Psychol. 15:113-118, 1960.

Selective Attention

Attentiveness to stimuli or situational context is selective and involves multiple interdependent activated neuronal populations. Each area may make its own contribution to the stimulus or situation analysis. Because real-world stimuli and situations comprise much more information than the nervous system can accommodate, the brain adapts to this sensory overload by selectively attending at any given moment to only a small subset of the stimulus/situation and then distributing an appropriate abstracted output to appropriate output targets. Past learning helps the nervous system to "decide" which features of the stimulus or situation are most salient.

Explanation

The nervous system is confronted with sensory overload. There is vastly more information in the real world than the brain can accommodate. As the "Information Pyramid" in this section's introduction diagram shows, there is a progressive loss in information-carrying capacity as one moves up the different levels of the pyramid. The world supplies more information than a given sensory system can faithfully register, which in turn is more than the sensory pathways can conduct into the brain, which in turn is more than the brain can attend to, which in turn is more than can be deposited and retrieved from memory.

The brain's solution to this dilemma is to sample only selected aspects of the sensory world at any one time, i.e., to abstract the stimulus. Because these multiple sensory samples activate multiple areas within the brain, each of which makes its own contribution to the sensory registration and processing, there must be some mechanism by which these areas are linked and interdependent.

The expression of this selective attention occurs first in simultaneous processing in multiple brain areas. There may be progressive recruitment over time of other brain areas. Attention amplifies the intensity of processing. As time progresses, the process may be sustained, presumably by re-entrant inputs back into the originally activated areas from other areas to which they had projected an output. Finally, this process is plastic, subject to the influence of hormones, drugs, and learning.

One modern theory for selective attention is that of Crick and Koch, who propose that single cells focus attention by "temporal tagging." That is, attention modulates the firing pattern of neuronal discharge; the overall rate may stay the same. Such tagging may be mediated by synchronized oscillatory neural activity in the 25-60 wave/sec range that has been observed in the visual system of the cat and the monkey.

Examples

Look at a large picture of somebody's face. Fixate on one part, such as the nose. What do you see? A nose, and maybe not much else. The point is that the picture has more information content than your sensory systems and brain can accommodate at one time. You can only assimilate all that information by scanning the scene, taking in the information in small steps.

Look at someone's eyes while they are reading. The eyes fixate, then jump to another fixation point, and continue this process across and down the page. Nobody sees the whole page in one fixation.

In the brain, the stimulus features that are detected are parcelled out as abstracted fragments. Visual scenes, for example, are registered by separate neurons, each specifically tuned to respond to certain colors, edge orientation, position, and dominant eye.

Related Principles

Conscious Awareness
Cortical Columns (Information Processing)
Readiness Response
Rhythmicity and Synchronicity (Information Processing)

References

Barlow, J.S. 1993. The Electroencephalogram: Its Patterns and Origins. MIT Press, Cambridge, Massachusetts.

Cohen, R.A. 1993. The Neuropsychology of Attention. Critical Issues in Neuropsychology. Plenum Press, New York.

Crick, F. and Koch, C. 1990. Towards a neurobiological theory of consciousness. Seminars in the Neurosciences. 2:263-275.

Gray, C.M. and Singer, W. 1989. Stimulus-specific neuronal oscillations in orientation columns of cat visual cortex. Proc. Nat. Acad. Sci. U.S.A. 86:1698-1702.

Meyer, D.E. and Kornblum, S. (eds.) 1993. Attention and Performance XIV. Synergies in Experimental Psychology, Artificial Intelligence and Cognitive Neuroscience. A Science Jubilee. MIT Press Cambridge, Massachusetts.

Naatanen, R. 1991. Attention & Brain Function. Lawrence Erlbaum Assoc., Hillsdale, New Jersey.

Niebur, E., Koch, C., and Rosin, C. 1993. An oscillation-based model for the neuronal basis of attention. Vision Res. 33:2789-2802.

Posner, M.I. (ed.) 1989. Foundations of Cognitive Science. MIT Press, Cambridge, Massachusetts.

Posner, M.I. and Peterson, S.E. 1990. The attention system of the human brain. Ann. Rev. Neurosci. 13:25-42.

Citation Classic

Hubel, D.H. and Wiesel, T.N. 1959. Receptive fields of single neurones in the cat's striate cortex. J. Physiol. 148:574-591.

Cognition

Cognition (which we may loosely regard as thinking) occurs in multiple brain areas, wherein each area makes its contribution to the analysis or task, yet each area is interdependent on the others. The cognitive process involves local computation in parallel, distributed sites, followed by re-entrant or delayed inputs back into those local process areas. The process depends on conscious awareness, is amplified by selective attention, is plastic, and involves progressive recruitment of other local areas over time .

Explanation

Cognition is more of a whole-brain function than mere registration of sensory information. As explained elsewhere, specific sensory modalities are registered in specifically located clusters of cells in the cerebral cortex. However, these signals have to be interpreted in the context of other simultaneously occurring sensory input and in the context of prior experience. To do this, other brain areas have to be activated to participate in evaluation. For these other brain areas to influence activity in the primary receiving site, the various brain areas need to be reciprocally connected—which they are, not only within one hemisphere but across hemispheres via a major fiber tract known as the corpus callosum.

Because many of the sensory sites are topographically mapped, they project a mapped output to other brain regions, which may also be topographically mapped for the output that they receive indirectly from primary sensory sites. Thus, what we have are interactive maps. This mapped approach to cognition seems to be nature's parsimonious way to take bits and pieces of diverse sensory information and bind them together into a coherent whole.

Examples

One of the most obvious ways to realize how dependent our thinking is on sensory context is to look at any of a number of well-known ambiguous figures. One is the well-known figure designed by Gaetano Kaniza, which in reality has 3 "pac man" figures and three Vs. Yet when viewed, most people see two large triangles, with one large white one, without borders, sitting on top of the other triangle and on top of three black circles. When viewing such a figure, cognitive processes impose features that are not really there. Presumably, when mapped regions receive this kind of input, they impose their previous learning in a way that says what the maps "think they saw" (Figure 5-5).

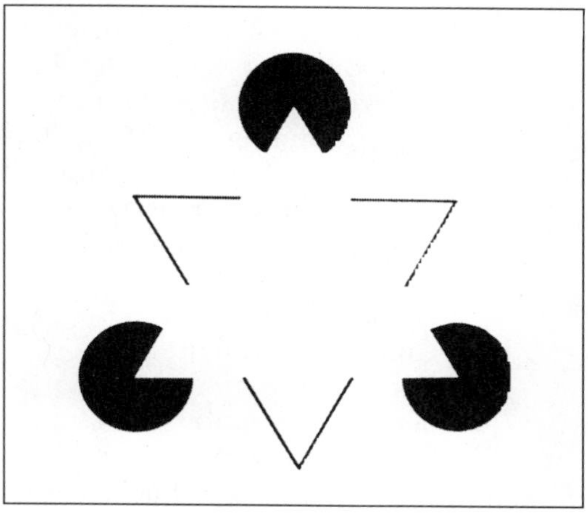

Fig. 5-5 Use of the famous Kaniza ambiguous figure to illustrate the difference between seeing and thinking. What we think when viewing this figure depends on context and memory. We may see three black circles and a black-outlined triangle, with a superimposed white, borderless triangle. Or we may see three "pac man" figures and three V-shaped figures.

Local computation during a thought process can actually be visualized by positron emission spectroscopy (PET) scans or topographical maps of the electroencephalogram. PET scans, which are images of regional changes in blood flow in the cortex, provide a picture of differential metabolic activity of neurons. Those neurons that are most active in a particular type of task have the most blood flow, and thus they "light-up" in color-coded PET scans. Such scans show that a given task is being processed in local sites, typically several sites simultaneously. Researchers have demonstrated that certain mental operations have local processing, that there are different brain control networks for different cognitive processes, and that sensory input and mental imagery can converge in the same place. When, for example, subjects are presented with different visual stimuli, the responses indicate that there are separate visual cortex areas for color, motion, and the visual form of words. In word-reading studies, words activate specific posterior visual areas. When subjects are required to indicate if the word is a noun or classify it in some other way, other specific areas light up in the PET scan; namely frontal, temporal-parietal, and certain midline areas of cortex light up. After learning sets in, the blood flow in the various areas ceases to increase.

EEG maps have the advantage that they can be created with very fine time resolution, using the electrical evoked responses in many given brain regions on a millisecond time scale. For example, a given cognitive stimulus, such as the meaning of a word presented visually, will evoke an electrical response in selected brain areas that is characterized by a series of two or three successive electrical waves of opposite polarity. The loci at which this "signature" appears may spread over time, as new local processing units are recruited. Other interesting EEG studies have shown that brain processes are distinctly different when imagining a visual or tactile object than when actually seeing or touching it.

The influence of focused attention can be seen as an expanding number of the active neurons as attention is enhanced. This can manifest as an expanding area of increased blood flow or a magnification of certain components in the evoked EEG response.

Processing of reciprocally-connected map information is suggested because during attention, the amplification of evoked EEG processing shows up after a substantial delay (300 msecs or more; recall that the "transit time" across a single synapse is on the order of 1 msec).

Related Principles

Conscious Awareness
Reciprocal Action (Information Processing)
Selective Attention
Topographical Mapping (Overview)
Rhythmicity and Synchronicity (Information Processing)

References

Baars, B.J. 1986. The Cognitive Revolution in Psychology. The Guilford Press, New York.
Flanagan, O. 1991. The Science of the Mind. 2nd ed. MIT Press, Cambridge, Massachusetts.
Gevins, A. 1989. Dynamic functional topography of cognitive tasks. Brain Topography. 2:37-56.
Humphrey, N. 1992. A History of the Mind. Evolution and the Birth of Consciousness. Simon & Schuster.
Posner, M.I., Petersen, S.E., Fox, P.T., and Raichle, M.E. 1988. Localization of cognitive operations in the human brain. Science. 240:1627-1631.
Posner, M.I. 1993. Seeing the mind. Science. 262:673-674.

Pain Perception

*Pain is a sensory **perception**, occurring in the consciousness. Several neural influences regulate pain threshold and neuronal connectivity ambiguities can distort the perception of pain.*

Explanation

The fact that anesthesia simultaneously produces unconsciousness and alleviates pain is no coincidence. Pain cannot be perceived without consciousness.

Any stimulus, if of sufficient intensity, can be painful. Nonetheless, certain sensory systems are more likely to mediate pain. Painful information is carried into the brain from spinal and cranial nerves in at least two fiber bundles. Some potential for altering transmission along pain pathways exists at several levels, including synapses in the spinal cord, the brainstem, thalamus, and sensory cortex.

Examples

The effects of anesthesia are well known. What may not be well known are studies showing that noxious stimuli that normally would be painful are still able to activate responses in spinal, thalamic, and cortical pathways during anesthesia that may be as robust as responses without anesthesia. In other words, under anesthesia, pain-producing information is *received* without being *perceived*.

What about pain-killing drugs that do NOT alter consciousness? These may act peripherally, in the mode of aspirin for example, to reduce the irritation and inflammatory causes of pain. Local anesthetics actually block transmission of action potentials along nerve fibers. Other drugs, such as codeine and other opiates, act on endogenous pain-

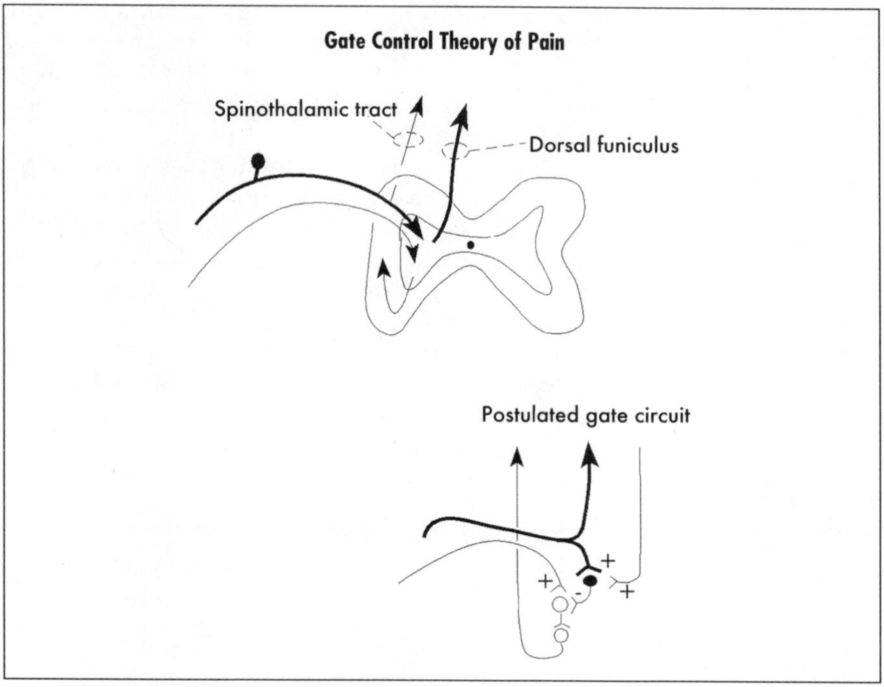

Fig. 5-6 Diagram of the "gate-control" theory of pain, originally espoused by Melzck and Wall. Stimuli that cause pain (nociceptive stimuli) flow in pathways indicated by the narrow line from the periphery into the dorsal part of the spinal cord. From there, the nociceptive information is sent in fiber tracts (spinothalamic and other tracts not shown) to reach the thalamus and neocortex. The spinal nerves that contain nociceptive fibers also contain larger fibers (darker lines) that convey touch and pressure information, which also converge in the same dorsal part of the spinal cord. A postulated gating circuit in that region of the spinal cord is shown below wherein touch/pressure inputs that occur simultaneously with pain may activate inhibitory neurons that shut down information flow along nociceptive pathways. These same inhibitory neurons may also be activated from descending fibers coming from the brain, thus providing a way for the brain to influence the perception of pain.

alleviating systems that have specific postsynaptic receptors for opiates (which coincidentally bind the endogenous opiate peptide neurotransmitters that are concentrated in certain brain areas that have been implicated in mediating pain).

One of the enigmas of pain is that the threshold for pain differs greatly among people. Hindu ascetics can perform amazing feats of walking on coals and lying on nails that would be excruciating to other people. Even within the same person, the pain threshold varies, depending on coincident circumstances. Soldiers in battle, for example, have been known to continue fighting without apparent awareness of pain from horrible wounds, only to feel the pain when the mind is not "distracted" by the heat of battle.

We also know that certain kinds of cutaneous stimulation relieves pain, such as liniments and even special electrical stimulators. One unifying explanation for these diverse modulatory influences on the perception of pain is a theory that routing of stimuli that cause pain (nociceptive stimuli) can be blocked at the spinal cord level, by either descending influences from the brain or simultaneously occurring sensory information of other kinds (Figure 5-6).

So-called "referred pain" is an example of how connectivity ambiguities can cause confused conscious interpretation of sensation. It is well known that heart attack victims, for example, commonly feel extreme pain in their left arm. The mind thinks that the arm hurts, when in actuality the pain is arising from heart muscle that is dying because of lack of oxygen. The reason for this confusion is that the neurons in the spinal cord that relay sensation from the left arm also get convergent sensory input from the heart. Since the heart is not represented by neurons in the cortex, while the arm is, the cortical receiving area for arm input "concludes" that the pain is coming from the arm.

TERMS

Endorphins — Endogenous opiate peptides; certain proteins or protein components (peptides) that act on the same receptors that coincidentally bind opiate drugs. The endorphins produce effects akin to those caused by opiates.

Nociceptive Stimuli — Stimuli that are damaging and which in humans would be perceived as painful.

Perception — As opposed to the receipt and registration of sensory input at the level of non-awareness, perception is a conscious awareness of the sensory input. A wide variety of sensory input can be consciously perceived (sight, sound, touch, temperature, pressure, taste, smell, etc.). However, some stimuli, such as the tone of muscles and various visceral sensations such as blood pressure and heart rate, may never be consciously perceived.

Related Principles

Circuit Design (Overview)
Conscious Awareness
Neurotransmitters (Cell Biology)
Selective Attention

References

Basbaum, A.I. and Fields, H.L. 1984. Endogenous pain control systems: brainstem spinal pathways and endorphin circuitry. Ann. Rev. Neurosci. 7:309-338.

Dubner, R. and Bennett, G.J. 1983. Spinal and trigeminal mechanisms of nociception. Ann. Rev. Neurosci. 6:381-418.

Kerr, F.W.L. and Wilson, P.R. 1978. Pain. Ann. Rev. Neurosci. 1:83-102.

Klemm, W.R. 1992. Are there EEG correlates of mental states in animals? Neuropsychobiology. 26:151-165.

Wall, P.D. and Jones, M. 1991. Defeating Pain, The War Against a Silent Epidemic. Plenum Press, New York.

Willis, W.D. 1985. Pain and Headache, Vol. 8, S. Karger, Farmington, Connecticut.

Citation Classics

D'amour F.E. and Smith D.L. 1941. A method for determining loss of pain sensation. J. Pharmacol. Exp. Ther. 72:74-79.

Fields H.L. and Basbaum A.I. 1978. Brain stem control of spinal pain transmission neurons. Ann. Rev. Physiol. 40:217-248.

Janssen P.A.J., Neimegeerss C.J.E., and Dony J.G.H. 1963. The inhibitory effect of fentanyl and other morphine-like analgesics on the warm water induced tail withdrawal reflex in rats. Arzneim.-Forsch.-Drug Res. 13:502-507.

Mayer D.J., Wolfle T.L., Akil H., Carder B., and Liebeskind J.C. 1971. Analgesia from electrical stimulation in the brainstem of the rat. Science 174:1351-1354.

Melzack R. and Wall P.D. 1965. Pain mechanisms: a new theory. Science 150:971-979.

Tsou K. and Jang C.S. 1964. Studies on the site of analgesic action of morphine by intracerebral micro-injection. Sci. Sinica 13:1099-1109.

Yaksh T.L. and Rudy T.A. 1976. Analgesia mediated by a direct spinal action of narcotics. Science 192:1357-1358.

Sleep

Sleep is a behavioral quiescence, generally presumed to produce rest. In higher animals, sleep is more than just quiescence, with special neural functions that generate and sustain it, along with characteristic neurophysiological changes.

Explanation

All vertebrates, even fish, show periods of behavioral quiescence that we call sleep. Humans know sleep as a spontaneous lapse into unconsciousness in which the brain disengages itself and the body from interaction with the environment. Sleep is recognized most obviously as a reversible unconsciousness and a relative immobility of an animal. However, because we cannot know the extent to which lower animals such as fish are conscious, we cannot equate their quiescence with sleep.

Behavior during sleep includes such characteristics as (1) reduced ability to analyze and respond to changes in the environment, (2) increased threshold to sensory stimulation, and (3) relative muscular inactivity. In all animals except birds, the tonic proprioceptive, labyrinthine, and visual reflexes that are responsible for righting the body and maintaining normal posture no longer operate.

In the brain, sleep is characterized by relatively synchronized activity of billions of neurons, especially those located in the thalamus and the cortex. Synchronization of cortical activity produces summated postsynaptic potentials, whose extracellular currents create relatively large, long-duration voltage waveforms in the electroencephalogram (EEG). Thus, typical sleep is sometimes called slow-wave sleep. Also conspicuous in the light stages of sleep are cortical EEG spindles, which are bursts of waves in the frequency range of 7-14 per sec. However, by themselves these electrical correlates are not specific indicators of sleep, for they can also occur in certain drug sedation states, during natural drowsiness, during anesthesia, and in certain disease conditions.

The slow-wave stage of sleep can be thought of as *deactivated sleep (DS)* to contrast it from another, qualitatively different, stage of sleep (activated sleep) that is discussed as the Dreaming principle. In DS, the flow of sensory information from the thalamus to the cortex is impaired by inhibitory processes in the thalamus. Spinal reflexes are also depressed. Absence of movement is conspicuous during DS, and muscle tone is generally reduced.

During DS, the dominance of the parasympathetic nervous system typically depresses visceral functions. Heart rate slows, and blood pressure decreases. Pupils constrict (subtle changes in pupil size can be monitored as an index of the degree of drowsiness). Breathing movements are slower, alveolar CO_2 levels drop, blood pH becomes slightly more acid, total metabolic rate decreases, and body temperature drops. However, no parallel signs of metabolic depression in the brain occur. Cerebral blood flow can even increase during sleep. Moreover, some neurons become more active during DS than during wakefulness.

Mechanisms of Deactivated Sleep

Many theories have been proposed to explain sleep, but no one theory seems sufficient by itself. Some insight is provided by the normal requirements for inducing sleep.

Sleep is best induced by limiting movement and by restricting excessive sensory stimulation, which can be aided by a continuous monotonous sound. Reducing the proprioceptive drive with muscle relaxants also promotes sleep. Under natural conditions, fatigue of postural tonus markedly decreases the number of proprioceptive impulses. The highest cerebral centers, thus partly isolated from input stimulation, are able to lapse into inactivity, resulting in sleep. Although fatigue is the usual cause of muscle relaxation, it is not necessary to induce sleep. Sometimes, conditioned responses control sleep and wakefulness cycles.

Among the several humoral theories of sleep is the shared premise that certain body chemicals, usually metabolic end-products, accumulate and excite the neurons that cause sleep. This possibility has prompted testing of dialysates and extracts of fluids from sleep-deprived subjects. Such studies reveal the existence of several naturally occurring peptides that promote DS.

Immunological mechanisms may be involved in the genesis of sleep. The leukocytic interleukin I, which also occurs in astrocytes, promotes sleep.

Several other, even more compelling, lines of evidence, suggest that serotonin is involved in sleep control: (1) brain levels of serotonin are high in sleep and low during behavioral activity; (2) intracarotid or intravertebral injections of serotonin induce EEG and ocular signs of sleep, as does topical application of serotonin to an area on the floor of the fourth ventricle; (3) depletion of serotonin interferes with sleep; (4) in such depleted animals, injection of the serotonin precursor (5-hydroxytryptophan) tends to restore normal sleeping; and (5) lesions of the serotonergic neurons in the midline region of the brainstem interfere with sleep.

Another possibility, not mutually exclusive of the serotonin theory, is the possible role of acetylcholine. Cholinergic substances injected in parts of the limbic system, preoptic region, parts of the thalamus, and the central core of the pons in the brainstem also can induce sleep.

The parasympathetic nervous system dominates during DS. This, however, does not prove a parasympathetic cause of sleep. For example, during extreme alertness, there is activation of the sympathetic system; but sympathetic activity results from, rather than causes, the alertness.

Since decorticate animals sleep, the basic sleep-wakefulness mechanisms must be subcortical. These subcortical mechanisms include the activating role of the central core of the brainstem (see the Readiness Response principle) and the deactivating role of the thalamus. Diminution of brainstem-arousing activity is an important predisposing cause of sleep. The basic question of why brainstem activity diminishes naturally has not been satisfactorily answered. However, impulse discharge rate in the BSRF and the level of behavioral arousal do correlate. Reduction of brainstem arousing activity, either normal or abnormal (by barbiturates, tranquilizers, experimental injury, or disease), can lead to various degrees of sedation, sleep-like states, and coma.

But sleep is more than a passive phenomenon. Sleep also appears to be promoted by activity in such areas as the preoptic region of the hypothalamus, the midline region of the thalamus, certain raphe nuclei in the midbrain, and the solitary tract region of the medulla.

What is the purpose of sleep? It is not clear. *Rest* can be achieved without sleep. No one has been able to associate sleep with any specific life-maintaining function related to measurable physiological variables. Human experiments seem to suggest that sleep is needed for the brain but not for the body. One possibility is that sleep is nature's way to provide the body with large amounts of free-radical scavenger compound: melatonin. Melatonin is produced in large amounts during nighttime sleep, and this compound has an unusually potent ability to scavenge free radical molecules that are normally produced as a byproduct of oxygen metabolism. Free radicals promote aging and can cause cancer, because they can bind and disrupt the genome.

Sleep is necessary for health, and animals deprived of sleep will eventually become ill. Forcing animals to stay awake can make them irritable and they may engage in vicious fights. Sleep deprivation apparently alters certain immune functions.

One purpose of sleep may be to enable the activated phase of sleep, for which some purposes have been hypothesized (see the Dreaming principle).

Examples

Most fish show periodic episodes of what could be called sleep. However, some fish, such as tuna, mackerel, and certain shark species seem to always be active. But the sleep of fish does not seem to be the same as that in higher animals; for example, they show no shift toward large, slow waves in the EEG. Amphibians also show quiescent periods, but again, show no EEG changes like those seen in the higher animals. Nocturnally active tree frogs exhibit behavioral sleep in the daytime, including being hyporesponsive to stimuli. However, behavioral signs can be misleading. The bullfrog, *Rana catesbiana*, apparently is fully responsive to external stimuli at all times, even when behaviorally withdrawn and quiet. With reptiles, some physiological signs of sleep appear in certain species. One study revealed EEG slow waves in young crocodiles when they became behaviorally asleep. Lizards commonly show high-voltage EEG spikes when they sleep. Birds and mammals generally show both behavioral and EEG signs of sleep. The EEG of mammals differ in that they have both slow EEG waves and spindles.

Scientists interested in the evolution of sleep have focussed on the Echidna (spiny anteater) because this is a primitive egg-laying mammal. These animals show typical slow-wave signs of sleep about 35% of each day, but there are no spindles. Also, they are one of the few terrestrial mammals that do not show physiological signs of activated sleep (dream sleep).

Sleeping marine mammals are an especially interesting case, because they cannot sleep in the same manner as other mammals, lest they drown. The dolphin apparently solves this problem by showing periodic EEG signs of sleep that are confined to ONE hemisphere at a time. The Northern fur seals also solve the problem by sleeping one hemisphere at a time; they can also sleep on their side with a flipper in motion to maintain a posture that keeps the nostrils above the water's surface.

TERMS

Deactivated Sleep (Slow-wave Sleep)	That phase of sleep when the brain is less active and the EEG is dominated by large, irregular, slow waves.
Activated Sleep ("Paradoxical" Sleep, "Rapid Eye Movement" Sleep)	That phase of sleep when the brain is intensely active, causing a low voltage, high frequency EEG and spastic twitches of muscles, including eye muscles. This stage of sleep is associated with dreaming.
Electroencephalogram (EEG)	A recording of the voltage changes from the brain, typically recorded from the scalp or skin overlying the cranial vault. It is actually a plot of voltage (microvolts) as a function of time. The EEG is created by multiple neuronal generators, particularly from the pools of neurons in the cortex, because they are closest to the surface.
Parasympathetic Nervous System	A division of the peripheral nervous system can help regulate visceral function. This system promotes normal resting functions of viscera.
Proprioception	Sensory information associated with muscle tone and tendon stretch that gives rise to sensations and unconscious awareness about body and limb position.
Sympathetic Nervous System	A division of the peripheral nervous system that helps regulate visceral function. This system activates the visceral functions that contribute to appropriate responses under emergency or stressful conditions and suppresses the visceral functions that do not contribute to appropriate responses.

Related Principles

Conscious Awareness
Dreaming
Neurotransmitters (Cell Biology)
Readiness Response
Rhythmicity and Synchronicity (Information Processing)

References

Antrobus, J. and Bertini, M. (eds.) 1992. Neuropsychology of Sleep and Dreaming. Lawrence Erlbaum Associates. Hillsdale, New Jersey.

Ayala-Guerrero, F. and Hitron-Resendiz, S. 1991. Sleep patterns in the lizard, *Ctenosaura pectinata*. Physiol. Behav. 49:1305-1307.

Buchet, C., Deswasnes, G., and LeMaho, Y. 1986. An electrophysiological study of sleep in emperor penguin under natural ambient conditions. Physiol. Behav. 38:331-335.

Flanigan, W.F. Jr., Wilcox, R.H., and Rechtschaffen, A. 1973. The EEG and behavioral continuum of the crocodilian, Caiman sclerops. Electroenceph. Clin. Neurophysiol. 34:521-538.

Hobson, J.A. 1989. Sleep. W.H. Freeman, New York.

Karmanova, I.G. and Lazarev. S.G. 1980. Neurophysiological characteristics of primary sleep in fish and amphibian, p. 437-442. In: Sleep 1978, edited by L. Popoviciu, B. Asgian, and G. Badiu. S. Karger, Basel.

Mancia, M. and Marini, G. 1990. The Diencephalon and Sleep. Raven Press, New York.

Meglasson, M.D. and Huggins, S.E. 1979. Sleep in a crocodilian, *Caiman sclerops*. Comp. Biochem. Physiol. 63A. 561-567.

Oleksenko, A.I., Mukhametov, L.M., Polyakova, I.G., Supin, A.Y., and Kovalzon, V.M. 1992. Unihemispheric sleep deprivation in bottle-nosed dolphins. J. Sleep Res. 1:40-44.

Steriade, M. and McCarley, R.W. 1990. Brainstem Control of Wakefulness and Sleep. Plenum Press, New York.

Steriade, M., McCormick, D.A., and Sejnowski, T.J. 1993. Thalamoccortical oscillations in the sleeping and aroused brain. Science. 262:679-684.

Wauquier, A. et al. 1989. Slow Wave Sleep: Physiological, Pathophysiological, and Functional Aspects. Raven Press, New York.

Weitzman, E.D. 1981. Sleep and its disorders. Ann. Rev. Neurosci. 4:381-417.

Citation Classics

Allison, T. Van Twyver, H. Van, and Goff, W.R. 1972. Electrophysiological studies of the echidna, *Tachyglossus aculeatus*. Arch. ital. Biol. 110:145-184.

Dement W. and Kleitman N. 1957. Cyclic variations in EEG during sleep and their relation to eye movements, body motility, and dreaming. EEG Clin. Neurol. 9:673-690.

Hobson J.A., Lydic R., and Baghdoyan H.A. 1986. Evolving concepts of sleep cycle generation-from brain centers to neuronal populations. Behav. Brain. Sci. 9:371-400.

Jouvet, M. 1969. Biogenic amines and the state of sleep. Science. 163:32-41.
(Reviewed in Current Contents, April 25, 1983)

Koella, W.P., Feldstein, A., and Czicman, J.S. 1968. The effect of parachlorophenylalanine on the sleep of cats. Electroenceph. clin. Neurophysiol. 25:481-490.
(Reviewed in Current Contents, October 26, 1981).

Dreaming

Dreaming is a unique stage of sleep in which the brain creates its own inner consciousness that is disconnected from awareness of events in the external world. The brain is activated and produces physiological effects that are quite distinct from ordinary sleep.

Explanation

In 1957, EEG studies in humans revealed that during sleep there were alternating periods of EEG activity in which the waves were not the large, slow waves typical of sleep but were actually small, high-frequency waves that were accompanied by stereotyped rapid-eye movements. When people were awakened in the midst of these episodes, they usually remembered a dream being interrupted. Subsequently, these same physiological changes were observed to occur periodically during the sleep of other mammalian species.

The dream stage of sleep is qualitatively different from deactivated sleep (DS). This is considered as an *activated sleep (AS),* except that the activation is not expressed behaviorally. Inhibition of motor activity during AS is one of its cardinal features. One of the paradoxes of this state (it is often called paradoxical sleep) is that as the activity in the brainstem arousal system increases, there is a corresponding increase in activity of motor inhibitory systems, particularly those in the medulla.

In AS, there is activation of the cortical EEG, hippocampal theta rhythm, spikelike EEG waves in several vision-related brain areas (pons, lateral geniculate nuclei, and occipital cortex), bursts of rapid eye movements (REM), and postural muscle atonia with super-imposed phasic twitching. During AS, both monosynaptic and polysynaptic reflexes are markedly depressed, except for those periods when phasic twitching occurs.

Another unusual feature is that AS is a *deep* stage of sleep, in that the threshold for arousing stimuli is greater than during DS.

The phylogenetic distribution of AS indicates that it is a relatively recent evolutionary development. AS does not occur in fish or amphibians and is poorly developed in reptiles, birds, and primitive mammals.

AS represents an ontogenetically earlier condition than slow-wave sleep. In week-old kittens, AS constitutes 90 percent of the total sleep time. During the second week, more variability appears in the EEG, with some arousal during wakefulness and some spindles during DS. At 3 weeks, the EEG resembles that of the adult.

Mechanisms of Activated Sleep

The brain area that seems to be most concerned with producing AS is the pons. Neither surgical removal of the cerebellum nor complete transections of the brainstem at the midbrain level prevent the peripheral physiological signs of AS. Similarly, transection of the brainstem caudal to the pons fails to prevent EEG signs of AS. Physiological signs of AS are abolished by lesions in the pons of intact animals, and low-level electrical stimulation of the midbrain or pontine reticular formation during DS can trigger AS. Intense stimulation during DS causes awakening.

"Executive" structures for AS are located in the pontine reticular formation. The system causing postural atonia is located in the medial part of the norepinephrine-containing neurons in the locus coeruleus and its immediate vicinity. These neurons have a descending pathway to the n. magnocellularis in the medulla, which is a zone that ultimately depresses muscle tone at the spinal level. Gigantocellular field units fire in correlation with movements, both in AS and in the awake state. They are thus not critical to the AS state and cannot be AS "executive neurons." The executive neurons appear to be located in the mediodorsal part of the pons, near the locus coeruleus, and the ventromedial part of the caudal pons and rostral medulla. In both regions, tonically activated neurons can be located during AS. What activates these neurons during AS? One possibility is input from noradrenergic and serotonergic neurons that are located nearby and that cease firing during AS (releasing the executive neurons from inhibition).

Lesion studies indicate that only the activity of certain neurons in the dorsolateral pons is crucial for the generation of AS. Total brainstem transections at the pontomedullary junction reveal that neither the pontine reticulum nor the medullary reticulum were sufficient for generating the indices of AS; both zones must interact to coordinate the various activities associated with AS.

We do not know why AS is necessary, but evidence exists for several possibilities, namely, that AS (1) serves as an endogenous source of stimulation to promote maturation, (2) is needed to establish neuronal pathways serving binocular vision, (3) is an internal reward mechanism, (4) promotes consolidation of memories for recently learned events, (5) is essential for maintaining emotional stability, and (6) is required for sustaining norepinephrine and dopamine neurotransmitter systems. All of these are consistent with another possibility; namely, that AS enables "readiness rehearsal" for the forthcoming day's awake situations and experiences. It may be detrimental for the brain to sustain long periods of depression, as in DS. Accordingly, one would expect that AS deprivation would make animals less ready to respond appropriately to biologically meaningful stimuli and situations during their wakefulness periods. This view predicts that such deprived animals would show decreased reaction times, lessened orienting, decreased adrenal stress responses, lessened EEG activation responses to stimuli, and decreased performance on various conditioning paradigms.

Examples

Humans have dreams that they can sometimes remember in their awake consciousness. Indeed, physiological monitoring reveals that humans dream much more than they think they do because the process of awakening interferes with the consolidation of memory (see the Memory Consolidation principle). Our introspection about the dreams we do remember tells us that our brain creates episodes of a virtual reality in our sleep.

The assertion that dreaming is a state of conscious awareness does not contradict our earlier claim that consciousness is inevitably linked to sensation. The difference here is that in dreaming, where there is usually no physical stimulation from the external world, the brain generates its own representation or "reminders" of stimulation. For example, just as the visual cortex becomes active in an awake person who is viewing scenes, it has been experimentally demonstrated that the visual cortex becomes activated during dream imagery.

Since AS is associated with dreaming in humans, the question arises whether a similar association exists in animals. Much anecdotal evidence seems to indicate that animals do dream. Sleeping pet animals such as dogs and cats provide the best opportunity to observe the signs of dreaming in animals. Dogs, for example, will paddle their feet, bark, twitch their noses, and show REM, as if they were engaged in a virtual reality chase of another animal (perhaps a rabbit or a cat). All of the physiological and behavioral signs of dreaming in people can also be confirmed in higher animals (mammals and birds).

Cats have been the most studied species, but physiological signs of dream sleep have been seen in most mammalian species studied. Birds show most of the physiological signs of AS, but they apparently do not have ponto-geniculo-occipital waves or hippocampal theta activity. Also, most birds do not show the postural collapse exhibited by mammals.

The tonic depression of muscle activity probably arises in the brainstem reticulum. A dramatic phenomenon of AS without such depressed motor activity can be demonstrated by producing small, bilateral lesions in the dorsolateral part of the pontine reticular formation. When sleeping cats with such lesions enter the AS stage, they raise their heads, make body-righting movements, alternately move their limbs, and even try to stand. Although these cats do not respond to stimuli, they act as though they are startled or searching, and sometimes they even attack nearby objects. During alert wakefulness these cats behave almost normally.

The basic age relationship is that young animals sleep more than older ones of the same species. Moreover, most of the sleep time of the young is spent in AS, and amount of AS time gradually decreases with maturation. The high incidence of AS in the young probably begins in the fetal stage. Fetal calves within 30 days of parturition have almost half of their total sleep time in AS; EEG signs of nonsleeping were present only about 15 percent of the time.

Some insight into AS mechanisms can be gleaned from the demonstration of a biological need for AS. This has been shown in cats, for example, by awakening them every time they exhibited physiological signs of AS (but not DS). On successive days of deprivation, an increasing number of awakenings was needed. On the first day of recovery after deprivation, when sleep was not interrupted, the AS phase increased to 53 percent of the total sleep time. Memory consolidation also seems to be promoted during AS. In one study of human learning of a visual discrimination task, the performance improved after a normal night's sleep. Selective disruption of the dream stage of sleep prevented the improvement normally found after a normal night's sleep. Disruption of the slow-wave stages of sleep did not affect subsequent performance. Disruption of dream sleep has no effect on a similar, but previously learned, task. Thus, REM sleep seems to be crucial for consolidation of memory for this learning situation.

TERMS

Hippocampus A part of the primitive cortex that in higher animals has become linked into a system of structures (the "limbic system") that regulates emotions. The hippocampus also has a special role in the formation (consolidation) of long-term memories.

Ontogenetic Refers to developmental changes in the young.

Phylogenetic Refers to the relationship among species.

Ponto-geniculo-occipital Waves Bursts of rhythmic electrical activity that occur during dream sleep in certain linked areas of the pons, geniculate body of the thalamus, and in the occipital cortex. These waves correlate with rapid eye movements and are thought to reflect the activity of the neurons that actually cause the eye movements.

Theta Rhythm A rhythmic brain-wave pattern, of 4-7 waves/sec, that is generated by the hippocampus.

Related Principles

Conscious Awareness
Memory Consolidation (Learning and Memory)
Readiness Response
Sleep

References

Gackenbach, J. and LaBarge, S. 1988. Conscious Mind, Sleeping Brain. New Perspectives on Lucid Dreaming. Plenum Press, New York.
Hobson, J.A., McCarley, R.W., and Wyzinski, P.W. 1975. Sleep cycle oscillation: reciprocal discharge by two brainstem neuronal groups. Science. 189:55-58.
Hobson, J.A., McCarley, R.W., and Nelson, J.P. 1983. Location and spike-train characteristics of cells in anterodorsal pons having selective decreases in firing rate during desynchronized sleep. J. Neurophysiol. 50:770-783.
Hobson, J.A. 1988. The Dreaming Brain. Basic Books, New York.
Jouvet, M. 1967. The states of sleep. Scientific American. 216:62-72.
Karni, A., Tanne, D., Rubenstein, B.S., Askenasy, J.J.M., and Sagi, D. 1994. Dependence on REM sleep of overnight improvement of a perceptual skill. Science. 265:679-682.
Winson, J. 1985. Brain and Psyche. The Biology of the Unconscious. Anchor Press/Doubleday, Garden City, New York.

Citation Classics

Aserinsky, E. and Kleitman, N. 1953. Regularly occurring periods of eye motility, and concomitant phenomena, during sleep. Science 118:273-274.
Dement, W.C. and Kleitman, N. 1957. Cyclic variations in EEG dur-

ing sleep and their relation to eye movement, body motility, and dreaming. Electroenceph. Clin. Neurophysiol. 9:673-690.

Dement, W.C. 1958. The occurrence of low voltage, fast electroencephalogram patterns during behavioral sleep in the cat. Electrophysiol. Clin. Neurophysiol. 10:291-296.

Jouvet, M. 1963. The rhombencephalic phase of sleep, pp. 406-424. In Brain Mechanisms, Progress in Brain Res. Vol. 1., ed. by G. Moruzzi, A. Fessard, and H.H. Jasper, Elsevier, Amsterdam.

States of Consciousness: Study Questions

1. List and explain each of the principles in this category.
2. For each of the principles, provide an example *that is not mentioned in this text.*
3. What is the significance of the fact that the brainstem reticular formation has so many neurons and synapses?
4. Why are the functions of the brainstem reticular formation considered to mediate a behavioral "readiness" response? Summarize the components of that response.
5. Why is consciousness considered to be an emergent property?
6. What areas of brain are most involved in consciousness? Why?
7. Are animals conscious to the same extent as humans? Why or why not?
8. Why is selective attention a useful capability?
9. Why do we say that cognition is more of a whole-brain function than is sensation?
10. How does the Kaniza figure illustrate the relationship between cognition and sensation?
11. What kinds of evidence indicate that cognition occurs in parallel, distributed networks?
12. Why do we say that pain can only occur as a conscious perception?
13. Explain referred pain.
14. How do we know if or when animals feel pain?
15. Why must there be some kind of sensory gating mechanism for pain?
16. Why do we say that sleep is an active—and not solely passive—process?
17. How do you interpret the fact that during sleep there seems to be an increase in the amount of synchronization among neuronal populations in the cortex?
18. What is the relationship between sleep and visceral control?
19. How do the physiological signs of activated sleep differ from those of deactivated sleep?
20. Discuss the various possible explanations for why higher animals need to have activated sleep.
21. Defend the argument that the purpose and biological significance of sleep is to enable dreaming.
22. As in the question above, what would then be the significance of the fact that lower animals do not sleep in the same way that mammals do.

CHAPTER 6

Emotions

And men should know that from nothing else but from the brain come joys, laughter and jests, and sorrows, griefs, despondency and lamentations. And by this, in an especial manner, we acquire wisdom and knowledge, and we see and hear and know what are foul and what are fair, what sweet and what unsavory... and by the same organ we become mad and delirious and fears and terrors assail us...

— Hippocrates, 5th century B.C.

Emotions are generated at the subconscious level by a diverse set of structures that are collectively called the limbic system. Components of the limbic system are highly interconnected. In addition, there are many access and egress routes of this system from the brainstem and cerebral cortex. In ways that are not yet understood, this **Neural Origin of Emotions** creates internal drives or **Motivation,** that guide animals toward goal-directed behaviors. These drives create positive or negative emotions, depending on the nature of the **Reinforcement** received from such goal-directed behaviors. In turn, the positive or negative emotions help determine whether the animal or person seeks or avoids situations that are associated with the reinforcement (Figure 6-1).

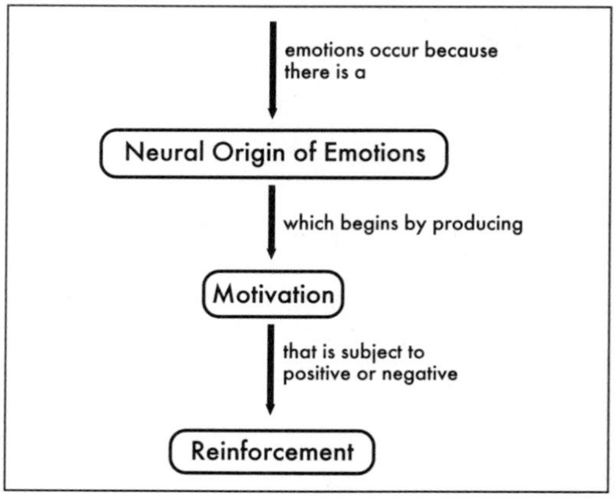

Fig. 6-1 Concept map for principles of emotion.

List of Principles

Neural Origin of Emotions	Neural systems generate subjective "feelings" and "emotions." These affective states seem to be regulated by a systems process involving a highly interconnected network of subcortical structures, collectively called the limbic system. Affective states help to drive the rest of the brain to engage in certain operations and suppress alternatives.
Motivation	Animals have internal drives that produce goal-directed behaviors. These drives motivate animals to produce specific avoidance or approach behaviors, depending on situational context.
Reinforcement	The brain has systems for both positive and negative reinforcement. That is, these systems promote behaviors that are either repeated or are avoided, as a consequence of the feedback from the first behavioral episodes.

Neural Origin of Emotions

Neural systems generate subjective "feelings" and "emotions." These affective states seem to be regulated by a systems process involving a highly interconnected network of subcortical structures, collectively called the limbic system. Affective states help to drive the rest of the brain to engage in certain operations and suppress alternatives.

Explanation

There is a system of brain areas that has been specifically linked in the genesis of emotional states. These brain areas include the septum, hippocampus, entorhinal cortex, amygdala, cingulate cortex, hypothalamus, and selected parts of the thalamus. Disease or experimental manipulation of any one part of these highly interconnected areas can lead to emotional disturbances. The whole gamut of emotions seems to be modulated by this system: rage, fear, joy, aggression, euphoria, dysphoria, etc.

The "systems" property of this collection of brain areas is indicated by the fact that no simple, one-to-one relationship can be found for a given part of the system and a given emotion. Note that people are not always consciously aware of their emotional state.

Examples

The first definitive implication of the limbic system in emotions was probably the observation that dogs with rabies ("mad dogs") had viral inclusion bodies in neurons of the hippocampal part of the brain. A study of the anatomy of the hippocampus revealed that it had extensive, reciprocal connections with such areas as the septum, amygdala, hypothalamus, and entorhinal cortex, and that the hippocampus could be considered as a key structure in a highly interconnected system of subcortical brain areas.

However, the hippocampus is not a "rage center," because lesions of it—or electrical stimulation of it—do not necessarily cause rage. This diversity of functional involvement is perhaps most evident in the amygdala, where lesions or electrical stimulation cause various affective behaviors in animals.

Emotions such as anxiety or fear are likewise generated in unknown ways from system-level operations of the limbic system. However, we do know that certain drugs (tranquilizers) reduce anxiety, suggesting that there are anti-anxiety receptors in the limbic system. Such receptors have been found, extracted, and purified. This suggests that the brain has one or more endogenous tranquilizer neurotransmitters, a presumption that is further justified by an earlier discovery that the brain's ability to respond to opiate narcotics was subserved by the pre-existence of an endogenous opioid neurotransmitter system, complete with a family of opioid transmitters and receptors.

Related Principles

Emergent Properties (Information Processing)
Motivation
Nodal Point (Cell Biology)
Parallel, Multi-level Processing (Information Processing)
Reinforcement

References

Archer, J. 1988. Behavioral Biology of Aggression. Cambridge University Press, New York.
Briley, M. and File, S. (eds.) 1991. New Concepts in Anxiety. CRC Press, Inc. Boca Raton, Florida.
Davis, M. 1992. The role of the amygdala in fear and anxiety. Ann. Rev. Neurosci. 15:353-75.
Doane, Benjamin K., and Livingston, K. (eds.) 1986. The Limbic System: Functional Organization and Clinical Disorders. Raven Press, New York.
Hammer, R.P. 1992 Neurobiology of Opiates. CRC Press, Boca Raton, Florida.
Horton, R. and Katona, C. (eds.) 1991. Biological Aspects of Affective Disorders. Neuroscience Perspectives Series. Academic Press. New York.
Mhatre, M., Mehta, A.K., and Ticku, M.K. 1988. Chronic ethanol administration increases the binding of the benzodiazepine inverse agonist and alcohol antagonist R015-4513 in rat brain. Europ. J. Pharmacol. 153:141-145.
Mohler, H., Schoch, P., and Richards, J.G. 1986. The GABA/benzodi-

azepine receptor complex: function, structure and location, pp. 91-96, In Molecular Aspects of Neurobiology, edited by R. Levi Montalcini, P. Calissano, E.R. Kandel, and A. Maggi. Springer-Verlag, Berlin.

Sachar, E.J. and Baron, M. 1979. The biology of affective disorders. Ann. Rev. Neurosci. 2:505-18.

Tallman, J.F., Paul, S.M., Skolnick, P., and Gallager, D.W. 1980. Receptors for the age of anxiety: pharmacology of the benzodiazepines. Science 207:274-281.

Citation Classics

Coppen A. 1967. The biochemistry of the affective disorders. Brit. J. Psychiatry 113:1237-64.

Mandler G. and Sarason S.B. 1952. A study of anxiety and learning. J. Abnormal and Soc. Psychol. 47:166-73.

Mohler, H. and Okada, T. 1977. Benzodiazepine receptor: demonstration in the central nervous system. Science 198:849-851.

Motivation

Animals have internal drives that produce goal-directed behaviors. These drives motivate animals to produce specific avoidance or approach behaviors that are contingent on situational context.

Explanation

Animals are motivated by internal drive states that increase the probabilities for behaviors that selectively promote eating, drinking, reproduction, offspring nurturing, aggression, and other behaviors that have adaptive value under specified circumstances. Often the value lies in promoting normal homeostatic regulation, as in eating and drinking when the body needs energy and liquid.

The neural systems that energize these drives involve different neural circuitry that tends to be specific for given drives. Some of these neural systems are influenced by circulating hormones. Most of this function is attributable to the limbic system.

The relationship of biological drives to emotions is reflected in the extent to which those drives are satisfied. Positive emotions ensue from satisfying biological drives, and negative emotions result when basic biological drives are thwarted.

The same limbic structures that produce emotions are also crucial for producing biological drives. Just as drives influence emotions, emotions can influence drives.

Examples

When an animal has gone without food, it has an internal drive to eat. In this case, there are "appetitostat" neurons in the hypothalamus that generate the drive to eat. Likewise, there are "satiety" neurons, in a different region of the hypothalamus, which, when active, cause the animal to stop seeking food. Other neuronal groups control drinking in a reciprocal fashion.

Many sexual behaviors are influenced by daylength cycles and associated changes in sexual hormones. In males, for example, testosterone promotes libido as well as aggression. In female animals, the build-up of estrogens as ovarian follicles mature causes them to engage in courtship behaviors.

Even relatively mild shifts in emotion can exert appreciable effects on behavior. Young-adult humans were studied while performing tasks in the presence of environmental sources of pleasant odors. Participants exposed to pleasant scents set higher goals on a clerical coding task and were more likely to adopt an efficient strategy for performing this task than control subjects. The odor-treated subjects also set higher monetary goals and made more concessions during face-to-face negotiations with an accomplice. They also reported weaker preferences for handling future conflicts with the accomplice through avoidance and competition.

Related Principles

Homeostatic Regulation (Overview)
Learning and Habituation (Learning and Memory)
Reciprocal Action (Information Processing)
Reinforcement

References

Baron, R.A. 1990. Environmentally induced positive affect: its impact on self-efficacy, task performance, negotiation, and conflict. J. app. Soc. Psychol. 20:368-384.
Satinoff, E. and Teitelbaum, P. 1983. Motivation. Plenum Press, New York.

Reinforcement

The brain has systems for both positive and negative reinforcement. That is, these systems promote behaviors that are either repeated or are avoided, as a consequence of the feedback from the first behavioral episodes.

Explanation

Several areas in the brainstem seem to mediate unpleasant sensations. An animal or person who experiences stimulus conditions that activate these brainstem areas will try to avoid such stimulus conditions and/or behaviors.

Other areas in the limbic system, particularly in a fiber bundle that connects many of the components, mediates reinforcement behavior. That is, stimulus conditions that activate these structures will cause animals or people to seek out such stimuli and indulge in behaviors that create those stimulus conditions.

Reinforcement has been linked to two neurotransmitter systems, norepinephrine and dopamine. The "mesolimbic" projection of dopaminergic fibers from the ventral tegmental area in the brainstem to several limbic system target structures seems to be especially important.

Examples

Painful stimuli are good examples of aversive stimuli. Brain areas that mediate painful sensations can initiate avoidance behaviors to get away from the painful stimulus. Experimentally, if one stimulates certain brainstem areas with electric current via implanted electrodes, an animal will exhibit discomfort and act as if it is trying to run away from or avoid something unpleasant. These areas include many sites in the anterior part of the brainstem reticular formation and parts of the central grey area of the midbrain. The avoidance behavior can be made explicit in animals that are trained to bar press in order to control stimulus presentation. If they are taught to discover that a bar press will stop delivery of electrical stimulation to such areas of the brain, they will press vigorously and often to protect themselves from such stimulation, which they apparently regard as unwanted and unpleasant.

On the other hand, there are other brain areas that animals like to have stimulated with electrical current. They will bar press to get the stimulus delivered to such sites via implanted electrodes. This self-stimulation behavior can be elicited from many sites within the limbic system, but the most robust response is often obtained from a major fiber tract, called the medial forebrain bundle. This is a tract of fibers that courses through the lateral hypothalamus. It contains axons projecting in both anterior and posterior directions, axons that connect such structures as the brainstem reticular formation areas with several parts of the limbic system. Also contained in this fiber bundle are the axons from major cell groups in the brainstem that make dopamine and serotonin as neurotransmitters (Figure 6-2).

Clinicians have long known the tendency of drug- or alcohol-free addicts to return to those settings associated with their addictive behavior, even when they are counseled that exposure to these settings is reinforcing for the drug or alcohol addiction. Presumably drug taking in these settings is more rewarding than would otherwise be the case. This behavior is analagous to place-seeking behavior in rodents. The experiment is conducted as follows: a rat with electrodes implanted in the medial forebrain bundle of the brain is electrically stimulated only when it accidentally happens to be located in a certain place in an open field. After a few such experiences, the rat starts spending more and more time in the place where the reward stimulation is given.

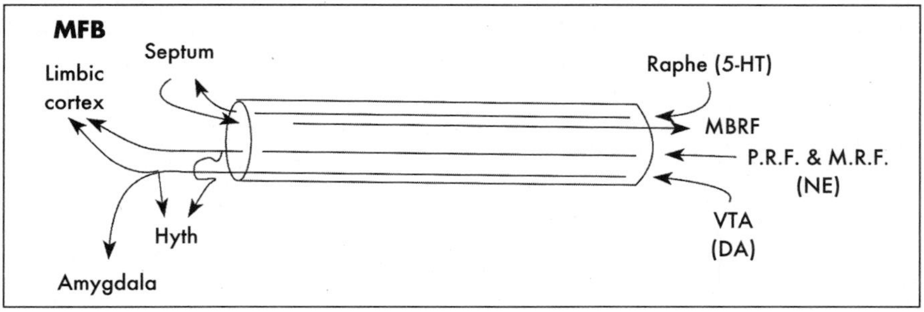

Fig. 6-2 Diagram of the origin of the fibers in the medial forebrain bundle, a fiber tract in the lateral part of the hypothalamus that seems to be crucial for mediating reinforcement and reward processes. Abbreviations: DA = dopamine neurotransmitter; Hyth. = hypothalamus; MBRF = midbrain reticular formation; NE = norepinephrine neurotransmitter; P.R.F. and M.R.F. = pontine and medullary reticular formation; VTA = ventral tegmental area.

TERMS

Aversion A physiological, psychological, or behavioral response that is sufficiently undesirable that the subject tries to avoid a repetition of the conditions that caused it.

Brainstem Reticular Formation The central core of the brainstem, which contains numerous small neurons that do not have extensive projections, plus some extremely large neurons that project extensively, either in anterior directions or posterior directions into the spinal cord.

Central Grey A group of neurons that surround the "canal" in the midbrain that connects the fluid-filled cavities in the forebrain (lateral and third ventricles) with a similar cavity (forth ventricle) underneath the cerebellum.

Dopaminergic Cells or fibers that release dopamine as a transmitter.

Limbic System A group of highly interconnected neuronal cell groups that collectively have a major influence on emotions. Such areas include the hypothalamus, amgydala, septum, hippocampus, cingulate cortex, anterior nucleus of the thalamus.

Mesolimbic The coupling between certain areas in the midbrain with those in the limbic system.

Reinforcement A physiological, psychological, or behavioral response that is sufficiently desirable or rewarding that the subject tries to repeat the conditions that caused it.

Related Principles

Motivation
Neurotransmitters (Cell Biology)
Neural Origin of Emotions

References

Engel, J. et al. 1988. Brain Reward Systems and Abuse. Proc. 7th International Berzelius Symposium. Raven Press, New York.

Meyer, R.E. 1994. Toward a comprehensive theory of alcoholism, pp. 238-250. In Types of Alcoholics. Edited by T.F. Babor et al. Ann. N.Y. Acad. Sci. Vol. 708. New York Acad. Sci. New York.

Netto, C.A. and Izquierdo, I. 1985. On how passive is inhibitory avoidance. Behav. Neural Biol. 43:327-330.

Samson, H.W. 1992. The function of brain dopamine in ethanol reinforcement, pp. 91-107. In Alcohol and Neurobiology. Receptors, Membranes, and Channels, edited by R.R. Watson, CRC Press, Boca Raton.

Schneider, L.H., Cooper. S.J., and Halmi, K.A. 1989. The Psychobiology of Human Eating Disorders. N.Y. Acad. Science, New York.

Citation Classics

Amand B.K. and Brobeck J.R. 1951. Hypothalamic control of food intake in rats and cats. Yale J. Biol. Med. 24:123-40.

Bolles R.C. 1970. Species-specific defense reactions and avoidance learning. Psychol. Rev. 71:32-48.

Crow T.J. 1972. Catecholamine-containing neurones and electrical self-stimulation: 1. A review of some data. Psychol. Med. 2:414-21.

Deneau G., Yanagita T., and Seevers M.H. 1969. Self-administration of psychoactive substances by the monkey: a measure of psychological dependence. Psychopharmacologia 16:30-48.

Ernst A.M. 1967. Mode of action of apomorphine and dexamphetamine on gnawing compulsion in rats. Psychopharmacologia 10:316-23.

Lenaerts F.M. 1967. Is it possible to predict the clinical effects of neuroleptic drugs (major tranquilizers) from animal data? Part IV: an improved experimental design for measuring the inhibitory effects of neuroleptic drugs on amphetamine- or apomorphine-induced "chewing" and "agitation" in rats. Arzneim. Forsch. Drug Res. 17:841-5.

Morley J.E. 1980. The neuroendocrine control of appetite: the role of the endogenous opiates, cholecystokinin, TRH, gamma-amino-butyric-acid and the diazepam receptor. Life Sci. 27:355-68.

Olds, J. and Milner, P. 1954. Positive reinforcement produced by electrical stimulation of septal area and other regions of rat brain. J. Comp. Physiol. 47:419-427.

Randrup A. and Munkvad I. 1967. Stereotyped activities produced by amphetamine in several animal species and man. Psychopharmacologia 11:300-10.

Sidman M. 1953. Avoidance conditioning with brief shock and no exteroceptive warning signal. Science 118:157-8.

Emotions: Study Questions

1. List and explain each of the principles in this category.
2. For each of the principles, provide an example *that is not mentioned in this text.*
3. Justify the premise that emotions are an emergent property of the limbic system.
4. What are the implications for the mechanisms of other emotions from the discovery of anti-anxiety receptors in the brain?
5. Why is it not a coincidence that both drives and emotions are controlled by the limbic system?
6. What are the relationships between biological drives and emotions.
7. Explain how emotions affect and are affected by behavior.
8. What is the relationship between reinforcement and conditioned learning (see Learning and Habituation principle).
9. What does it mean to say that there are "pleasure" and "punishment" centers in the brain.
10. What is the role of feedback in reinforcement?
11. How is the probability of exhibiting a given behavior affected by reinforcement contingencies?

CHAPTER 7

Learning and Memory

Our brains learn to view new information within the context of previous learning. Thus, past learning is like a pair of sunglasses, filtering the input so that only a portion is registered. The more limited the past learning has been, the more restricted the outlook will be. In a large sense, recognition of this filtering effect and dealing with it in oneself is a hallmark of maturity and education.

— W. R. Klemm

Animals learn the biological significance of novel stimuli. Without biological significance, animals often learn to ignore (habituate to) such stimuli. **Learning and Habituation** are associated with memory of the learned experience. That memory may be a "working memory" of very short duration, such as the memory of a telephone number that you have just looked up and are dialing. In such situations, the memory process is contained and represented by a spatio-temporal pattern of action potentials in widely distributed circuitry. Certain features of the memory may be contained in certain parts of the circuitry, in somewhat modular fashion, which then are reassembled and bound together as a system-level operation (Figure 7-1).

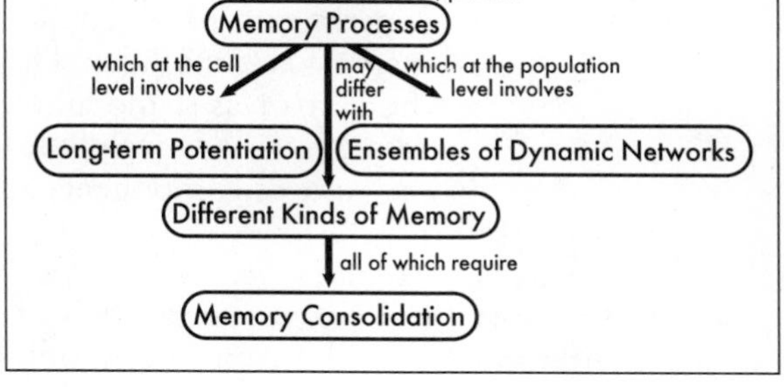

Fig. 7-1 Concept map for the mechanisms of learning and memory.

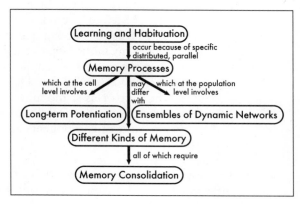

At the single-cell level, repeated presentations of a learning event can cause a **Long-term Postsynaptic Potentiation** that, if repeated, can cause a postsynaptic potentiation that outlasts the stimulus by hours or even days.

At the population level, memory typically encodes in large **Ensembles of Dynamic Networks** that operate as parallel, distributed circuits. The spatio-temporal pattern of firing in these ensembles is specific to each learning event. If the ensembles can be made to recreate a given spatio-temporal firing pattern, the associated memory is reproduced.

Memory Kinds Reflect Memory Processes, as reflected in whether memory operates at the conscious level ("declarative memory") or as some unconscious automatism, such as well-learned motor skills ("procedural memory"). Declarative memories require a normally functioning hippocampus, whereas procedural memories require normally functioning cerebellum and movement control circuits.

Regardless of the kind of memory, **Memory Consolidation** processes convert short-term, working memories into longer-term and perhaps permanent memories. Consolidation requires the learning event to be clear and significant and progresses with time (on the order of a few hours) and with rehearsal. Consolidation creates an active metabolic and structural change in synapses. Conflicting or distracting events that occur shortly after a learning event may interfere with consolidation by perturbing the set of spatio-temporal action potential patterns that have encoded the neural representation of the learned event.

List of Principles

Learning and Habituation Neural systems can recall a learning experience long after it has terminated. They may also "learn to unlearn" or ignore experiences that lose biological significance. This is an intrinsic capability that underlies the capacity for adaptive change based on experience.

Memory Processes One well-known quote says a lot about memory processes: "Memory is not a thing in a place but a process in a population." Memory, even for a single event, does not seem to be confined to a specific brain area, but rather is distributed throughout many widely separated parts of the brain. The most recent experiments have led to a concept that certain aspects of memory are modular. That is, each distributed site might be thought of as a module that encodes a specific aspect of a memory. Collectively, the processing in these modules binds the various components and reconstructs the learned event.

Long-term Post-tetanic Potentiation Memory occurs at the cell level and is manifest in changed responsiveness at the synapse level. One prominent feature of such memory is a long-lasting potentiation of postsynaptic responsivity that occurs on repetition of the same stimulus. In brief, the postsynaptic response to a given sensory input becomes magnified on subsequent repetitions. Because this potentiation outlasts the stimulus, it is a form of memory of that stimulus. A related phenomenon of postsynaptic depression is also known.

Ensembles of Dynamic Neural Networks Networks of functionally linked neurons are the building blocks of higher brain functions. These neural assemblies are triggered into event-specific activity that has an association with spatio-temporal pattern. When this input is repeated, the synaptic connections in the assembly are strengthened. When reactivated, these assemblies produce appropriate spatio-temporal output patterns that can "play back" the memory of the original event.

Memory Kinds Reflect Memory Mechanisms Behavioral expression of memory reveals that there are different kinds of memory, each of which has an associated set of separate yet interacting neural systems and subsystems. The two most basic kinds are declarative memory, operating in the consciousness, and procedural memory, often operating subconsciously.

Memory Consolidation The formation of lasting memories is both time and event dependent. Learning experiences, depending on their intensity and biological significance and, depending on sufficient time for normal neural functions, convert from being short-term to intermediate- and long-term memory.

Learning and Habituation

Neural systems can recall a learning experience long after it has terminated. These systems can also "learn to unlearn"; that is, ignore experiences that lose biological significance. This is an intrinsic capability that underlies the capacity for adaptive change based on experience.

Explanation

Acquisition of new capability from repeated experience is a property that we commonly call learning. This property can be exhibited at the level of small neural circuits and even single neurons, although the most obvious form of learning is that which affects large segments of the nervous system and is expressed as behavioral change. The learning may be associative or non-associative. In non-associative learning, there is a single type of stimulus, which when repeated may cause the neural system to become more responsive to it (i.e., sensitized) or less responsive (i.e., habituated). In associative learning, the organism makes an association between different sets of stimuli to produce a response capability that was not there before. In this case, typically, a stimulus that is neutral acquires the ability to cause a new effect simply by virtue of its pairing, or association, with another stimulus that naturally causes a distinct response.

The world and its objects are initially unlabeled in the brain. The nervous system "evolves" a categorization scheme for new stimuli based on the pattern of neural activity that is triggered within the "hard-wired" circuits that can respond to the stimulus. As these circuits are sculpted by experience, the brain learns to recognize previously experienced input patterns.

Examples

The best-known example of sensitization is the gill withdrawal reflex of the marine mollusc, *Aplysia*, which offers the great experimental advantage of having a simpler nervous system that makes it easier to discern basic principles of learning and memory. Most studies of learning in *Aplysia* have focused on the abdominal ganglion, which contains about 2000 neuronal cell bodies that control movement of the gills and the siphon which pulls water across the gills. When the head or tail of this mollusc is weakly stimulated by touch or electrical current, the gills withdraw, serving to protect this delicate tissue from potentially damaging stimuli. This reflex response, controlled by about 13 motor neurons, can become larger and last longer as stimuli are repeated. The mechanism is attributable to the progressive recruitment of facilitatory neurons that develop excitatory synapses with the presynaptic terminals of the sensory neurons. This causes the sensory neurons to release more than the usual amount of neurotransmitter when they are activated by stimulus. There apparently is also enhanced responsiveness of the postsynaptic receptors on the motor neurons, as well as the recruitment of other excitatory neurons to act on motorneurons (Figure 7-2).

Aplysia also provides a good illustration of habituation, which can be considered as learning not to learn. If the siphon is lightly touched, it and the gill withdraw. But with repetition of such stimuli at short intervals, the response may abate, i.e., become habituated. The basis for this habituation is a combination of depressed synaptic transmission between sensory cells and interneurons that connect with motor neurons, and between sensory neurons and motor neurons. The size and duration of excitatory postsynaptic potentials (EPSPs) in these pathways becomes diminished, causing less discharge of action potentials in the motor neurons and a diminished motor response. The synaptic depression is not a result of motor neurons becoming "tired." Direct electrical stimulation of these motor neurons at rates in which behavioral stimuli cause habituation are NOT accompanied by habituation. Since stimulation of the sensory nerves of the siphon causes habituation, the effect must result from some processing reactions within the abdominal ganglion. This habitua-

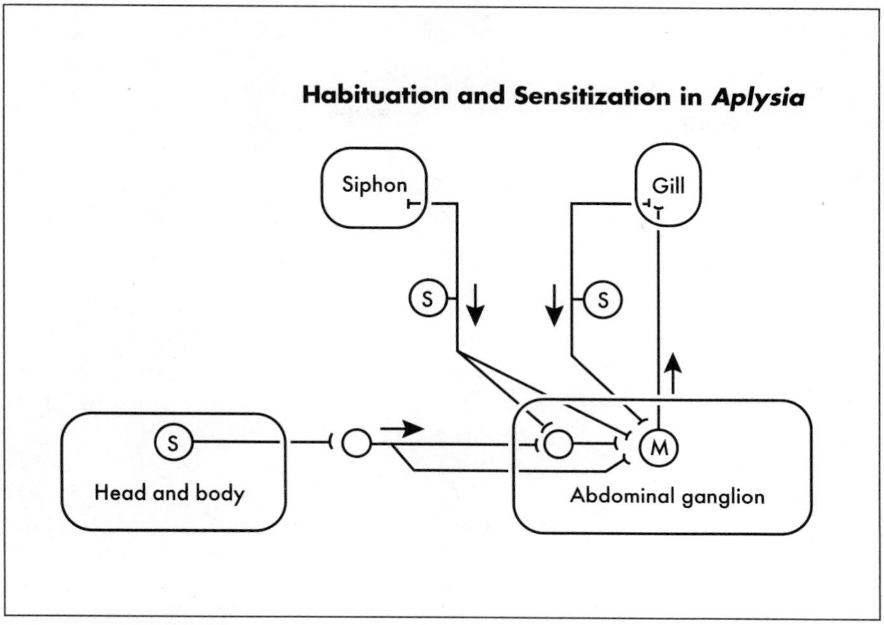

Fig. 7-2 Diagram of the circuitry that mediates sensitization and habituation of the visceral withdrawal reflex in *Aplysia*. The basic reflex is initiated by stimulation of the siphon and/or gill, which causes a reflex motor response that withdraws the gill and protects it. Sensory neurons (S) excite motor neurons (M) in the abdominal ganglion, which in turn activate the muscles that cause gill withdrawal. There are also interneurons in the ganglion (not shown) that contribute to reinforcing this reflex. Sensory neurons (S) in the head and body also feed into this circuit and can sensitize the system to become more responsive to the stimuli that activate the reflex. This action occurs by enhancing neurotransmitter release from presynaptic terminals of the siphon and gill sensory nerves.

tion can be seen in a depression of the monosynaptic response of motor neurons after successive stimulation of siphon sensory nerves.

Dishabituation can be demonstrated by stimulation of the head/neck region in a habituated animal. Under these conditions, the habituation is reversed, accompanied by a restoration of full-sized EPSPs in the ganglion motor neurons. Other studies have shown that dishabituation arises from presynaptic facilitation of the sensory neuron, which increases the amount of neurotransmitter that they release. Receptor sensitivity does not seem to be a factor.

Associative, or conditioned, learning is best known from the famous studies of "Pavlov's dogs." Pavlov noted that dogs salivate and their stomach secretions increase when they are shown tasty food. This is an unlearned response that occurs spontaneously as part of nature's way to prepare the digestive tract for eating. But Pavlov also noted that a second stimulus that is normally unrelated to food can acquire new meaning if it is repeatedly presented in combination with food. Ringing a bell, for example, and then presenting food will eventually cause the dog to salivate and secrete stomach juices on hearing the bell, *before* food is presented. Thus, the previously neutral stimulus of the bell has acquired a new meaning by virtue of its association with food. The bell is said to have been converted to a "conditioned" stimulus, in contrast to the unconditioned stimulus of the food. The dog's response to food alone is an unconditioned response, but that same response to the bell is a conditioned response.

Similar associative learning can be demonstrated at the level of simple circuits and single neurons. In *Aplysia,* for example, a tactile stimulus to the siphon or the mantle causes an unconditioned, *weak* response of gill withdrawal; electrical stimulation of the tail produces a *strong* gill withdrawal response. If that mantle stimulus, for example, is paired with a succeeding electrical stimulus to the tail, the gill withdrawal gradually grows into a strong response after stimulation of the mantle—even if there is no longer a tail stimulus. As a control, an experimenter can show that magnified gill withdrawal never occurs from random delivery of mantle or siphon stimuli that are not time-locked to tail stimulation (Figure 7-3).

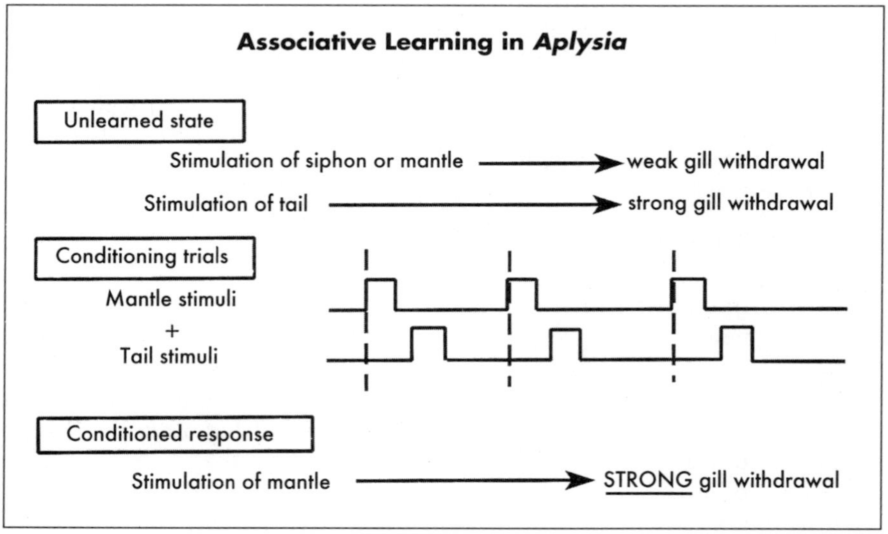

Fig. 7-3 Diagram of the process of associative learning, using *Aplysia* as an example. In the unlearned state, stimulation of either the siphon or the mantle leads to a weak reflex withdrawal. Stimulation of the tail evokes a much stronger reflex withdrawal. If one conditions this system by pairing a stimulation of the mantle with a delayed and time-locked stimulation of the tail, the organism eventually learns to attribute the properties of tail stimulation to mantle stimulation. That is, a strong gill withdrawal response develops to stimulation of the mantle, which in the unconditioned state only evokes a weak response.

TERMS

Classical Conditioning	A training situation in which an originally neutral stimulus acquires the capability to produce a reflex response because it has been repeatedly paired with a stimulus that typically elicits this same response without training.
Dishabituation	A reversal of the habituated state by presentation of another, usually stronger, stimulus. If given under non-habituated conditions, such a stimulus can sensitize the system and make it more responsive than usual.
Monosyaptic Pathway	One that involves only one synapse.

Related Principles

Action Potentials (Cell Biology)
Long-term Postsynaptic Potentiation
Nodal Point (Cell Biology)
Neurochemical transmission (Cell Biology)
Reflex Action (Information Processing)

References

Brown, T.H., Kairiss, E.W., and Keenan, C.L. 1990. Hebbian synapses: biophysical mechanisms and algorithms. Ann Rev. Neurosci. 13:475-511.

Kandel, E.R. 1976. Cellular Basis of Behavior. Freeman, San Francisco.

Rescorla, R.A. 1988. Behavioral studies of pavlovian conditioning. Ann. Rev. Neurosci. 11:329-52.

Wu, J.-Y., Cohen, L.B., and Falk, C.X. 1994. Neuronal activity during different behaviors in *Aplysia:* a distributed organization? Science 263:820-823.

Citation Classics

Bolles R.C. 1970. Species-specific defense reactions and avoidance learning. Psychol. Rev. 71:32-48.

Castellucci, V. and Kandel, E.R. 1976. Presynaptic facilitation as a mechanism for behavioral sensitization in *Aplysia.* Science 194:1176-1178.

Jensen A.R. 1969. How much can we boost IQ and scholastic achievement? Harvard Educ. 39:1-123.

Mandler G. and Sarason S.B. 1952. A study of anxiety and learning. J. Abnormal & Soc. Psychol. 47:166-173.

Pavlov, I.P. 1927. Conditioned Reflexes: An Investigation of the Physiological Activity of the Cerebral Cortex. G.V. Anrep (trans.). London: Oxford University Press.

Rescorla R.A. and Soloman R.L. 1967. Two-process learning theory: relationships between Pavlovian conditioning and instrumental training. Psychol. Rev. 74:151-183.

Sidman M. 1953. Avoidance conditioning with brief shock and no exteroceptive warning signal. Science 118:157-158.

Memory Processes

One well-known quote says a lot about memory processes: "Memory is not a thing in a place but a process in a population." Memory, even for a signal event, does not seem to be confined to a specific brain area, but rather is distributed throughout many widely separated parts of the brain. The most recent experiments have led to a concept that certain aspects of memory are modular. That is, each distributed site might be thought of as a module that encodes a specific aspect of a memory. Collectively, the processing in these modules binds the various components and reconstructs the learned event.

Explanation

Learning is a late form of neuronal differentiation that continues throughout a lifetime. Learning processes are widely distributed throughout the brain. Although the hippocampus is crucial for *forming,* i.e., consolidating memories, once they are formed their recall does not seem to depend on the hippocampus nor usually on other specific brain regions. Memories for coordinated movements do, however, depend on normal function of the motor cortex and cerebellum—otherwise, learned movements cannot be properly expressed. Neurons in a wide variety of widely disbursed areas of the brain are electrically responsive during the formation and recall of a given memory. Moreover, specific features about a memory may be encoded in different areas of the brain.

The key unresolved issue is how all the components of a memory, as in an image, are "put together" in memory. Is there some central "memory manager" that integrates the activity in all the distributed regions to produce a complete reconstruction? For now, there is no evidence for such a manager. The only alternative theory for how a reconstructed whole can be produced from the parallel, distributed processes that encode memory is that certain high-frequency electrical rhythms link the various modules into coherent population processing.

Examples

Experiments in the 1920s by Karl Lashley showed that it was difficult to find where in the brain memory was located. When rats were taught a task, subsequent experimental damage to the neocortex had little effect on the recall until the total area of damage became extensive. That is, regional lesions did not impair the memory, no matter what region of the cortex was damaged.

In 1861, Broca reported from autopsy observations that two people who were known to have serious speech impairments had lesions in the inferior frontal convolution in the left hemisphere. Later, Wernicke in 1874 observed a case with a brain lesion of the superior temporal gyrus of the left hemisphere of a person who had been known to have problems in comprehending speech. Clinicians in the following years identified other parts of the brain that seemed to be associated with memories for certain functions: vision, sound, touch, and motor control, and even more specific functions such as word blindness, word deafness, writing skills, making logical propositions and naming things. This led for some time to a "localizationalist" concept in which specific memories were thought to be "found" in specific places. We now know that for many memories this is an oversimplification and misleading.

Later and more sophisticated experiments used electrophysiological recordings to try to localize areas of the brain where memory resided by finding its electrical "signature." Electrically evoked responses associated with any given memory could be found in many widely scattered areas, both cortical and subcortical. Similar results have since been obtained by other topographical imaging techniques such as PET scan. This led to the obvious conclusion that memory is a parallel process distributed throughout many neural networks. How these diverse locales are orchestrated to reproduce the learning is not known, except that there is evidence for coherent electrical rhythms in widely scattered loci.

Much study of memory has involved visual memories, and experiments with this sense have led to a modular concept of memory. Within the visual cortex, there are specialized zones, wherein some cells respond to color, some to movement, some to shape, and some to line orientation and to other features of an image. Neurons in these areas hold visual representations "on line" in working memory for a brief time after the stimulus is withdrawn. This activity is a neuronal correlate of working memory. The memories associated with such abstracted features of an image appear to reside not only in their initial registration sites in the visual cortex, *but also elsewhere in the brain.* For example, neurons in the prefrontal area of the neocortex, near its rostral pole, respond to different features of a visual stimulus. One prefrontal region seems to respond to spatial features of a visual stimulus, while neurons in another prefrontal region respond to specific features of a stimulus. In other words, some neurons respond to "what" an object is, and other neurons respond to "where" it is located (Figure 7-4).

TERMS

Hippocampus A region of "primitive" cortex that is folded underneath the cerebral cortex ("neocortex"). The hippocampus mediates the formation of long-lasting memories, but does not seem to be instrumental in recall.

Parallel Process The neural activity associated with a given function or behavior that occurs more or less simultaneously in parallel circuits.

PET (positron emission tomography) A topographical mapping technique for brain function based on using a radioactive analog of glucose that accumulates in very active neurons.

Visual Cortex That part of the neocortex at the caudal pole of the brain. Neurons in this region register specific features of a visual scene.

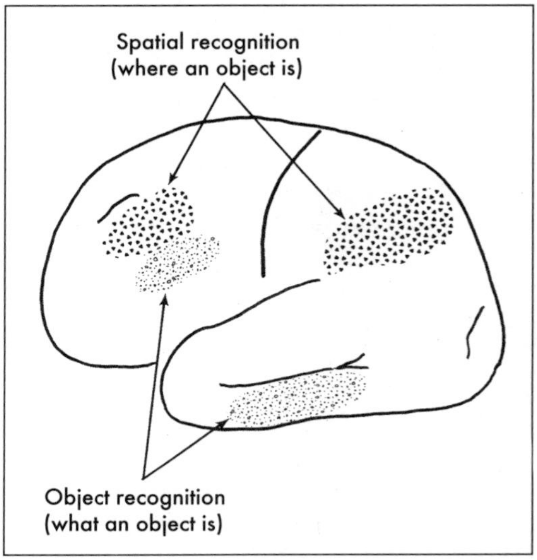

Fig. 7-4 Diagram of regions in primate neocortex that participate in registration and storage of various components of a visual memory.

Related Principles

Ensembles of Dynamic Neural Networks
Memory Consolidation
Parallel, Multi-level Processing (Information Processing)
Rhythmicity (Overview)
Synchronicity (Information Processing)

References

Abraham, W. et al. (eds.) 1990. Memory Mechanisms. A Tribute to G. V. Goddard. Lawrence Erlbaum Associates, Hillsdale, New Jersey.

Alkon, D.L. 1988. Memory Traces in the Brain. Cambridge University Press, New York.

Brown, T.H., Kairiss, E.W., and Keenan, C.L. 1990. Hebbian synapses: biophysical mechanisms and algorithms. Ann. Rev. Neurosci. 13:475-511.

Cohen, N. and Eichenbaum, H. 1993. Memory, Amnesia, and the Hippocampal System. MIT Press, Cambridge, Massachusetts.

Gazzaniga, M. (ed.) 1988. Perspectives in Memory Research. MIT Press, Cambridge, Massachusetts.

McGaugh, J.L. 1989. Involvement of hormonal and neuromodulatory systems in the regulation of memory storage. Ann. Rev. Neurosci. 12:255-87.

Schacter, D.L., Chiu, C.-Y.P., and Ochsner, K.N. 1993. Implicit memory: a selective review. Ann Rev. Neurosci. 16:159-82.
Schwartz, J.H. and Greenberg, S.M. 1987. Molecular mechanisms for memory: second-messenger induced modifications of protein kinases in nerve cells. Ann. Rev. Neurosci. 10:459-76.
Squire, L.R. 1982. The neuropsychology of human memory. Ann. Rev. Neurosci. 5:241-273.
Thatcher, R.W. and John, E.R. 1977. Foundations of Cognitive Processes. Lawrence Erlbaum Associates. New York.
Thompson, R.F. and Krupa, D.J. 1994. Organization of memory traces in the mammalian brain. Ann. Rev. Neuroscience. 17:519-550.
Thompson, R.F., Berger, T.W., and Madden, J., IV. 1983. Cellular processes of learning and memory in the mammalian CNS. Ann. Rev. Neurosci. 6:447-91.
Wilson, F.A.W. Scalaidhe, S.P., and Goldman-Rakic, P.S. 1993. Dissociation of object and spatial processing domains in primate prefrontal cortex. Science. 260:1955-1958.
Zola-Morgan, S. and Squire, L.R. 1993. Neuroanatomy of Memory. Ann. Rev. Neurosci. 16:547-63.
Zucker, R.S. 1989. Short-term synaptic plasticity. Ann. Rev. Neurosci. 12:13-31.

Citation Classics

Lashley, K.S. 1950. In search of the engram. Symp. Soc. Exp. Biol. 4:454-482.
Underwood B.J. 1969. Attributes of memory. Psychol. Rev. 76:559-73.

Long-term Post-tetanic Synaptic Memories

Memory occurs at the cell level and is manifest in changed responsiveness at the synapse level. One prominent feature of such memory is a long-lasting potentiation of postsynaptic responsivity that occurs on repetition of the same stimulus. In brief, the postsynaptic response to a given sensory input becomes magnified on subsequent repetitions. Because this potentiation outlasts the stimulus, it is a form of memory of that stimulus. A related phenomenon of postsynaptic depression is also known.

Explanation

An electronic neural network loses its memory whenever its power is turned off unless provision is made for permanent storage. The nervous system has a permanent storage mechanism, but it is not well understood. Evidence to date is that learning causes lasting biochemical and structural changes in the synapses that participated in the learning experience. Much debate and controversy exists over the nature of the biochemical changes in synapses that are associated with memory. What is generally agreed on is that these changes bias certain synapses and pathways so that they reconstruct the response to an earlier stimulus condition; that is, the memory of a learned event is encoded in electrical activity that resembles that which was originally activated during learning.

The relevant electrical properties that lead to biochemical change are those than occur in the synapses, namely, postsynaptic responses, either IPSPs or EPSPs. If these potentials are evoked often enough at high-enough rates, then they come potentiated.

In addition to long-term potentiation, there is a counterpart phenomenon of long-term depression that has been demonstrated to be due to a long-term decrease in release of presynaptic neurotransmitter.

The molecular basis for long-term postsynaptic potentiation (LTPP) is not known, although it is assumed that semi-permanent changes in neurotransmitter and receptor functions must take place to provide a sustained bias in synaptic transmission. An even more enigmatic issue is how the molecular representation of learning is preserved over long spans of time, given that all molecules of synaptic membrane are continually being broken down and reconstructed. No one knows how epigenetic influences produce long-term changes in inducible genetic mechanisms to preserve the learned information.

Examples

LTPP has been most widely studied in a trisynaptic pathway in the hippocampus, in the large pyramidal cells in the region known as CA1. This potentiation seems to be mediated by the *N*-methyl-D-aspartate receptor, which typically responds to the transmitter, glutamate.

A typical procedure is to record from a target neuron and observe the magnitude of its extracellular field potential in response to each pulsed activation of an input fiber from an adjacent cortical region called entorhinal cortex. When a tetanizing stimulus is used at the appropriate frequency for a few seconds, the magnitude of the field potential gets larger than before the high-frequency stimulation. Likewise, a single-pulse stimulus yields a field potential that is much larger than it would have been if that single-pulse stimulus had not followed a tentanizing stimulus. Such potentiated responses can be demonstrated for days or even weeks after the tetanizing stimulus.

In addition, certain patterns of stimulation can produce a long-term depression of the target neurons. This has been demonstrated in both the hippocampus and in the neocortex. Long-term depression has been shown to depend on postsynaptic calcium ion entry through voltage-gated ion channels paired with activation of glutamate receptors (Figure 7-5).

Some scientists suspect that LTPP involves the release of retrograde messengers from postsynaptic cells that act on the presynaptic terminals to enhance release of neurotransmitter. One candidate retrograde chemical signal is nitric oxide (NO), a membrane-permeant gas. Injecting mice with drugs that inhibit NO synthesis also blocks the electrical signs of LTPP in the hippocampus.

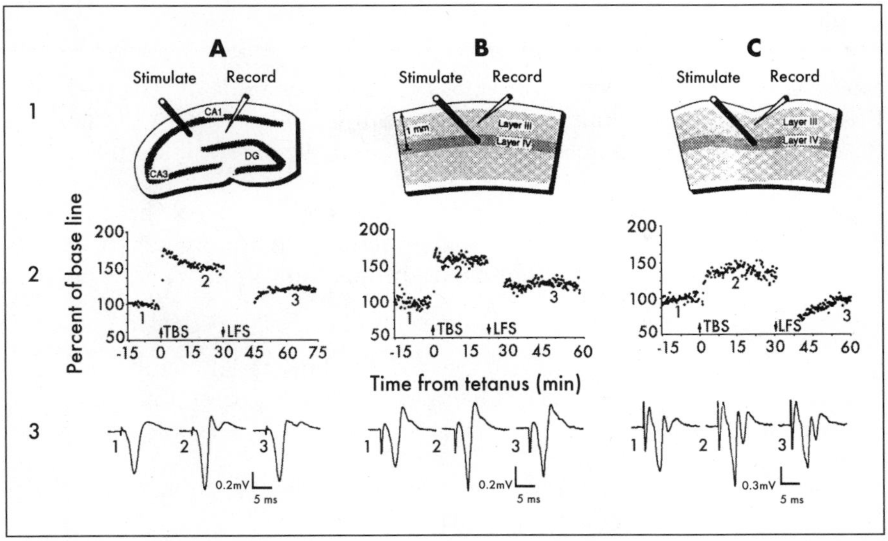

Fig. 7-5 Similar signs of long-term postsynaptic potentiation in three experimental situations: A – rat hippocampus, B – rat visual cortex, and C – cat visual cortex. Row 1 shows the location of stimulus and recording electrodes in the three conditions. Row 2 shows the extracellular field-potential response, as a percent change from no-stimulus baseline, in postsynaptic target cells. Experiments were conducted in vitro, with brain slices kept alive in warm, nutrient-laded fluid. When the stimulus was a short burst of fluctuating current at the "theta" rhythm frequency of 5-7/sec (TBS) delivered to axons of the CA3 region that send collaterals to the cells in the CA1 region, the response to a single-pulse test stimulus (each data point) was clearly augmented for up to 30 min or more after the TBS conditioning (section 2 of the data in column A). Similar results occurred in similar experiments in rat and cat visual cortex, where experimenters observed augmented responses to single-pulse stimuli given to cells in layer IV that supply input to layer III cells. The enhanced response seemed to depend on NMDA receptors. When input pathways were conditioned by low-frequency stimulation (LFS), at 1/sec, the response to test stimulus was suppressed, in the hippocampus and in cortical tissue (section 3 of data). Raw electrophysiological data are shown in row 3. These curves are the averages of four evoked responses taken before (1), during (2), and after (3) conditioning stimulation with TBS or LPS.

As an example of molecular change associated with learning, we can cite the case of habituation in *Aplysia.* Habituation occurs because of sustained inactivation of calcium channels in excitatory presynaptic terminals, which in turn suppress the release of excitatory neurotransmitter.

Second messenger activation occurs in *Aplysia* sensitization and gill reflex conditioning. Intracellular cyclic AMP levels increase, which phosphorylates and activates protein kinase, which in turn activates certain presynaptic potassium channels. Because potassium efflux terminates action potentials, this prolongs the action potentials. In turn calcium influx is prolonged, and there is greater release of neurotransmitter.

TERMS

Excitatory Postsynaptic Potentials (EPSPs)	Depolarizations of the neuronal membrane that move the membrane potential toward the threshold for discharging action potentials; that is, EPSPs make it more likely that a neuron will fire.
Hippocampus	A primitive part of the cerebral cortex that in higher mammals is folded inward underneath the more advanced neocortex; it is a key component of the limbic system and is crucial to memory formation.
Inhibitory Postsynaptic Potentials (IPSPs)	Hyperpolarizations of the neuronal membrane that make the neuron less likely to discharge action potentials.
Neocortex	The most advanced kind of cortex that is dominant in higher animals, especially humans.
Tetanizing Stimulus	Repetitive stimulation at a frequency that elicits maximal response.

Related Principles

Action Potentials (Cell Biology)
Calcium and Transmitter Release (Cell Biology)
Ion Channels (Cell Biology)
Neurotransmitters (Cell Biology)
Nodal Point (Cell Biology)
Membrane Receptors (Cell Biology)
Memory Consolidation
Memory Processes
Second Messengers (Cell Biology)

References

Baudry, M. and Davis, J. (eds.) 1991. Long-term Potentiation: A Debate of Current Issues. MIT Press, Cambridge, Massachusetts.

Baudry, M., Thompson, R.F., and Davis, J.L., (eds.) 1993. Synaptic Plasticity: Molecular, Cellular, and Functional Aspects. MIT Press, Cambridge, Massachusetts.

Bolshakov, V.Y. and Siegelbaum, S.A. 1994. Postsynaptic induction and presynaptic expression of hippocampal long-term depression. Science. 264:1148-1152.

Kirkwood, A., Dudek, S.M., Gold, J.T., Aizenman, C.D., and Bear, M.F. 1993. Common forms of synaptic plasticity in the hippocampus and neocortex in vitro. Science. 260:1518-1521.

O'Dell, T.J., Huang, P.L., Dawson, T.M., Dinerman, J.L., Snyder, S.H., Kandel, E.R., and Fishman, M.C. 1994. Endothelial NOS and the blockade of LTP by NOS inhibitors in mice lacking neuronal NOS. Science. 265:542-546.

Madison, D.V., Malenka, R.C., and Nicoll, R.A. 1991. Mechanisms underlying long-term potentiation of synaptic transmission. Ann. Rev. Neurosci. 14:379-97.
Teyler, T.J. and DiScenna, P. 1987. Long-term potentiation. Ann. Rev. Neurosci. 10:131-61.
Thompson, R.F., Berger, T.W., and Madden, J. 1983. Cellular processes of learning and memory. Ann. Rev. Neurosci. 6:447-491.
Zucker, R.S. 1989. Short-term synaptic plasticity. Ann. Rev. Neurosci. 12:13-31.

Citation Classics

Bliss, T.V.P. and Lømo, T. 1973. Long-lasting potentiation of synaptic transmission in the dentate area of the anaesthetized rabbit following stimulation of the perforant path. J. Physiol. (Lond.) 232:331-356.
Hebb, D.O. 1949. The Organization of Behavior: A Neuropsychological Theory. Wiley & Sons, New York.

Ensembles of Dynamic Neural Networks

Networks of functionally linked neurons are the building blocks of higher brain functions. These neural assemblies are triggered into event-specific activity that is associated with a spatio-temporal pattern. When this input is repeated, the synaptic connections in the assembly are strengthened. When re-activated, these assemblies produce appropriate spatio-temporal output patterns that can "play back" the memory of the original event.

Explanation

At birth, the neural circuits that are most functionally complete are those that subserve life sustaining functions, such as regulation of body temperature, respiration, and cardiovascular reflexes. Many other functions have to be learned and represented in a new circuit pattern. The newborn brain has developed in the very cloistered and relatively stimulus-free environment of the womb. It is as if it were an empty slate, waiting to be written upon.

Spatio-temporal input patterns program the newborn brain. This same phenomenon occurs throughout life, being only more evident in the young. This programming takes the form of "sculpting" associated circuitry that represents and acts on this pattern of input. As the pattern of input is repeated, the synapses in the associated circuit become facilitated and made more likely to respond when exposed again to this spatio-temporal pattern of input in a way that can represent the original learning event.

Experimental detection of linked activity within a functionally segregated circuit is not easy to demonstrate. The first requirement is to be able to record from many individual neurons simultaneously. Then there is the formidable problem of having an analytical method for evaluating temporal order, not only within the impulse train within a single neuron, but also for identifying correlated activity among many neurons. Many studies have now documented that there is a temporal ordering in impulse trains within a single neuron; that is, certain interval patterns occur at much higher than chance rates. Research efforts in the future are likely to focus on analytical techniques that can search across a population of neurons that are likely to be a functional assembly for an association between repeated impulse patterns and externally observable events (specific stimuli, behaviors, etc.).

Examples

At birth, even some of the more basic reflex functions are still undeveloped. One example can be seen with the reflex called the Babinski reflex. In the newborn, tickling the sole of the foot causes the toes to curl upward, rather than downward, as is the case in mature nervous systems. As the nervous system matures, the sensory world not only becomes encoded in various learned contexts but also actually helps to "sculpt" the neuronal circuitry to make it more effective at dealing with those kinds of sensory experience.

An example of how memories are encoded in parallel, distributed processes in widely scattered circuitry is that the electrical responses evoked by a stimulus for which a memory exists will appear in many different and widely separated brain areas. A brief novel stimulus may well elicit a short-latency cortical evoked response in a few sites. However, as this stimulus is repeated and learning occurs, long-latency components appear and the evoked response waveforms become more similar in many different brain areas. Another method that leads to a similar conclusion about the distributed nature of memory comes from studies wherein radioactive glucose is injected to produce brain maps of changes in glucose utilization as a function of memory. In humans receiving such injections, multiple foci of brain activation can be seen, and the topography of these foci differ depending on whether the stimuli were words or visual images. Time-dependent co-activation changes can be seen as learning progresses, which is especially prominent in limbic and associational cortices.

Working memory, i.e., short-term memory that is used during task performance, involves passive (automatic) and active (volitional) mechanisms that appear to operate in large parallel, distributed ensembles. For example, when monkeys are trained to hold a sample picture "in mind" so that they can signal when the picture was presented again, the response of widely separated neurons in visual and prefrontal cerebral cortex neurons fall into two classes. About half of the cells discharge a memory-specific pattern of action potential discharge when monkeys signal their recognition that a picture matched one they had recently seen. About 2/3 of the "memory cells" are generally suppressed by any subsequent picture, whether or not it matched the reference picture. The response of these neurons thus seems to be a passive and automated response to exposure to test pictures, irrespective of their content. The other memory neurons discharge at higher rates, but only when a picture matches the reference picture. The enhancement effect uniquely identified the one picture in a set of pictures that matched the original stimulus picture. This seems to reflect an active, working memory system.

Simple memories, especially procedural ones, may involve relatively small circuits that can actually be identified. For example, the circuitry that seems to encode the recognition of male odors (pheromones) by females has been identified to require the vomeronasal organ (VNO) and its associated neural connections with the accessory olfactory bulb. Selective destruction of the VNO prevents the ability of pheromones from strange males to block implantation of fertilized ova and pregnancy in mice. Presumably, in normal mice without such lesions, this mechanism promotes genetic heterogeneity, because impregnation by a strange male is made more likely (Figure 7-6).

The circuitry mediating this detection of strange-male pheromones relies on matching the stimulus with a recognition memory for pheromone from familiar males. The circuitry includes connections to the hippocampus, which is necessary for the consolidation of many memories. However,

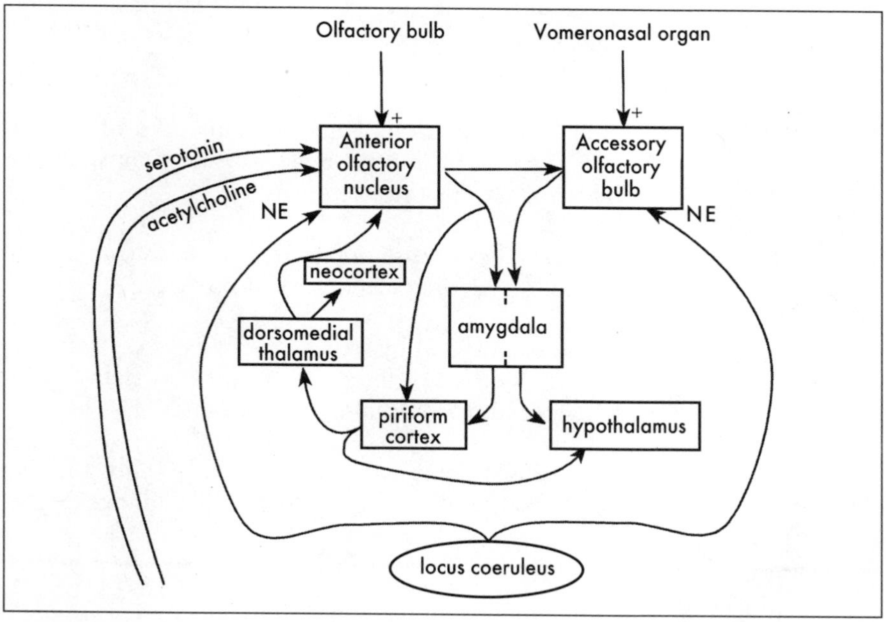

Fig. 7-6 Diagram to illustrate the multiple parallel and interacting regions of the brain that participate in the formation of primitive memories. This diagram shows the sensory pathway from the vomeronasal organ to the accessory olfactory bulb, to the medial amygdala, to several nuclei of the hypothalamus (anterior, arcuate, and ventromedial) and the bed nucleus of the stria terminalis. Multiple neurotransmitter influences include serotonin, acetylcholine, and norepinephrine (NE).

lesions of the hippocampus do not impair formation of pregnancy-block pheromone memories. This suggests that the more restricted circuitry involving the accessory olfactory bulb and its hypothalamic connections are necessary and sufficient. The memory may be residing in the earliest parts of the pathway. Injection of local anesthetic at the first two relays in the accessory olfactory bulb disrupts the pheromone memory system. Both the granule cell layer and the external plexiform layer of the anterior olfactory bulb receive a rich noradrenergic input from the locus coeruleus part of the brain. Formation of pregnancy-block pheromone memories depends on this noradrenergic input. Lesions of this system before mating prevent the formation of pheromone recognition memory. In intact anterior olfactory bulb, mating activates noradrenergic turnover for at least four hours, and this correlates with the length of time that pheromone must be present for recognition memory to develop.

TERMS

Hypothalamus The region of the brain that is bounded dorsally by the thalamus, ventrally by the pituitary, dorsally by the thalamus, anterioraly by the septum and posterioraly by the midbrain. It contains numerous nuclear groups that have several functions: controlling appetite, thirst, sexual cycles and behavior, and the autonomic nervous system.

Locus Coeruleus A cluster of neurons lying on both sides of the midline in the dorsal part of the pons region in the brainstem; these neurons release norepinephrine transmitter.

Noradrenergic Neurons that release norepinephrine as a transmitter (noradrenalin and norepinephrine are synonyms.

Pheromone A chemical signal by which members of the same species recognize each other; different chemicals may help to regulate sexual behavior and maternal-offspring bonding.

Vomeronasal Organ An accessory sensor that lies in a closed tube that lies embedded in the bone of the roof of the midline, alongside the base of the nasal septum.

Related Principles

Memory Consolidation
Memory Kinds
Memory Processes
Neurochemical Transmission (Cell Biology)
Topographical Mapping (Overview)

References

Abeles, M. and Gerstein, G.L. 1988. Detecting spatiotemporal firing patterns among simultaneously recorded single neurons. J. Neurophysiol. 60:909-924.

Brennan, P., Kaba, H., and Keverne, E.B. 1990. Olfactory recognition: a simple memory system. Science 250:1223-1226.

Friston, K.J., Frityh, C.D., and Frackowiak, R.S.J. 1993. Time-dependent changes in effective connectivity measured with PET. Human Brain Mapping. 1:69-79.

John, E.R. 1990. Representation of information in the brain, p. 27-56. In Machinery of the Mind, edited by E. R. John. Birkhäuser, Berlin.

Lang, M., Lang, W., Uhl, F., and Kornhuber, A. 1989. Patterns of event-related brain potentials in paired associative learning tasks: learning and directed attention, pp. 323-325. In Topographic Brain Mapping of EEG and Evoked Potentials, edited by K. Maurer, Springer-Verlag, Berlin.

Miller, E.K. and Desimone, R. 1994. Parallel neuronal mechanisms for short-term memory. Science. 263:520-522.

Peterson, S.E. Fox, P.T., Posner, M.I., Mintun, M., and Raichle, M.E. 1988. Positron emission tomographic studies of cortical anatomy of single-word processing. nature 331:585-589.

Thatcher, R.W. and John, E.R. 1977. Functional Neuroscience. Vol. 1. Foundations of Cognitive Processes. Lawrence Erlbaum, Hillsdale, New Jersey.

Citation Classic

Hebb, D.O. 1949. The Organization of Behavior: A Neuropsychological Theory. Wiley & Sons, New York.

Memory Kinds Reflect Memory Mechanisms

Behavioral expression of memory reveals that there are different kinds of memory, each of which has an associated set of separate yet interacting neural systems and subsystems. The two most basic kinds are memories that are declarative, operating in the consciousness, and procedural, often operating subconsciously.

Explanation

Many authorities in the field of human learning and memory conclude that there are two distinct kinds of memory: declarative and procedural. Declarative memories are constructed consciously, whereas procedural memories are commonly processed as automated and subconscious motor behaviors (Figure 7-7).

Conscious beings explicitly recall declarative memories. But procedural memories are implicit, involving information of skills acquired earlier but that are not accessible to conscious recall.

Studies involving surgical lesions in animals and by observation of naturally occurring disease in humans show that the important neural

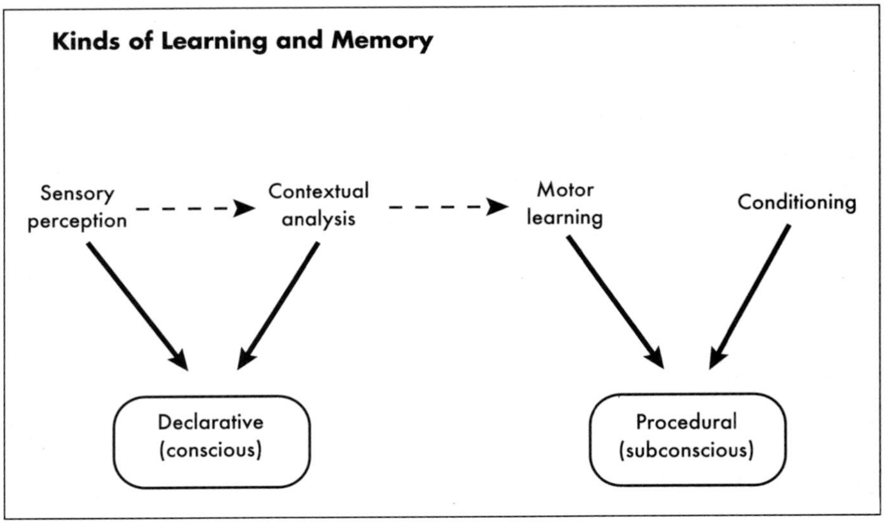

Fig. 7-7 Diagram of some kinds of learning and their relationships to the two main kinds of memory. Note that conditioning can occur from sensory stimulation that does not necessarily involve conscious perception. Other procedural memories may evolve from repetition and rehearsal of declarative memory processes.

structures for explicit memory involve the medial temporal lobe, especially the hippocampus and the midline diencephalon. Key structures in the temporal lobe, in addition to the hippocampus, include the anatomically related entorhinal, peri-rhinal, and parahippocampal cortices. The amygdala is not a part of this system. The important structures in the diencephalon are the anterior thalamic nucleus and the mediodorsal thalamic nucleus. Presumably, these structures and their connections with the neocortex establish lasting explicit memories by binding together the multiple areas of neocortex that collectively subserve perception and short-term memory of whole events. Gradually, the neocortex acquires the capacity to support long-term memory without the continued involvement of medial temporal lobe and midline diencephalic structures.

Examples

Examples of declarative memories include memorization of concepts and facts, or of rules of procedure, or of the words and notes of a song, or any number of things that we recall while being aware of their recall.

Examples of procedural memories include motor skills that develop over long-term practice, such as components of piano playing, touch typing, or playing basketball. Less obvious, perhaps, are stereotyped behavioral responses.

In a typical demonstration of explicit memory, a person may be shown a series of words, pictures, or some other set of items to be remembered. Then, at a later time, the person is tested for recall. In a typical demonstration of implicit memory, the impact of prior learning experiences is examined in performance tasks that do not require recollection of the early learning experiences. Examples include completing a fragment of a word, choosing which of two stimuli are preferred, or reading inverted text. Many motor skills, such as the precision performance of well-trained athletes fall into this category of memory.

Humans who have wounds or a disease of the hippocampal or temporal parts of the brain have severe deficits in the formation of declarative memories, but relatively little impairment of procedural memories. Conversely, people with damage to the cerebellum or other motor-control parts of the brain will have severe limitations in acquiring procedural memories, while their declarative memories may remain relatively intact.

A classic example of implicit learning is the tragic case of a severely epileptic patient, "H.M.," whose hippocampus had to be surgically removed to control the epilepsy. This patient could no longer consolidate new learning experiences into memory, although he did retain memory of many of his life experiences before the surgery. After surgery, learning experiences only lasted as short-term memory, for a few seconds, and then they were lost forever. Many attempts were made to teach this patient a game, but H.M. never learned it, insisting that each time the game was played, he had never seen it before. Nonetheless, H.M. took less and less time to relearn the game each time it was played. There was thus some residual implicit memory that facilitated new learning. Because the hippocampus was destroyed, this neural system must not play a large role in implicit memory, although its role in consolidation of explicit memories is crucial.

TERMS

Declarative Memory "Knowing that," characterized by memories that are often consciously recalled and can be "declared" verbally or in some other equivalent way. These memories can typically be manipulated, recalled, reorganized, and used in novel contexts.

Diencephalon That part of the brain that includes the pineal gland, habenula, thalamus, and hypothalamus. It generally lies along the midline, underneath the cortex.

Hippocampus A segment of primitive cortex that is prominent in primitive animal species and is intimately linked with olfaction. In higher species, the hippocampus is contiguous with and folded underneath the much more prominent neocortex.

Procedural Memory "Knowing how," characterized by memories that are not consciously evident and are relatively inflexible. These often develop after repeated training. In many cases, these memories involve complex motor acts, such as playing the piano, high-skill sports, etc. Brain systems supporting these memories seem to be specifically dedicated and unavailable to other processing systems.

Related Principles

Ensembles of Dynamic Neural Networks
Memory Consolidation
Memory Processes

References

Milner, B., Corkin, S., Teuber, H.L. 1968. Further analysis of the hippocampal amnesic syndrome: Fourteen year follow-up study of H.M. Neuropsychologia. 6:215-234.
Schacter, D.L., Chiu, C.-Y. Peter, and Ochsner, K.N. 1993. Implicit memory: a selective review. Ann. Rev. Neurosci. 16:159-182.
Squire, L.R. and Butters, N. (eds.). 1984. Neuropsychology of Memory. Guilford Press, New York.
Zola-Morgan, S. and Squire, L.R. 1993. Neuroanatomy of memory. Ann. Rev. Neurosci. 16:547-563.

Citation Classic

Nauta, W.J.H. 1958. Hippocampal projections and related neural pathways to the mid-brain in the cat. Brain 81:319-340.

Memory Consolidation

The formation of lasting memories is both time and event dependent. Learning experiences, depending on their intensity and biological significance and depending on sufficient time for normal neural functions, convert from being short-term to intermediate- and long-term memory. Synaptic structoral change accompanies long-term memory.

Explanation

Consolidation is the process whereby newly formed memories are made permanent. Without consolidation, a new experience may be recalled only for a few seconds or minutes. A still poorly understood set of biochemical processes is needed to convert short-term memories, which are held in a spatially distributed pattern of electrical activity, into a form that can outlast that particular pattern of electrical activity (Figure 7-8).

One place in the brain, the hippocampus, is crucial for the creation of long-term memories, but memory itself is apparently not stored there. Also, many studies have shown that forgetting is often a matter of faulty retrieval: the memories are in storage, but just not accessible in the absence of the right cues.

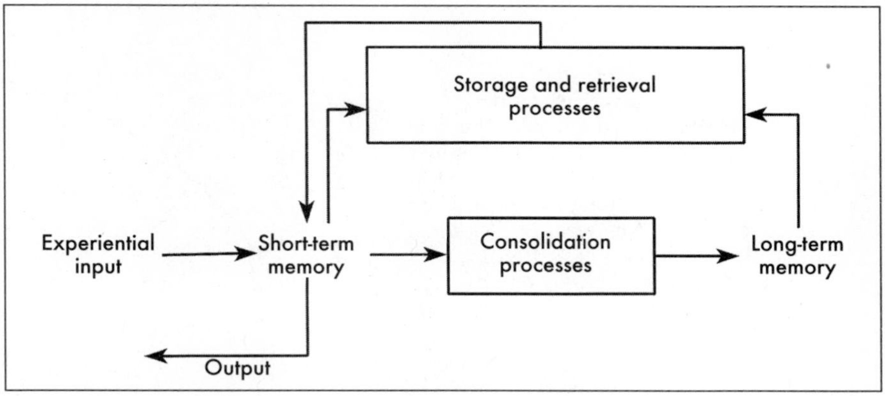

Fig. 7-8 Diagram of the memory consolidation process. Learning stimuli are initially held in the nervous system in a short-term memory form, available for recall only for a short time after the experience. After a certain amount of time, the learned experience may consolidate into a permanent form that can be retrieved from memory stores many days, months, or even years after the initial experience.

Memory consolidation is not quite as simple as it seems, in part because testing for it is always confounded with other variables such as stimulus registration and recall processes. Likewise, the central role of the hippocampus has some apparent exceptions. Hippocampal damage does not impair consolidation of some tasks.

Long-term memories are associated with visible changes in the extent of axon terminal length, branching, and varicosities. The number of synapses and vesicles also increases.

Examples

A common experience that illustrates the consolidation idea is when a person dials a telephone number that has just been found in the telephone directory. The dialer remembers the phone number long enough to dial it, but typically, does not retain memory of that number more than a few minutes. Sometimes, if you are not concentrating, you might not keep the number in your "working memory" long enough to get it dialed, and it has to be looked up again. Moreover, consolidation has failed. That is, the information is learned and remembered only for a short while; it is not consolidated into more permanent form.

In experimental animals, consolidation processes are typically studied in learning situations that involve only one trial, so as to control for variables associated with repeated learning trials. For instance, if a rat is placed on an elevated platform, it will normally step off and move to the nearest wall. However, if the floor of the test chamber is electrified, and a strong-enough foot shock is received when stepping down, a rat usually learns in that one-time experience never to step down from the platform. If removed from the chamber and placed on the same platform the next day, the rat will refuse to step down. However, whether or not this one-trial experience is remembered

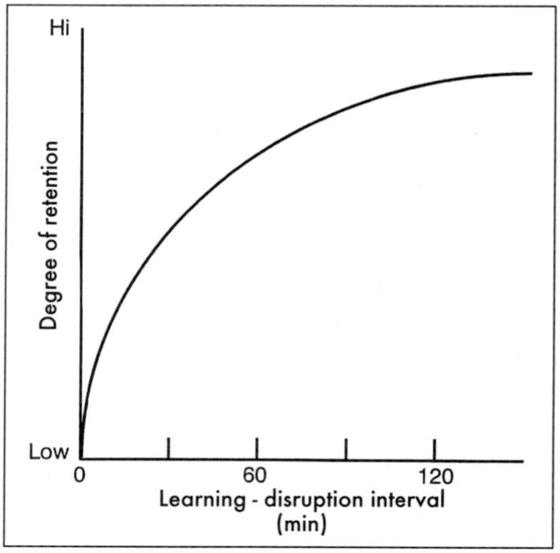

Fig. 7-9 Illustration of experimental results that can be obtained when one tests the time dependence of consolidation in experimental animals exposed to one-trial learning situations. The degree of long-term retention (tested several days after initial experience) depends on how much "uninterrupted" time occurs between the initial experience and the imposition of some new state (such as a set of powerful new stimuli, injection of a sedative drug, or electroconvulsive shock).

(consolidated) depends on allowing the rat time, on the order of minutes, for the brain to convert the learning experience into a long-term memory. If shortly after learning, the rat is subjected to intense distractions, or brain-disrupting treatments such as drugs or electroconvulsive shock, the learning experience may not get consolidated. The key is how much time elapses between the initial learning and exposure to the disruption (drug, shock, etc.). Consolidation is also event dependent, in the sense that the intensity of the learning experience can increase the probability that long-term memory ensues rapidly (Figure 7-9).

Formation of permanent memories requires structural change. The original evidence that new protein synthesis had to occur has been followed by many microanatomical demonstrations of increased axon terminal arborization and electron micrograph evidence of increased synaptic vesicles.

TERMS

Working Memory The kind of memory that is actively retrieved and being used at the moment.

Related Principles

Memory Kinds Reflect Memory Mechanisms
Memory Processes
Motivation (Emotions)
Reinforcement (Emotions)

References

Agranoff, B.W. 1984. Current biochemical approaches to memory formation, p. 353-358. In Neurobiology of Learning and Memory, edited by G. Lynch et al. The Guilford Press, New York.

Carpenter, G.A. and Grossberg, S. 1988. Neural dynamics of category learning and recognition: attention, memory consolidation, and amnesia, p. 233-283. In Brain Structure, Learning, and Memory, edited by J.L. Davis et al. Amer. Assoc. Adv. Science, Washington.

Changeux, J.P. and Konishi, M. (eds.). 1987. The Neural and Molecular Bases of Learning. Wiley & Sons, New York.

Cohen, N.J. and Eichenbaum, H. 1993. Memory, Amnesia, and the Hippocampal System. MIT Press. Cambridge, Massachusetts.

Mahut, H. and M. Moss. 1984. Consolidation of memory: the hippocampus revisited, pp. 297-316. In Neuropsychology of Memory, edited by L. R. Squire and N. Butters. The Guilford Press, New York.

McGaugh, J.L. 1989. Involvement of hormonal and neuromodulatory systems in the regulation of memory storage. Ann. Rev. Neurosci. 12:255-288.

McPhail, E.M. 1993. The Neuroscience of Animal Intelligence. From the Seahare to the Seahorse. Columbia Univ. Press., New York.

Rosenzweig, M.R. and Bennett, E.L. 1984. Basic processes and modulatory influences in the stages of memory formation, pp. 263-288. In Neurobiology of Learning and Memory, edited by G. Lynch et al. The Guilford Press, New York.

Schwartz, J.H. and Greenberg, S.M. 1987. Molecular mechanisms for memory: second messenger induced modification of protein kinases in nerve cells. Ann. Rev. Neurosci. 10:459-476.

Citation Classic

McGaugh J.L. 1966. Time-dependent processes in memory storage. Science 153:1351-8.

Learning and Memory: Study Questions

1. List and explain each of the principles in this category.
2. For each of the principles, provide an example *that is not mentioned in this text.*
3. Explain sensitization, habituation, associative learning.
4. Distinguish classical and operant conditioning.
5. Defend and illustrate the premise that some memories have identifiable locations, whereas others clearly involve widespread, distributed processes.
6. What is the electrical manifestation of LTPP?
7. What do we know (and not know) about the molecular bases of memory?
8. Why do we think that memory is registered in large ensembles of distributed, parallel, and dynamic neural networks?
9. What is memory consolidation and why is it important?
10. What are some things—not mentioned in the text—that can disrupt memory consolidation?
11. Why is a one-trial test paradigm so useful for studying memory consolidation?
12. What is a common disease that is characterized by progressive loss of memory?

CHAPTER 8

Motor Output and Control

I guess, doctor, that your electricity is stronger than my will.

— patient of José Delgado as he tried to prevent a hand-clenching movement produced by electrical stimulation of the motor cortex.

This quote is not meant to convey the notion that the motor cortex is the seat of will for movement. The cortex is an executive agent for movement and not, by itself, the seat of motor will. In fact we do not know how the various motor centers in the brain are orchestrated in developing motor will.

We do know that movement occurs because of antecedent **Motor Preparation,** whether conscious or unconscious. This preparation takes into account the existing state of muscle tone, the existing position of limbs and joints, and the required (or desired) future position of the limbs and joints. Motor commands are ultimately expressed through a **Final Common Path** that terminates in the neurons that make direct contact with muscle or glands. Some of these neural targets are smooth muscle cells or glands, providing a basis for **Visceral Control** (Figure 8-1).

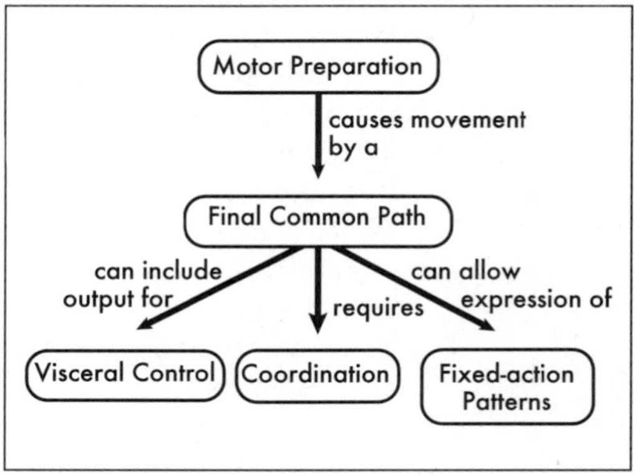

Fig. 8-1 Concept map for motor output and control.

For skeletal muscle targets, neural circuits must have a high degree of interaction and feedback information about the precision of movement so that there is appropriate **Coordination**. Many elemental movements are more or less pre-programmed and can result in complex behaviors that are nonetheless **Fixed-action Patterns** of movement.

List of Principles

Motor Preparation Brain events drive motor output. Electrical changes in the brain precede—and cause—movement.

Final Common Path In the spinal cord (and brainstem) there are "final common pathway" neurons that receive converging motor command signals and, if excitatory threshold is reached, produce direct excitatory effects on skeletal muscle.

Visceral Control Certain neural systems are dedicated to control of visceral movements and glandular secretions. These act in concert with hormones to achieve visceral homeostasis.

Coordination Neural systems that control bodily movements map the body in space and excite specific muscle groups in specific sequence patterns that do not cause conflicting behaviors but rather assure that the motion is appropriate to the stimulus or situation that triggered it. Servo-regulation lies at the heart of this capability.

Fixed-action Patterns Each species has a repertoire of motor-activity patterns. These are generated by central pattern-generator circuitry in response to rather specific stimuli or environmental contingencies. Fixed-action patterns may underlie and subserve more complex (and less reflexive) patterns of motor activity.

Motor Preparation

Mental events drive motor output. Electrical changes in the brain precede—and cause—movement.

Explanation

Certain neurons in the brain can drive lower motor neurons in the spinal cord or motor neurons in cranial nerves to cause muscle contractions that subserve postural tone and body movements. That is, when higher neural circuits "make a decision" that certain body movements should be made, the output of those circuits serve as pre-programmed instructions to create a specified pattern of muscle contractions. In short, movements are created not only by reflex action but by intent. Many of these "intentional" neurons reveal directional coding. For example, when a hand moves in different directions, neuronal activity is greatest for a certain direction.

These descending motor influences can be quite complex, due in large measure to the population of interneurons in the spinal cord. A single interneuron can receive input from more than ten different and diverse regions of the brain.

Examples

Willful intent to move is self-evident to humans. It has been clearly demonstrated in animals as well. In an experiment that demonstrated the phenomenon in animals, monkeys were trained to move a lever in a certain direction when they received two successive cues. The first cue told the monkey the direction to move the lever, and thus served as a signal to the brain to "plan" a future motor act. After a certain delay, the second cue told the monkey to produce the movement. Extracellular recordings from neurons in the motor cortex showed that some of these cells became active after the first cue, before the second cue was presented. These neurons are thought to be generating the motor-planning program that is triggered into action by other neurons ("go" neurons) that respond to the second cue.

Similar indexes of motor intent can be seen from strategically placed EEG electrodes or by PET scans of cerebral blood flow.

Motor activity originating in the spinal cord is modulated not only by the actions of interneurons, as described above, but also by direct projections from the motor cortex. In monkeys, for example, stimulation of each cell leads to a discrete movement, such as flexion of the contralateral thumb. These same cells are activated by passive movement of the appropriate joint of that thumb and can also be activated by anticipated movement of that thumb.

The neurons in the motor cortex's topographical map are modulated during development of implicit (unconscious) and explicit (conscious) learning. This has been demonstrated in human subjects sitting in front of a computer screen who responded to a screen presentation of the number 1, 2, 3, or 4 with a corresponding button press by an assigned finger on the right hand (for example, the ring finger was to press button 3). In the subjects that received stimuli, with a repeated sequence of cues, response times progressively decreased with practice, reflecting implicit learning. Transcranial magnetic stimulation was used to map the motor cortex output to specific finger muscles. The size of these maps became larger as response times decreased in the subjects as they acquired implicit knowledge of the cue sequence. Motor maps do not change in controls who received cues presented randomly. With continued experience, test subjects gradually acquired explicit knowledge of the cue sequence, and their reaction times continued to decrease because they could anticipate the next cue instead of reacting to it. Surprisingly, at this point, the size of the motor maps returned to pre-training level. It would seem that a large motor map supports implicit learning but is not needed for explicit learning because other brain systems take over task execution.

Related Principles

Memory Kinds Reflect Memory Mechanisms (Learning and Memory)
Plasticity (Development)
Reflex Action (Information Processing)
Topographic Mapping (Overview)

References

Czarkowska, J., Jankowska, E., and Sybirska, E. 1981. Common interneurones in reflex pathways from group 1A and 1B afferents of knee flexors and extensors in the cat. J. Physiol. 310:367-380.

Deecke, L., Scheid, P., and Kornhuber, H.H. 1969. Distribution of readiness potential, pre-motor positivity, and motor potential of the human cerebral cortex preceding voluntary finger movements. Exp. Brain Res. 7:158-168.

Evarts, E.V. 1968. Relation of pyramidal tract activity to force exerted during voluntary movement. J. Neurophysiol. 31:14-27.

Georgopoulos, A.P. et al. 1983. Spatial coding of movement: a hypothesis concerning the coding of movement direction by motor cortical populations. Exp. Brain Res. Suppl. 7, 327-336.

Citation Classics

Evarts, E.V. 1966. Pyramidal tract activity associated with a conditioned hand movement in the monkey. J. Neurophysiol. 29:1011-1027.

Evarts, E.V., and Tanji, J. 1976. Reflex and intended responses in motor cortex pyramidal tract neurons of monkey. J. Neurophysiol. 39:1069-1080.

Final Common Path

Output of the nervous system occurs in "final common pathway" neurons that integrate converging motor command signals and, if excitatory threshold is reached, produce direct excitatory effects on skeletal muscle.

Explanation

The last neuron in a chain of neurons that connects to muscle is called a motor neuron. Such neurons are the last way station in the chain of neurons in the spinal cord and brain that can lead to muscle excitation and contraction. In other words, these neurons mediate motor commands from all other places in the nervous system (Figure 8-2).

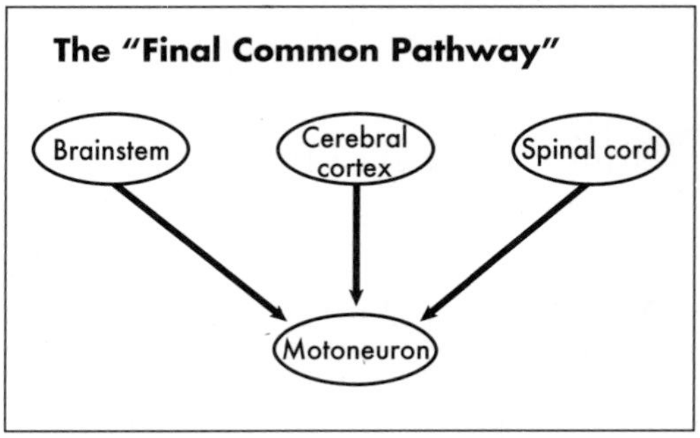

Fig. 8-2 Overview of the relationship to motoneurons to inputs from other places within the nervous system.

In the spinal cord, these motor neurons are organized by body segment; that is, when firing, the neurons activate muscles in the corresponding body segment. These neurons receive inputs from several sources. These include sensory neurons from the same body segment, which can activate spinal reflexes. Sensory neurons from adjacent spinal segments also supply input. There are also inputs that descend from neurons in the brainstem and higher centers of the brain. Finally, there are neurons that are part of a negative feedback loop within the same segment.

A similar principle applies to the cranial nerve nuclei in the brainstem that have motor neurons for muscles of the face, head, and neck.

Contraction of muscle fibers is coupled to depolarization of axon terminals of the motor neurons. The process opens calcium channels in muscle cell membrane, and the calcium influx activates muscle-cell proteins that initiate contraction. The rate of impulse activity in motor neurons determines whether the muscle response is a single twitch or a sustained contraction.

Examples

The best example of the final common pathway concept is the neuronal circuit that underlies spinal reflexes, as described earlier under "Reflex Action." The neuron that initiates the muscle contraction response to a stimulus receives inputs not only from the sensory neuron, but also inputs from adjacent spinal segments, contralateral sensor neurons, descending inputs from the brainstem and higher brain centers, and recurrent feedback loops involving so-called Renshaw cells. Since all of these inputs converge on a final common target, this principle is appropriately called the final common pathway for skeletal motor activity (Figure 8-3).

A similar idea applies to the control of smooth muscles, but here the segmental organization is less evident and the target muscle commonly is also under the influence of circulating hormones.

Complex premotor networks in the brain interact with simpler spinal reflexes. Premotor networks must have complex distributed outputs in order to satisfy the requirement for coordination of simultaneously active muscles in an orchestrated program of excitation of certain motor neurons and inhibition of others.

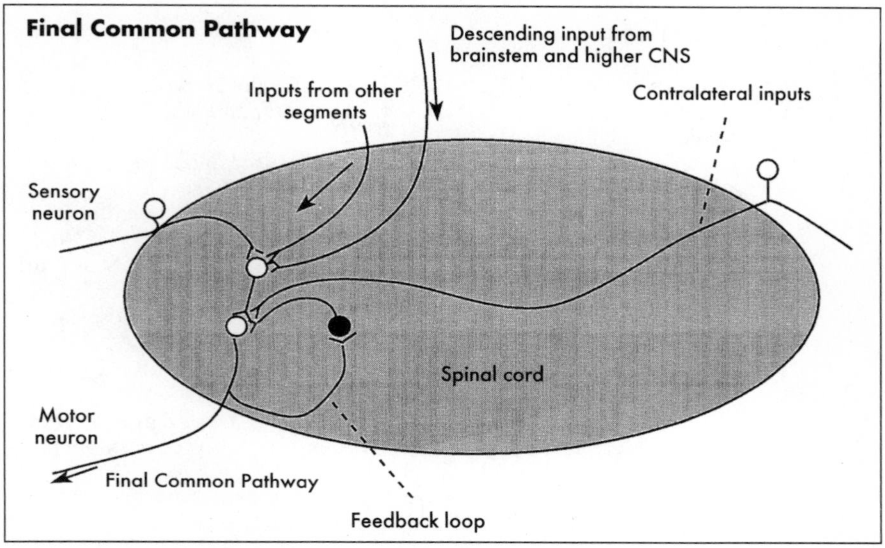

Fig. 8-3 Circuit diagram of the final common pathway motor neuron in the spinal cord. The last neuron in the chain that can activate skeletal muscle receives converging input from multiple sources, such as ipsilateral and contralateral sensory neurons, descending inputs from adjacent spinal segments, the brainstem, the higher brain centers, and from recurrent feedback neurons.

TERMS

Body (Spinal) Segment	A specific region of the body and/or spinal cord that is linked by common innervation. For example, the arms have sensory and motor nerves that are found in one region of the spinal cord (caudal neck and rostral thorax areas), while the legs have nerves found in another region of the cord (lumbar and sacral regions).
Contralateral	On the opposite side of the body.
Cranial Nerve Nuclei	Clusters of neurons in the brainstem that give rise to the nerves of the face and head.
Motoneurons	Synonym for motor neurons.
Negative Feedback Loop	A pathway in which some of the output is fed back to suppress excitatory input.

Related Principles

Reflex Action (Information Processing)
Topographic Mapping (Overview)

References

Alexander, G.E. and DeLong, M.R. 1986. Organization of supraspinal motor systems, pp. 352-369. In Diseases of the Nervous System, Vol. 1, ed. by A.K. Asbury, G.M. McKhann, and W.I. McDonald. Saunders, Philadelphia.

Armstrong, D.M. 1988. Review lecture: the supraspinal control of mammalian locomotion. J. Physiol. (London) 405:1-37.

Dennis, M.J. 1981. Development of the neuromuscular junction: inductive interactions between cells. Ann. Rev. Neurosci. 4:43-68.

Hoyle, G. 1983. Muscles and Their Neural Control. Wiley, New York.

Lundberg, A. 1975. Control of spinal mechanisms from the brain, pp. 253-265. In The Nervous System, Vol. 1: The Basic Neurosciences, ed. by D.B. Tower. Raven Press, New York.

Pearson, K.G. 1993. Common principles of motor control in vertebrates and invertebrates. Ann. Rev. Neurosci. 16:265-297.

Shinoda, Y., Yokota, J.-I., and Futami, T. 1981. Divergent projection of individual corticospinal axons to motor neurons of multiple muscles in the monkey. Neurosci. letter. 23:7-12.

Stein, P.S.G. 1978. Motor systems, with specific reference to the control of locomotion. Ann. Rev. Neurosci. 1:61-81.

Citation Classics

Dubner R., Sessle B.J., and Storey A.T. 1978. The Neural Basis of Oral and Facial Function. New York: Plenum Press.

Eccles, J.C. 1964. The Physiology of Synapses. Springer. Berlin.

Jackson, J.H. 1932. Selected Writings of John Hughlings Jackson, Vol. II ed. by J. Taylor. Hodder and Stoughton, London.

Lloyd, D.P.C. 1946. Integrative pattern of excitation and inhibition in two-neuron reflex arcs. J. Neurophysiol. 9:439-444.

Penfield W. and Jasper H. 1954. Epilepsy and functional anatomy of the human brain. Boston, MA: Little, Brown.

Renshaw, B. 1941. Influence of discharge upon excitation of neighboring motoneurons. J. Neurophysiol. 4:167-183.

Sherrington, C. 1947. The Integrative Action of the Nervous System, 2nd ed. Yale University Press, New Haven.

Taxi J. 1965. Contribution a l'etude des connexions des neurones moteurs du systeme nerveux autonome. (Contribution to the study of the connections of motor neurons in the autonomic nervous system.) Ann. Sci. Natur. Zool. 7:413-674.

Visceral Control

Certain neural systems are dedicated to control of visceral movements and glandular secretions. These act in concert with hormones to achieve visceral homeostasis.

Explanation

The role of visceral muscles and glands is quite different from that of skeletal muscles. The purpose of skeletal muscle is to provide selective action, in specified temporal sequences, to produce coordinated, purposeful movements. To be most adaptive, these movements should have a control system that can allow movements to be quick, specific, quickly terminated, and ordered in an appropriate sequence with other joint movements. No such requirements exist for the control of visceral muscle and glands. In fact, the requirements are often just the opposite.

Neural control of visceral muscle needs to be slower to act, more undifferentiated, more divergent, and longer lasting than neural control of skeletal muscle. In higher animals, visceral organs such as the heart and digestive tract must have intrinsic contractions and secretions that are modulated by neural controls. Most visceral organs receive a dual and antagonistic innervation. This arrangement allows viscera to be activated or deactivated, and the nervous system controls normally operate to produce homeostatic balance of activity. Both smooth muscle and glands of viscera are subject to this control.

Visceral control is accomplished unconsciously by control centers located mostly in the hypothalamus and brainstem. However, a significant portion of the system lies outside the brain and spinal cord. There are two divisions, called sympathetic and parasympathetic. As a simplification, they can be regarded to have mutually opposing action, with the sympathetic division designed to mobilize the body for emergency, life-threatening conditions, and the parasympathetic division designed to support nurturing, regenerative physiological processes.

Neurons of the sympathetic system are clustered together in ganglia that lie outside the brain and spinal cord. Input to these ganglia come from certain cranial nerves and from nerves arising out of the thoraco-lumbar part of the spinal cord. The ganglion cells (called postsynaptic because they are the last ones in the output chain, receiving presynaptic cholinergic input) send axons that distribute widely to most of the visceral organs: heart, lungs, digestive tract, urogenital tract. The axon terminals dump neurotransmitter (norepinephrine) at the junctions with smooth muscle cells and glands. The adrenal gland is an interesting exception in that the postsynaptic neurons constitute the adrenal medulla itself and the secretory product is epinephrine, which is released into the bloodstream (and thus is a hormone).

In the parasympathetic division of the autonomic nervous system, the postganglionic neurons are located in the visceral organs themselves. Why is this an advantage? For tubular organs, such as gut and bladder, the postganglionic neurons form more or less continuously circumferential layers, so that the whole gut can be activated more coherently, as the neurons can spread their excitation of muscle and gland secretions uniformly as the neurons excite each other at the same time. This would be harder to do if the neurons were located in the brain or spinal cord.

In both systems there are feedback signals, neuronal and hormonal, that inform both brain and local reflex circuits of the consequences of the autonomic output (Figure 8-4).

Because of these controls, animals can adjust rapidly and automatically to environmental conditions, mobilizing for "fight or flight" when

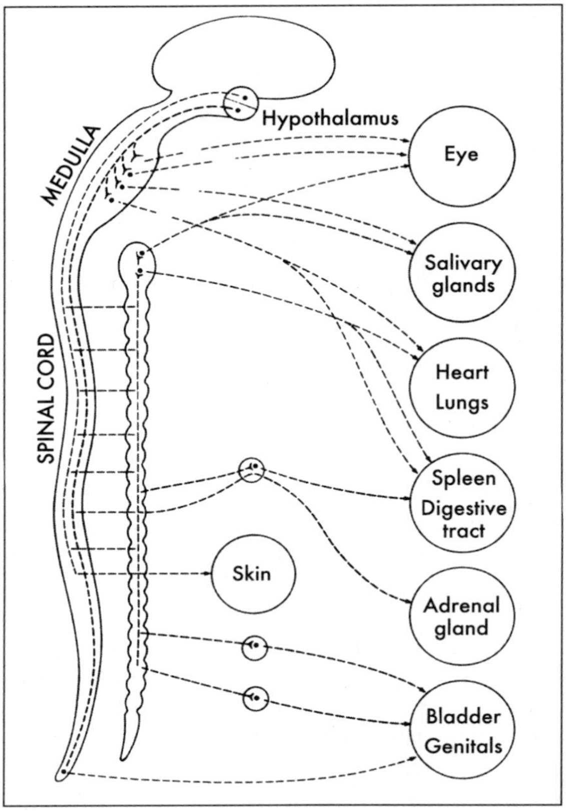

Fig. 8-4 Neural control systems for visceral function. Note that sympathetic pathways (emerging from the middle region of the spinal cord) have synaptic junctions outside the brain and spinal cord, either in a series of ganglia (cell clusters) lying alongside the spinal cord or in certain ganglia in the thoracic and abdominal cavities. The adrenal gland is an exception, in that it is nervous tissue and is the functional equivalent of a group of postsynaptic neurons. Parasympathetic neurons (emerging from the brainstem and the caudal end of the spinal cord) have their synapses outside the brain, located in the target organs themselves. (Reprinted with permission from Klemm, 1972.)

needed or relaxing for bodily maintenance functions. Some of the specific effects on viscera of the two regulatory systems are outlined in Table 1.

The output pathway from the central nervous system has two synaptic relays. The first uses acetylcholine as the transmitter. The second relay in the parasympathetic division also uses acetylcholine, but the second relay in the sympathetic division uses norepinephrine, a derivative of epinephrine (adrenalin). In many synapses, there is also co-release of peptide transmitters that can modulate the kinetics, duration, and strength of action on target muscles and glands.

Examples

Would your body need to do the same things when you are calmly eating a steak and when you recognize that a wild animal was about to attack you? Of course, when under attack you would need sympathetic activation of such visceral functions as increased heart rate, blood pressure, air flow through the lungs, and epinephrine release from the adrenal gland. Note from the table of functions that appropriate blood flow changes also occur. Flow increases in the heart and skeletal muscle, where it is needed, and decreases in the skin, digestive tract, and spleen, where it is not needed. Note also that the hormone, epinephrine, has many of the same functions as the sympathetic division. It thus acts as a reinforcer to prolong and intensify the effects triggered in the nervous system. Have you noticed how long it takes you to calm down after being frightened or after an emergency?

Conversely, during non-emergency situations, parasympathetic influences dominate, promoting such maintenance functions as digestion, rest for the heart, sexual activity, and appropriate redistribution of blood.

TABLE 8-1 SUMMARY OF AUTONOMIC FUNCTIONS

Sympathetic (Emergency)		Parasympathetic (Routine)
Salivation of mucus		Salivation, watery
Blood vessels constricted	Digestive	—
Peristalsis decreased	Organs	Peristalsis increased
Sphincters contracted		Sphincters relaxed
Secretions decreased		Secretions increased
Rate increased		Rate decreased
Contractile force increased	Heart	Contractile force decreased
Coronary arteries dilated		—
Blood vessels dilated	Skeletal Muscle	—
Blood vessels constricted		—
Hair erected	Skin	—
Sweat, from palms		Sweat, general
Bronchioles dilated	Lungs	Bronchioles constricted
Contracted	Spleen	Dilated
Wall dilated	Bladder	Wall contracted
Sphincter contracted		Sphincter relaxed
Ejaculation	Genitals	Erection
Pupils dilated	Eyes	Pupils constricted
Epinephrine released	Adrenal Gland	—

Unlike the case with skeletal muscle control, there are no particular advantages in having millisecond speed, because smooth muscle and glands can't respond that fast anyway. Second, the action needs to involve diffuse targets. For example, biological emergencies often call for blood to be diverted away from the digestive tract to skeletal muscles; it makes little sense to divert blood away from just one region of the digestive tract. To be most effective, the action needs to occur all at once, not spread out in sequence over time. Finally, this is the kind of response that you don't want to have turned off immediately. Most biological emergencies last longer than milliseconds. In the case of skeletal muscle action, for example, there may be a need for a continuously changing pattern of vigorous muscle contractions, which can occur only on a background of a sustained availability of increase in blood supply.

TERMS

Autonomic Nervous System	That part of the nervous system that controls viscera (smooth muscle and glands).
Parasympathetic Nervous System	That part of the autonomic nervous system that promotes routine, restorative kinds of functions (lower blood pressure, slower pulse, enhanced digestion.
Sympathetic Nervous System	That part of the autonomic nervous system that promotes mobilization of those bodily functions that contribute to emergency conditions (such as increased heart rate and blood pressure, dilation of bronchioles for exchange air in the lungs, diversion of blood from gut to skeletal muscle).

Related Principles

Hierarchical Control (Overview)
Homeostatic Regulation (Overview)
Neurotransmission (Cell Biology)
Readiness Response (States of Consciousness)
Reciprocal Action (Information Processing)
System Modulation (Information Processing)

References

Barman, S.M. 1990. Brainstem control of cardiovascular function, pp. 353-382. In Brainstem Mechanisms of Behavior, ed. by W.R. Klemm and R.P. Vertes. Wiley & Sons, New York.

Dennis, M.J., Harris, A.J., and Kuffler, S.W. 1971. Synaptic transmission and its duplication by focally applied acetylcholine in parasympathetic neurons in the heart of the frog. Proc. Roy. Soc. Lond. 117:509-539.

Elfin, L.G. (ed.) 1983. Autonomic Ganglia. Wiley, New York.

Furness, J.B. and Costa, M. 1980. Types of nerves in the enteric nervous system. Neuroscience. 5:1-20.

Klemm, W.R. 1972. Science, The Brain, and Our Future. Bobbs-Merrill, New York.

Orem, J. and Dick, T.E. 1990. Brainstem respiratory neurons and their control during various behaviors, pp. 383-406. In Brainstem Mechanisms of Behavior, ed. by W.R. Klemm and R.P. Vertes. Wiley & Sons, New York.

Rose, J.D. 1990. Brainstem influences on sexual behavior, p. 407-464. In Brainstem Mechanisms of Behavior, ed. by W.R. Klemm and R.P. Vertes. Wiley & Sons, New York.

Smith, O.A. and DeVito, J.L. 1984. Central neural integration for the control of autonomic responses associated with emotion. Ann. Rev. Neurosci. 7:43-65.

Swanson, L.W. and Sauchenkeo, P.E. 1983. Hypothalamic integration: Organization of the paraventricular and supraoptic nuclei. Ann. Rev. Neurosci. 6:269-324.

Citation Classics

Bard, P. 1928. A diencephalic mechanism for the expression of rage with special reference to the sympathetic nervous system. Amer. J. Physiol. 84:490-515.

Cannon, W.B. 1932. The Wisdom of the Body. Norton, New York.

Coordination

Neural systems that control bodily movements map the body in space and excite specific muscle groups in specific sequence patterns that do not cause conflicting behaviors but rather assure that the motion is appropriate to the stimulus or situation that triggered it. Servo-regulation lies at the heart of this capability.

Explanation

Whole-body and limb movements involve patterns of muscular contraction and relaxation that achieve movement of joints, either flexing or extending them. The pattern of joint movements, considered together, produces observable behavior. Since each joint tends to have opposing muscles (those that extend, those that flex when contracting), it is clear that no joint action occurs if both opposing sets of muscles are active at the same time. Thus, to produce extension, the extensors must contract while at the same time flexors must relax. To get a pattern of such joint movements over time, neural coordinating mechanisms are needed to produce a succession of joint actions that are appropriate to the stimulus situation.

Coordination is achieved through neuronal circuitry that not only issues commands for muscles to contract, but also is automatically informed of the consequences of those motor actions. This feedback allows corrective action.

Neural processing reactions may use computations that involve multiple solutions for the same result. For example, in motor control, the computation of spatial coordinates for desired movements defines a single set of target locations but permits multiple sets of direction, amplitude, and velocity of movements that can produce the same end. Vectorial combination of coarse postural maps in the spinal cord can yield a large repertoire of motor behaviors.

Example

A common way to illustrate servo-control is to consider a mosquito homing in on a human target. Mosquitos fly upstream in a plume of human body heat and odor. The insect cannot "know" where the target is, but as it flies toward it and accidentally exits the plume, continuous monitor sensors inform the motor control system to turn around in the

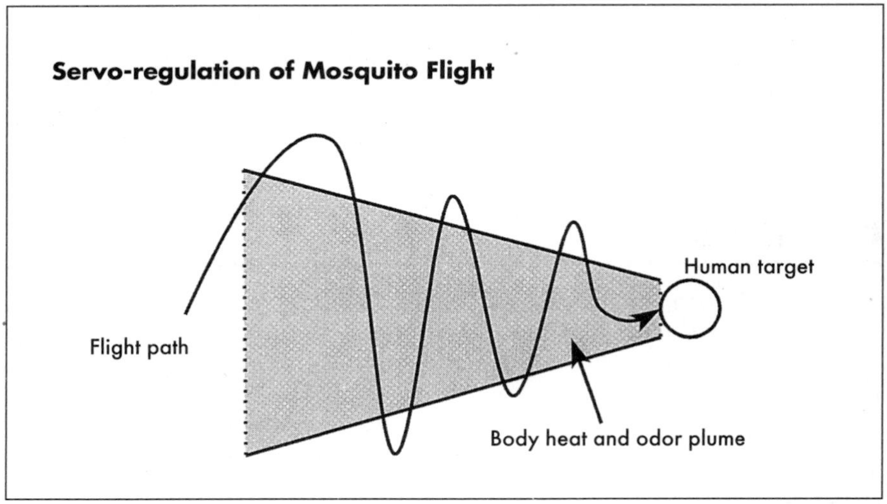

Fig. 8-5 Diagram of how a mosquito coordinates its flight movements to reach the target. The plume of body heat and odor from a human target causes the mosquito to fly upstream. Each time the insect leaves the plume, it reflexly corrects by right or left upstream movements until it gets back into the plume.

direction of the airstream. This automatically repositions the insect in the plume. A succession of such events keeps the insect progressing toward the target (Figure 8-5).

The mechanisms by which such servo-regulation occurs in higher animal brains is quite complex. We do know that motor output signals arise from certain discrete brain areas (motor cortex, red nucleus, vestibulospinal nuclei, reticulospinal nuclei, and others). But there are peripheral sensors in muscles and tendons that feed back information on the status of contraction/relaxation of muscles and tension in tendons into certain coordination centers in the brain. Chief among these is the cerebellum, which not only gets this feedback but also gets feed-forward input from cells in the motor cortex. In other words, when the motor cortex issues a command for certain muscles to contract, the cerebellum "knows" in advance about this command before action even occurs. Because of the feedback that the cerebellum gets, it is poised to compare intended action with consequence.

Other brain structures are also involved in servo-regulation, but in less clearly understood ways. These include such structures as the caudate, putamen, and globus pallidus.

TERM

Servo-regulation	Automatic control over output, achieved by feedback from sensors that inform command structures of the consequences of intended action.

Related Principles

Feedback and Re-entry (Information Processing)
From Input to Output (Information Processing)

References

Alexander, G.E., DeLong, M.R., and Strick, P.L. 1986. Parallel organization of functionally segregated circuits linking basal ganglia and cortex. Ann. Rev. Neurosci. 9:357-381.

Asanuma, H. 1989. The Motor Cortex. Raven Press, New York.

Bernardi, G. et al. 1991. The Basal Ganglia III. Advances in Behavioral Biology Series. Vol. 39.

Bizzi, E., Mussa-Ivaldi, F.A., Giszter, S. 1991. Computations underlying the execution of movement: a biological perspective. Science. 253:287-291.

Eccles, J. and Dimitrijevic, M.R. (eds.) 1985. Upper Motor Neuron Functions and Dysfunctions. Rec. Achieve Restorative Neurol. Vol. 1, S. Karger, Farmington, CT.

Franks, A.J., 1992. Function and dysfunction in the basal ganglia. Elsevier, Oxford.

Georgopoulos, A.P. 1986. On reaching. Ann. Rev. Neurosci. 9:147-170.

Georgopoulos, A.P. 1991. Higher order motor control. Ann. Rev. Neurossci. 14:361-377.

Gerfen, C.R. 1992. The neostriatal mosaic: multiple levels of compartmental organization in the basal ganglia. Ann. Rev. Neurosci. 15:285-320.

Thach, W.T., Goodkin, H.P., and Keating, J.G. 1992. The cerebellum and the adaptive coordination of movement. Ann. Rev. Neurosci. 15:403-442.

Ito, M. 1985. The Cerebellum and Neural Control. Raven, N.Y.

Kien, J. 1992. Neurobiology of motor programme selection. Elsevier, Oxford.

Pearson, K.G. 1993. Common principles of motor control in vertebrates and invertebrates. Ann. Rev. Neurosci. 16:265-297.

Penney, J.B. and Young, A.B. 1983. Speculations on the functional anatomy of basal ganglia disorders. Ann. Rev. Neurosci. 6:73-94.

Soechting, J.F. and Flanders, M. 1992. Moving in three-dimensional space: frames of reference, vectors, and coordinate systems. Ann. Rev. Neurosci. 15:167-191.

Stein, P.S.G. 1978. Motor systems, with specific reference to the control of locomotion. Ann. Rev. Neurosci. 1:61-81.

Tarsy, D. and Baldessarini, R.J. 1980. Dopamine and the pathophysiology of dyskinesias induced by antipsychotic drugs. Ann. Rev. Neurosci. 3:23-41.

Fixed-action Patterns

Each species has a repertoire of fixed motor activity patterns. These are generated by central pattern-generator circuitry in response to rather specific stimuli or environmental contingencies. Fixed-action patterns may underlie and subserve more complex (and less reflexive) patterns of motor activity.

Explanation

A fixed-action pattern resembles a reflex. It is an inborn, unlearned set of motor activities that subserve some adaptive behavior under very specific sensory or situational conditions. Indeed, a fixed-action pattern can be thought of as a complex reflex that is compounded from a series of elemental reflexes. Other ways that fixed-action patterns may differ from elemental reflexes is that they tend to be all-or-none, rather than increasing in expression with increasing stimulus strength. Also, fixed-action patterns may sometimes occur in the absence of any external stimulus; that is, they can be internally generated.

Certain fixed-action patterns that have clear adaptive value may be common to many species. Other patterns may be highly species specific.

Examples

Fixed-action pattern generators are particularly evident in invertebrates. Certain invertebrates have "command" neurons that elicit complex sequences of behavior, involving flying, walking, swimming, and feeding behavior. In the crayfish, for example, electrically stimulating a single command neuron will evoke a complex defensive response that involves dozens of different muscles. The output of command neurons is divergent, exciting some and inhibiting others of the follower neurons. The interconnecting circuitry of follower neurons governs the specific motor output pattern.

There are no clearly identified command neurons in higher vertebrates, but they do have specific groups of cells that trigger preprogrammed motor acts. Such acts include certain appetitive behaviors, swallowing, vomiting, yawning, a variety of courtship behaviors, and a locomotor pattern generator in the brainstem.

The orienting response that animals display to certain stimuli is an example of a complex fixed-action pattern. This orienting is a component of the "arousal response," wherein an animal turns its head toward a stimulus and attends to it, particularly with visual, auditory, and olfactory senses. The behavioral adaptiveness of such a response is to help an animal find goal objects (food, water, nesting material) or to avoid stimuli that signal danger.

Another example is the courtship behavior of bulls. Females will mate only one day a month, due to cyclic hormonal changes that regulate development of eggs in the ovary and that also regulate several aspects of physiology and behavior. As a cow approaches this mating day, she gives off an odor, called pheromone, that a bull can smell. The bull spends several days in close physical proximity to this female and "guards" her. The bull will try to ward off other bulls who have been attracted and they may even get into fights. A series of chained behaviors occurs when a bull detects female pheromone. These behaviors can be grouped into three sets involving attraction to the odor, detection of the nature of the chemical signal, and preparation for mating. These behaviors are interdepen-

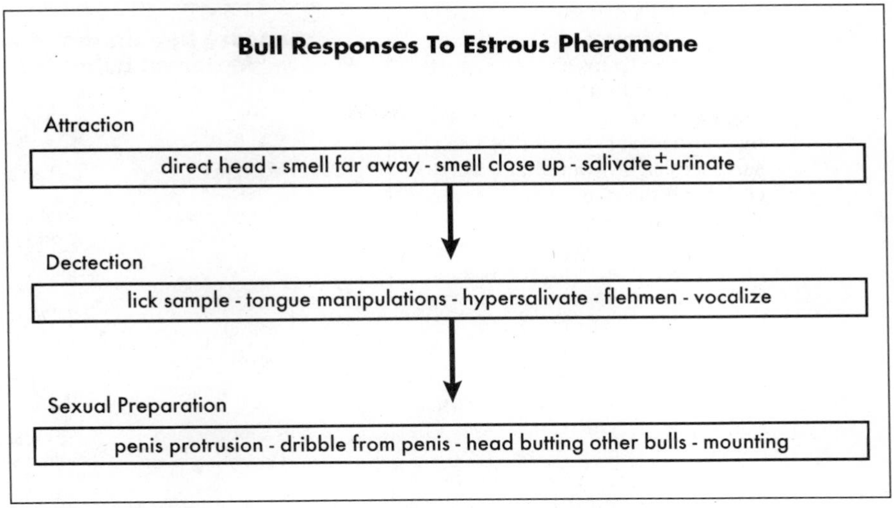

Fig. 8-6 Diagram of the set of chained behaviors that bulls exhibit when stimulated by pheromone from cows as they approach the day of mating and ovulation in the reproductive cycle. The stereotypical behaviors occur in left-to-right sequence. Generally, the probability of any one behavior occurring depends on the occurrence of the preceding behaviors.

dent. One behavior does not occur without certain specified preceding ones, and this can be demonstrated even when a cow is not physically present by presenting body fluids from a cow in an open dish. When first presented a body-fluid sample from a cow that is nearing the day of mating, the bull's attraction behaviors include—in sequence—turning the head toward the source of the odor, smelling it from a distance, being attracted to approach the sample, smelling it close up, salivation, and possibly urination. The detection behaviors, again in a prescribed sequence, include licking the sample, certain tongue movements that include stroking of the roof of the mouth with the tip of the tongue, hypersalivation, and a peculiar curling of the upper lip with head and neck extended upward. The latter is called flehmen behavior and seems to be associated with getting the liquid-borne odors sucked into an accessory olfactory organ known as the vomeronasal organ. On the basis of this sampling, the bull "decides" whether or not the pheromone is present. If not, the chain of sexual behaviors stops. If the pheromone is present, the bull then enters the sequence of sexual preparation behaviors that includes vocalization (bellowing), penis protrusion, dribbling from the penis, agonistic behaviors with any other nearby bulls, and mating of the female (Figure 8-6).

Even behaviors that are seemingly very complex can be little more than the sum of a chained series of rather simple acts. For example, the nurturing care that mother rats give to their young has been found to be a fixed-action pattern (Figure 8-7).

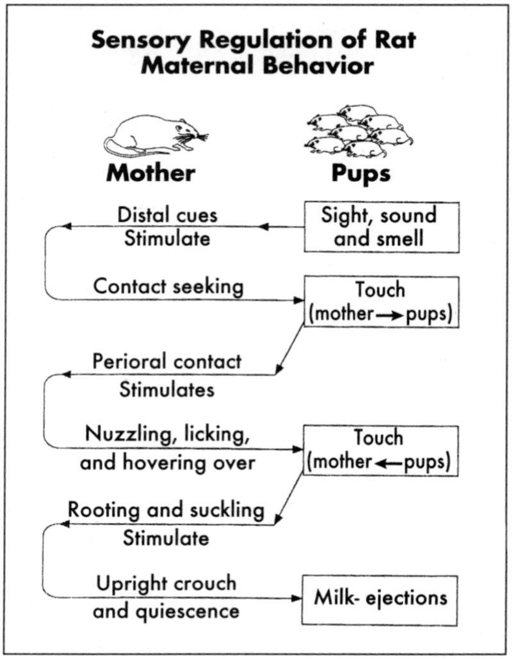

Fig. 8-7 Chained series of fixed-action patterns of mother rat and her newborn pups as they engage in stereotyped nursing and suckling behavior. Sex hormones of the mother bias the mother's sensitivity to sight, sound, and odor cues from the pups. This induces the mother to seek contact with the pups, gathering them together in the nest. The perioral sensations in the mother trigger her to nuzzle and lick the pups and hover over them. The contact sensations from the pups touching her mammae and underbelly skin trigger her to crouch over the pups in a splayed leg, immobile position. The tactile stimulation and the ventral skin cues from the mother trigger rooting and suckling responses from the pups, which in turn lead to nursing and milk let down. (After Stern and Johnson, 1989).

Fixed-action patterns are not exclusively devoted to causing movement. One notable example of inhibition of movement is the immobility reflex ("Animal Hypnosis"), in which certain stimuli cause a total arrest of behavior.

TERMS

Adaptive Behavior Behavior that is useful to an animal or to its species.

Immobility Reflex (Animal Hypnosis) A state of motor "waxy flexibility" that can ensue in certain species when they are placed in an awkward posture and restrained. It can also occur spontaneously in response to certain visual stimuli.

Related Principles

Coordination
Inhibitory Routing (Information Processing)
Reflex Action (Information Processing)

References

Bjursten, L.-M., Norrsell, K., and Norrsell, U. 1976. Behavioural repertory of cats without cerebral cortex from infancy. Exp. Brain Res. 25:115-130.

Cohen, A.H. et al. 1988. Neural Control of Rhythm and Movement in Vertebrates. Neurobiology Series. John Wiley & Sons, New York.
Delcomyn, F. 1980. Neural basis of rhythmic behavior in animals. Science. 210:492-498.
Grillner, S. and Wallen, P. 1985. Central pattern generators for locomotion, with special reference to vertebrates. Ann. Rev. Neurosci. 8:233-261.
Jankowska, E. Padel, Y., and Tanaka, R. 1976. Disynaptic inhibition of spinal motoneurones from the motor cortex in the monkey. J. Physiol. (London) 258:467-487.
Klemm, W.R. 1990. Behavioral inhibition, pp. 497-534. In Brainstem Mechanisms of Behavior, edited by W.R. Klemm and R.P. Vertes. Wiley & Sons, New York.
Kupfermann, I. and Weiss, K.R. 1978. The command neuron concept. Behav. Brain Sci. 1:3-39.
Pearson, K.G. 1993. Common principles of motor control in vertebrates and invertebrates. Ann. Rev. Neurosci. 16:265-297.
Rivard, G. and Klemm, W.R. 1989. Two body fluids containing bovine estrous pheromone(s). Chemical Senses. 14:273-279.
Roberts, T.D.M. 1979. Neurophysiology of Postural Mechanisms. Butterworths. London.
Shik, M.L. and Orlovsky, G.N. 1976. Neurophysiology of locomotor automatism. Physiol. Rev. 56:465-501.

Citation Classics

Darwin, C. 1872. The Expression of the Emotion in Man and Animals. Murray, London.
Grillner, S. and Shik, M.L. 1973. On the descending control of the lumbosacral spinal cord from the "mesencephalic locomotor region." Acta Physiol. Scand. 87:320-333.
Lorenz, K.Z. 1950. The comparative method in studying innate behaviour patterns. Symp. Soc. Exp. Biol. 4:221-268.
Magoun, H.W. and Rhines, R. 1946. An inhibitory mechanism in the bulbar reticular formation. J. Neurophysiol. 9:165-171.
Sherrington, C.S. 1898. Decerebrate rigidity, and reflex coordination of movements. J. Physiol. (London) 22:319-332.
Sherrington, C. 1947. The Integrative Action of the Nervous System. 2nd Ed. Yale Univ. Press, New Haven.
Stern, J.M. and Johnson, S.K. 1989. Perioral somatosensory determinants of nursing behavior in Norway rats (Rattus norvegicus). J. Comp. Phychol. 103:269-280.
Tinbergen, N. 1951. The Study of Instinct. Clarendon Press, Oxford.
Wiersma, C.A.G. 1938. Function of the giant fibers of the central nervous system of the crayfish. Proc. Soc. Exp. Biol. Med. 38:661-662.

Motor Output and Control: Study Questions

1. List and explain each of the principles in this category.
2. For each of the principles, provide an example *that is not mentioned in this text.*
3. What kinds of evidence can document that electrical changes in the brain precede—and cause—movement?
4. Where are final common pathway motor neurons located?
5. Why is the input to these final common pathway neurons convergent?
6. What criteria determine whether a given behavior is a fixed-action pattern?
7. What are the advantages and disadvantages of fixed-action patterns of behavior?
8. What is a "command" neuron? What are the advantages and disadvantages of having such command neurons?
9. What is the role of temporal sequencing in motor coordination?
10. What is the role of feedback in the coordination of motor behavior?
11. What are some primary differences in the mechanisms of visceral control and skeletal muscle control?
12. The sympathetic nervous system has a more extensively divergent output than the parasympathetic nervous system. What is the functional consequence of that difference?
13. What is it about the sympathetic and parasympathetic nervous systems that enable them to be important in visceral homeostasis?

CHAPTER 9

Development

The mass of neural tissue controlling a particular function is appropriate to the amount of information processing involved in performing the function.

— H.J. Jerison

Soon after an embryo forms, certain cell lines segregate that will ultimately create the nervous system, undergoing explosive **Neuron Division.** A super-abundance of new neurons leads to an **Early Death** of most neurons. Those that survive become "life's quota" of neurons, inasmuch as shortly after birth no new neurons will ever appear.

Formation of the nervous system is characterized by **Programmed Development** in which the genetic capabilities propel cells to divide and grow processes. But the physical and molecular environment in which development occurs creates **Epigenetic** forces that direct the **Migration** of

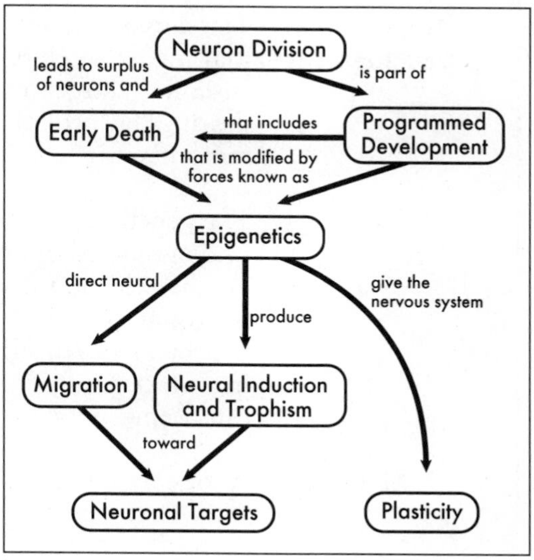

Fig. 9-1 Concept map of neural development processes.

certain neuronal populations to a final location, and produce neural **Induction and Trophism**. These forces dictate where and how growing processes of neurons will reach their **Neuronal Targets**. Collectively, these processes dictate the final location and differentiation of neuronal cell types. The circuitry of functional synapses is "sculpted" across space and time by the combination of genetic and epigenetic influences. Even after birth, the nervous system exhibits considerable **Plasticity**, in that environmental and experiential influences can continue to create new functional circuits.

List of Principles

Neuron Division Neurons have a narrowly defined time window in which to divide. During embryogenesis, each type of neuron is formed at a characteristic stage of development. Mature neurons do not divide. When they die, they are not replaced. Damaged neuronal processes, however, can regenerate.

Early Death Cell death gives birth to the nervous system. The developmental formation of the brain arises out of a competition of proliferating neurons that are "seeking" their final target contacts with other neurons and with muscles and glands. Many neurons seem to have a genetic program to kill themselves if the right target is not reached. In some systems, up to 70% die before the structure in the region is completed. Death of these neurons is accompanied by the sculpting of circuitry by the survivors.

Programmed Development The nervous system is created by a developmental program that involves cell division, migration and navigation, differentiation, target recognition, synapse formation, and cell death. The program is directed by genetic and epigenetic molecular signals that are spatially and temporally defined.

Epigenetics The genetic code does not entirely specify how a brain is to be constructed. It does specify a set of immutable constraints on the selection and competitive processes associated with neuronal division, proliferation of processes, sticking of neurons to each other, cell death, and selective strengthening of certain synapses. These competitive processes are thus "epigenetic."

Migration At one time or another in their lifetime, all neurons are gypsies—they migrate to their final positions and connections with other cells.

Neural Induction & Trophism	During embryonic development, certain chemicals act as inducers of neural differentiation and growth. These chemicals appear at different stages of development, and they regulate gene expression to dictate the fate of dividing neuronal precursor cells.
Neuronal Targets	The growing processes of neurons are guided to their targets by physical forces in their environment and by chemical signals "emitted" by the targets. Depriving a neuron of contact with its normal contact may cause the neuron to degenerate or die and its targets to become supersensitive to the chemical secretions of the missing neuron.
Plasticity	The nervous system can exhibit considerable plasticity. Neurons can grow new axonal and dendritic processes, and new synapses can be formed. Moreover, neural circuits can be recruited into existence and existing circuits can be re-organized.

Neuron Division

Neurons have a narrowly defined time window in which to divide. During embryogenesis, each type of neuron is formed at a characteristic phase of development. Mature neurons do not divide. When they die, they are not replaced. Damaged neuronal processes, however, can regenerate.

Explanation:

During embryogenesis, there is a staged sequence of neuronal division. Moreover, division occurs along spatial gradients, from anterior to posterior, and in vertebrates, also from ventral to dorsal. As a result, certain brain and spinal cord areas become differentiated before others. Most of the stem cells that give rise to neurons are originally located as a single cell layer on the inner (ventricular) surfaces of the neural tube. The newly formed neurons migrate outward from the stem cells. But shortly after birth, all this neuronal division stops. Neurons that die are never replaced.

We do not know why mature neurons do not divide. Other cell types in the body, including the glial cells that surround and support neurons, divide readily. All we can surmise is that there is some change in gene expression caused by biochemical signals associated with differentiation within neurons.

What these feedback signals are remain a mystery. It is not likely that the signal is associated with neurosecretion. Many other types of cells also secrete chemicals and hormones, yet they continue to divide throughout life.

One possibility is the pervasive, rapidly changing electrical environment in which neurons live. This may also explain why other electrically active cells, such as skeletal muscle, also do not divide after maturity.

Example

The time at which neuronal division stops varies with species. Some species are more mature at birth than others. Also, termination of neuronal division varies with brain area. In many species, cell division in cortical areas continues for a few weeks or months after division in other areas. This makes these dividing neurons and their emerging circuitry particularly vulnerable to epigenetic influences.

There are two notable exceptions to this principle. One is found in the olfactory mucosa, where the sensory cells (which at least should be considered as modified neurons) divide throughout life. Another apparent exception is in a group of neurons associated with singing in songbirds. Here, birth and death of neurons seem to change with seasons of the weather. In Spring, when birds are courting with songs and mating, the neurons in this nucleus proliferate noticeably, only to regress after the mating season.

TERMS

Differentiation	The process whereby cells reach their mature, final form as they develop from precursor (undifferentiated) cells.
Ventricle	A cavity inside the brain, filled with cerebrospinal fluid.
Neural Tube	A dorsally placed tube in the vertebrate embryo that is made up of neuronal precursor cells.

Related Principles

Epigenetics
Neuron Number and Types (Overview)
Programmed Development

References

Finger, S. et al. 1988. Brain Injury and Recovery. Plenum Press, New York.
Goodman, C.S. and Spitzer, N.C. 1979. Embryonic development of identified neurons: differentiation from neuroblast to neuron. Nature. 280:208-213.
Gorio, Alfredo, (ed.) 1993. Neuroregeneration. Raven Press, New York.
Keynes, R. and Lumsden, A. 1990. Segmentation and the origin of regional diversity in the vertebrate nervous system. Neuron. 2:1-9.
Rose, F.D. and Johnson, D.A. (eds.) 1993. Recovery from Brain Damage. Reflections and Directions. Advances in Experimental Medicine and Biology Series. Plenum Press, New York.
Williams, R.W. and Herrup, K. 1988. The control of neuron number. Ann. Rev. Neurobiol. 11:423-454.

Early Death

Cell death gives birth to the nervous system. The developmental formation of the brain arises out of a competition of proliferating neurons that are "seeking" their final target contacts with other neurons and with muscles and glands. Many neurons seem to have a genetic program to kill themselves if the right target is not reached. In some systems, up to 70% die before the structure in the region is completed. Death of these neurons is accompanied by the sculpting of circuitry by the survivors.

Explanation

Many types of neurons are produced in great excess during embryogenesis. Survival depends, at least in large part, by contact with cells that are to be targets in adulthood. The death of these neurons is thought to reflect the failure of these neurons to obtain adequate amounts of specific neurotrophic factors that are produced by the target cells and that are needed by neurons to survive. This mechanism is epigenetic. It does not seem likely that genes could specify creation of all cell-cell connections in exquisite detail. Neuronal death is a kind of natural selection process, akin to the natural selection that occurs at the evolutionary level of speciation. There is some evidence that the mechanism originates in a genetically controlled "cell suicide program," operating by default when a cell is deprived of appropriate neurotrophic signals.

Developing neurons do not depend exclusively on signals from their targets. Some neurons require signals from the neurons that innervate them. Other neurons require specific hormones. Signals from nearby glial cells may also be important.

Developing neurons also have certain surface molecules that promote cell-cell recognition. These so-called adhesion molecules dictate which cells develop a lasting interrelationship. The activation of genes that make these adhesion molecules alters the physical environment in which neurons develop and thus *physically* governs which neurons survive.

Although this may seem to be a wasteful use of energy, there are clear advantages to this kind of selection of survivors. First, this mechanism ensures that cells that accidentally project to inappropriate targets are eliminated. Also, by having a super-abundance of neurons, it helps to assure that all target cells become innervated. Finally, the selective reduction of excess neurons helps to achieve an appropriate match between neurons and their targets, minimizing superfluous innervation. Perhaps also there is a finite limit to active synapses that can be metabolically supported by a given target neuron.

Selective reduction of neurons also serves to sculpt circuitry. It is important to know more about the forces that determine how the brain eventually wires itself. The genetic code does not prescribe precisely how the brain wires itself. Genetic influences probably act indirectly via the production of adhesion molecules and neurotrophic factors. Epigenetic forces associated with the developmental history unique to each cell help to dictate how neurons are interconnected and, therefore, which neurons survive. These epigenetic forces are a process of natural selection for circuitry. Gerald Edelman calls the process "Neural Darwinism."

Example

The best studied example of programmed neuronal death is with sensory neurons and neurons in the sympathetic division of the nervous system. These neurons depend on Nerve Growth Factor (NGF) for their survival. If theses cells are treated with antibody to NGF, which effectively prevents the NGF trophic influence, the neurons die. If the cells are treated with NGF, they survive in excess.

If a target tissue is removed, the neurons that normally innervate it will die. In short, the programmed development of the nervous system produces a super-abundance of neurons that is eventually reduced by cell death to match the size of the targets.

The importance of target size is illustrated by experiments in which transplanting an extra limb bud in an embryo increases the number of neurons in the spinal ganglia that supply the limbs. Conversely, removing limb buds reduces the number of surviving neurons in the limb ganglia.

NGF is only one of a family of trophic substances that help keep developing neurons alive. Others include brain-derived neurotrophic factor (BDNF), neurotrophin-3 (NT-3), and neurotrophin-4/5 (NT-4/5).

Recently, BDNF has been shown to be associated with activity-dependent survival of peripheral neurons. In embryonic neuronal cell cultures, survival is enhanced by activation of voltage-dependent calcium channels, but not by activation of the ligand-dependent *N*-methyl-D-aspartate channels. The effect is mediated by increased expression of BDNF by the voltage-regulated calcium channels. Treatment of cultures with antibodies to BDNF reduces the survival of cultured neurons.

The programmed cell death extends also to certain glial cells as well as neurons. A large scale die-off occurs in the cells that form myelin coverings around the axonal processes of neurons. These glia, oligodendrocytes as they are called, wrap their membranes concentrically around neuronal processes. There is an optimal number needed to coat neurons, governed by the number and length of axons that require coating. The fact that the excess of glia die suggests that the glia normally depend on the axons for their survival.

Neuronal death is apparently inflicted by one or more specific proteins. Chemicals that inhibit RNA or protein synthesis can suppress or postpone cell death.

Neurons also die as a normal consequence of aging. Sometimes this is caused by medical conditions involving impaired blood supply or infection. Recent experiments indicate that neuronal death can result from excessive or persistent activation of ion channels that are regulated by the excitatory neurotransmitter, glutamate. A related and interacting mechanism involves the damaging effects of oxygen radicals and hydrogen peroxide, which are generated by oxidative metabolism. Note that the brain consumes more than 20% of the body's oxygen even though in humans the brain often weighs less than 2% of the body weight.

TERMS

Voltage-/ligand Dependent Ion Channels	Ion channels that open, respectively, to either a change in membrane potential or to specific neuro-transmitter-receptor binding.

Related Principles

Calcium and Transmitter Release (Cell Biology)
Epigenetics (Cell Biology)
Ion Channels (Cell Biology)
Neuron Numbers and Types (Overview)
Neuronal Targets
Programmed Development
Neural Induction and Trophism

References

Coyle, J.T. and Puttfarcken, P. 1993. Oxidative stress, glutamate, and neurodegenerative disorders. Science 262:689-700.

Cuello, A.C. 1993. Neuronal Death and Repair. Restorative Neurology. Vol. 6. Elsevier, Amsterdam.

Gagliardini, V., Fernandez, P. -A. Lee, R.K.K., Drexler, H.C.A., Rotello, R.J., Fishman, M.C. and Yuan, J. 1994. Prevention of vertebrate neuronal death by the crmA gene. Science 263:826-828.

Ghosh, A., Carnahan, J., and Greenberg, M.E. 1994. Requirement for BDNF in activity-dependent survival of cortical neurons. Science 263:1618-1622.

Johnson, E.M., Jr., and Deckwerth, T.L. 1993. Molecular mechanisms of developmental neuronal death. Ann. Rev. Neurosci. 16:31-46.

Levi-Montalcini, R. 1972. The morphological effects of immuno-sympathectomy, pp. 55-78. In G. Steiner and E. Schönbaum (eds.). Immunosympathectomy. Elsevier, Amsterdam.

Oppenheim, R.W. 1991. Cell death during development of the nervous system. Ann. Rev. Neurosci. 14:453-501.

Purves, D. and Lichtman, J.W. 1980. Elimination of synapses in the developing nervous system. Science 210:153-157.

Raff, M.C. Barres, B.A., Burne, J.F., Coles, H.S. Ishizaki, Y., and Jacobson, M.D. 1993. Programmed cell death and the control of cell survival: lessons from the nervous system. Science 262:695-700.

Ruitshauser, U., Acheson, A., Hall, A.K., mann, D.M. and Sunshine, J. 1988. The neural cell adhesion molecules (NCAM) as a regulator of cell-cell interactions. Science 240:53-57.

Sanes, J.R. 1989, Extracellular matrix molecules that influence neural development. Ann. Rev. Neurosci. 12:419-516.
Stent, G.S. 1981. Strength and weakness of the genetic approach to the development of the nervous system. Ann. Rev. Neurosci. 4:163-194.
Truman, J.W. 1983. Programmed cell death in the nervous system of an adult insect. J. Comp. Neurol. 216:445-452.
Williams, R.W. and Herrup, K. 1988. The control of neuron number. Ann. Rev. Neurosci. 11:423-53.

Citation Classics

Hamburger, V. 1939. Motor and sensory hyperplasia following limb-bud transplantations in chick embryos. Physiol. Zool. 12:268-284.
Hamburger, V. 1975. Cell death in the development of the lateral motor column of the chick embryo. J. Comp. Neurol. 160:535-546.
Hamburger, V. and Levi-Montalcini, R. 1949. Proliferation, differentiation and degeneration in the spinal ganglia of the chick embryo under normal and experimental conditions. J. Exp. Zool. 111:457-501.
Hamburger V., Brunso-Bechtold, J.K., and Yip, J.W. 1981. Neuronal death in the spinal ganglia of the chick embryo and its reduction by nerve growth factor. J. Neurosci. 1:60-71.
Yarmolinsky M.B. and de la Haba G.L. 1959 Inhibition by puromycin of amino acid incorporation into protein. Proc. Nat. Acad. Sci. USA 45:1721-1729.

Programmed Development

The nervous system is created by a developmental program that involves cell division, migration and navigation, differentiation, target recognition, synapse formation, and cell death. The program is directed by genetic and epigenetic molecular signals that are spatially and temporally defined.

Explanation

Development of the nervous system occurs in a specific, patterned, and progressive way. The sequence of events generally follows a pattern of large numbers of cell divisions. In general, the different types of neurons in the fully developed nervous system are not all born at the same time. Every cell type is formed during a characteristic phase in development.

Newborn neurons then migrate toward their targets, guided by physical and chemical interactions along the way. Some of the early migration and differentiation cues come from the chemicals released from the underlying germinal cell layer, the mesoderm. In the course of migration, most neurons lose the ability to divide, and they then differentiate into a specific cell type that can never revert back to the earlier undifferentiated stage. Neurons recognize their final target sites (muscles, glands, or other neurons) by chemical cues at the target sites. After recognition, synapse formation ensues. Neurons that do not accomplish a successful migration and target recognition then die. Many, perhaps the majority, of newborn neurons never survive the development process.

Differentiation is governed by two influences, genetic and epigenetic. The genetic influence is most obvious in the lineage of a given neuron. The history of previous divisions, and its "position in the family tree" has a large role in determining what kind of neuron will emerge. The epigenetic influence is seen in the role of the environment in which a neuron is born and migrates. Here, the spatial arrangement of cells and the interaction with neighbors provides physical and chemical cues that help to regulate genetic expression.

Examples

After a fertilized egg reaches about the 1,000 cells stage, a second stage occurs in which the surface of the spherical glob of cells invaginates at one of the poles, and, in most species, a three-layered embryo emerges. The outermost layer, called ectoderm will give rise to the nervous system. The first sign of the nervous system appears in the gastrula when a dorsally located groove appears on the ectodermal surface. The groove thickens and then curls upward and seals off at the top to form a tube of precursor neurons. At the same time, another separate group of neurons becomes isolated above the neural tube; these "neural crest" cells divide and migrate to create the peripheral nervous system.

As examples of different "birth dates" for different neuron types, the motor systems in the spinal cord become differentiated and functional before the sensory systems. Another example is that the deep nuclei of the cerebellar cortex form before the appearance of large numbers of "granule cells," which will ultimately form a major portion of the cerebellar cortex.

An example of the effect of time on neuronal function can be found in certain neurons that initially form action potentials that are created by calcium influx. Later, these same neurons switch to action potentials that are created by a combination of calcium and sodium influx. Later, these cells switch entirely to a sodium-influx mode for generating action potentials.

Related Principles

Early Death
Migration
Neuron Division
Neural Induction and Trophism
Neuron Numbers and Types (Overview)
Neuronal Targets
Topographical Mapping (Overview)

References

Brown, M.C. et al. 1991. Essentials of Neural Development. 2nd ed. Cambridge U. Press, New York.
Evrard, P. and Minkowski, A. 1989. Developmental Neurobiology. Raven Press, New York.
Hall, B. K. and Hörstadius, S. 1988. The Neural Crest. Oxford U. Press, London.
Hamburger, Viktor. 1990. Neuroembryology. The Selected Papers of Viktor Hamburger. Birkhauser, Boston, Massachusetts.
Ito M. 1989. Neural Programming. Taniguchi Symposium of Brain Science. No. 12. S. Karger, Farmington, Connecticut.
Jacobson, M. 1985. Clonal analysis and cell lineages of the vertebrate central nervous system. Ann. Rev. Neurosci. 8:71-102.
Jacobson, M. 1991. Developmental Neurobiology. 3rd ed. Plenum, New York.
Keynes, R. and Lumsden, A. 1990 Segmentation and the origin of regional diversity in the vertebrate central nervous system. Neuron. 2:1-9.
Patterson, P.H. and Purves, D. 1982. Readings in Developmental Neurobiology. Cold Spring Harbor, New York.
Purves, D. and Lichtman, J.W. 1985. Principles of Neural Development. Sinauer, Sunderland.
Sharma, S.C. (ed.) 1984. Organizing Principles of Neural Development. Plenum, New York.
Truman, J.W. 1983. Programmed cell death in the nervous system of an adult insect. J. Comp. Neurol. 216:445-452.
Udin, S.B. and Fawcett, J.W. 1988. Formation to topographical maps. Ann. Rev. Neurosci. 11:423-454.

Epigenetics

The genetic code does not entirely specify how a brain is to be constructed. It does specify a set of immutable constraints on the selection and competitive processes associated with neuronal division, proliferation of processes, sticking of neurons to each other, cell death, and selective strengthening of certain synapses. These competitive processes are thus "epigenetic."

Explanation

Specific neurons in one part of the nervous system preferentially form synapses with specific neurons in other parts. These key developmental events are not precisely prespecified by the genes. These events occur only if certain previous events have taken place. In short, the brain is to a large extent, self-organizing. Both extracellular environment and patterns of action potential activity direct the formation of synapses and functional circuitry.

In early stages of embryogenesis, most neurons migrate, sometimes long distances, to their final adult location. This migration is guided by the extracellular chemical and physical milieu. During development and regeneration, axons correctly recognize their appropriate targets. This recognition must depend on chemical signals in the external environment.

Examples

One of the best studied examples of epigenetic control on synapse formation is the development of synapses along the visual pathway. Retinal neurons make synapses with specific neurons in the brainstem, ignoring nearby neighbor cells. Transplantation experiments show that developing neurons form synapses with the correct targets even when the growing neurons have been moved from their original, normal position in the tissue. This does not prove an epigenetic cause, but it is hard to imagine how a "one gene-one synapse" mechanism could direct synaptogenesis when cells have been moved to abnormal locations. A more parsimonious explanation is that chemical cues in the target environment and cell-cell recognition phenomena create an epigenetic mechanism.

Muscle innervation provides another example of epigenetics. The innervation of muscle depends greatly on polysialic acid (PSA) in the membrane. Fast-twitch muscle, which has higher innervation density, also has a higher density of PSAs. Treating developing muscle with an enzyme that destroys PSAs reduces the innervation density. A related observation is that blockade of junctions between nerve and muscle with the toxin, curare, causes a coincident increase in PSA and increased axonal branching. The relationship of PSAs to axonal sprouting during development probably also holds for repair and plasticity phenomena in the mature nervous system.

Other examples of epigenetic influences can be found in several of the modules listed under Related Principles.

Related Principles

Neuron: The Operational Unit
Early Death
Neural Induction and Trophism
Migration
Neuronal Targets
Plasticity
Programmed Development

References

Brauth, S. et al. 1991. Plasticity of Development. MIT Press, Cambridge, Massachusetts.

Edelman, G.M. 1988. Topobiology: an introduction to molecular embryology. Basic Books, New York.

Edelman, G.M. 1987. Neural Darwinism. The Theory of Neuronal Group Selection. Basic Books, New York.

Evrard, P. and Minkowski, A. 1989. Developmental Neurobiology. Nestle Nutrition Workshop Series, Vol. 12. Raven Press, New York, New York.

Levi-Montalcini, R. 1982. Developmental neurobiology and the natural history of nerve growth factor. Ann. Rev. Neurosci. 5:341-362.

Michod, R.E. 1989. Darwinian selection in the brain. Evolution. 43:649-696.

O'Leary, D.D.M., Schlaggar, B.L., and Tuttle, R. 1994. Specification of neocortical areas and thalamocortical connections. Ann. Rev. Neurosci. 17:419-439.
Rutishauser, U. and Landmesser, L. 1991. Polysialic acid on the surface of axons regulates patterns of normal and activity-dependent innervation. Trends in Neurological Sciences. 12:528-532.
Sanes, J.R. 1989. Extracellular matrix molecules that influence neural development. Ann. Rev. Neurosci. 12:491-517.
Shatz, C.J. 1990. Impulse activity and patterning of connections during CNS development. Neuron. 5:745-756.
Steinman, L. Miller, A., Bernard, C.C.A., and Oksenberg, J.R. 1994. The epigenetics of multiple sclerosis: clues to etiology and a rationale for immune theory. Ann. Rev. Neuroscience. 17:247-266.

Citation Classics

Dennis, M.J. 1981. Development of the neuromuscular junction: inductive interactions between cells. Ann. Rev. Neurosci. 4:43-68.
Edelman, G.M. 1984. Modulation of cell adhesion during induction, histogenesis, and perinatal development of the nervous system. Ann. Rev. Neurosci. 7:339-77.
Hamburger, V. 1980. Trophic interactions in neurogenesis: a personal historical account. Ann. Rev. Neurosci. 3:269-78.
Weiss, P. 1936. Selectivity controlling the central-peripheral relations in the nervous system. Bio. Rev. 11:494-531.
Weiss, P. and Taylor, A.C. 1944. Further experimental evidence against "Neurotropism" in nerve regeneration. J. Exp. Zool. 95:233-257.

Migration

At one time or another in their lifetime, all neurons are gypsies—they migrate to their final positions and connections with other cells.

Explanation

During development, neurons divide, migrate, send out processes, release inducing chemical signals to other neurons, and stick to each other. These events depend on location (and which other cells are nearby) and time (when in the developmental sequence one event occurs in relation to another). Most notably, neurons differentiate; that is, they express different sets of genes at certain times and places.

Examples

After neurons are born, the chief early event in their life is to migrate. In the spinal cord, the migration is relatively simple. The first part of the cord to differentiate is the ventral (anterior in bipeds) horn of the grey matter. The grey-matter horns are formed simply by cells migrating from the ventricular zone of the neural tube where they were born to the adjacent region. Formation of the cerebral cortex is more complex. Newborn cells in the innermost ventricular zone migrate in an inside-out fashion to reach the surface. That is, the "first-born" neurons settle in an area near the external surface. Then, successive waves of newborn neurons migrate through this first layer and settle only when further outside, building up a succession of layers toward the outside. After all neurons have left their sites of origin, all that is left is the single-cell layer lining of the cerebral ventricles.

Migration in the cerebellum is still different. Here, all the cells originate from a small patch of neuroepithelium in the rostral part of the wall of the fourth ventricle. At least three waves of migration leave this area, with cells migrating outward toward what later becomes the surface of the cerebellum. In the first wave, a short migration of cells produces the so-called deep nuclei of the cerebellum. A second wave of cells moves on beyond the deep nuclei area and produces a temporary surface layer of the cerebellar cortex. The cells at this stage are unusual in that they still retain the capacity for division. Their progeny migrate back inward to create a layer of dense, small cells organized into what is called the granular layer of the cerebellar cortex. To create this layer, the precursor granule cells have to migrate inward through a layer of large, pyramid-shaped neurons. To do this, the granule cells assume a bipolar shape, and as they migrate inward, they leave their axons behind in the surface layer of pyramidal cells. Upon arriving at their final destination, the granule cells begin dendritic proliferation, which creates a dense "molecular" layer, formed in combination with pyramidal cell dendrites and small interneurons known as stellate and basket cells.

Cells of the neural crest migrate the furthest. They actually spread into other parts of the embryo, creating, for example, the ganglia of the sympathetic nervous system, the medullary cells of the adrenal gland, and melanocytes in the skin.

What controls migration is not known. One commonly accepted theory for development of the cerebral cortex is that the migration path is guided by physical contact with the processes of pre-existing neuroglial cells. In the young embryo, the wall of the neural tube is traversed by elongated basal processes of glial cells that are radially oriented. Migrating neurons can be seen to line up along these glial cells, and the assumption is that the neurons are "climbing" their way to the surface. Neurons do spontaneously exhibit "peristaltic contractions," as seen in time-lapse photography. After the migrations are complete, the radial glial cell scaffolding structure disappears.

TERMS

Neural Crest A region of undifferentiated neural precursor cells that is dorsal to the neural tube and give rise to the peripheral nervous system.

Neural Tube A central, tubular zone of cells, formed early in an embryo from ectoderm. Germinal cells in this tube give rise to the central nervous system.

Peristaltic Contractions Slow, undulating movements, which in this case help to move chemicals down the length of the axon by a squeezing action.

Related Principles

Cortical Columns (Information Processing)
Neural Induction and Trophism
Programmed Development

References

Bentley, D. and Caudy, M. 1983. Pioneer neurons lose directed growth after selective killing of guidepost cells. Nature 304:62-65.

LeDouarin, N.M., Smith, J., and LeLievre, C.S. 1982. From the neural crest to the ganglia of the peripheral nervous system. Ann. Rev. Physiol. 43:653-671.

Rakic, P. 1971. Neuron-glia relationship during granule cell migration in developing cerebellar cortex. A Golgi and electron-microscopic study in *Macacus rhesus*. J. Comp. Neurol. 141:283-312.

Rakic, P., 1972. Mode of cell migration to the superficial layers of the fetal monkey neocortex. J. Comp. Neurol. 141:61-83.

Rakic, P. 1977. Prenatal development of the visual system in rhesus monkey. Phil. Trans. R. Soc. Lond. B 278:245-260.

Sharma, S.C. (ed.) 1984. Organizing Principles of Neural Development. Plenum, New York.

Udin, S.B. and Fawcett, J.W. 1988. Formation of topographic maps. Ann. Rev. Neurosci. 11:289-423-454.

Walicke, P.A. 1989. Novel neurotrophic factors, receptors, and oncogenes. Ann. Rev. Neurosci. 12:103-26.

Neural Induction and Trophism

During embryonic development, certain chemicals act as inducers of neural differentiation and growth. These chemicals appear at different stages of development, and they regulate gene expression to dictate the fate of dividing neuronal precursor cells. In addition, other chemicals have a nurturing function for both embryonic and adult neurons.

Explanation

When a sex cell is fertilized, it begins to divide. The proliferating cells form into a ball of about 1,000 cells, presumably under the influence of some chemical signal that alters gene expression in a way that causes differential cell growth in certain areas. Next, a region of the ball invaginates (in the "gastrulation" stage of development), with cells streaming inward and dividing to form three layers. Cells in these layers respond to many chemical signals that tell them how to divide and differentiate—that is, tell them what kinds of tissues to become.

One chemical signal comes from the yolky half of the embryo, signaling to a band of cells at the embryo's equator to stream inward during gastrulation to form "mesoderm," which eventually forms the muscles, blood, and bones. Mesoderm, in turn, produces various other chemical signals. One zone of the mesoderm, called the "Spemann organizer" releases a chemical message that induces the formation of the back, including the vertebral canal and the adjacent part of the ectoderm that develops into neural tissue. Ectoderm responds to inducer chemicals to form a plate of tissue that later folds inward to form the spinal cord and brain (Figure 9-2).

Many chemical signals cause differentiation of already-formed neurons. Epigenetic influences control the fate of early neurons; i.e., chemical and physical signals in their environment have more to do with their fate than genetic endowment.

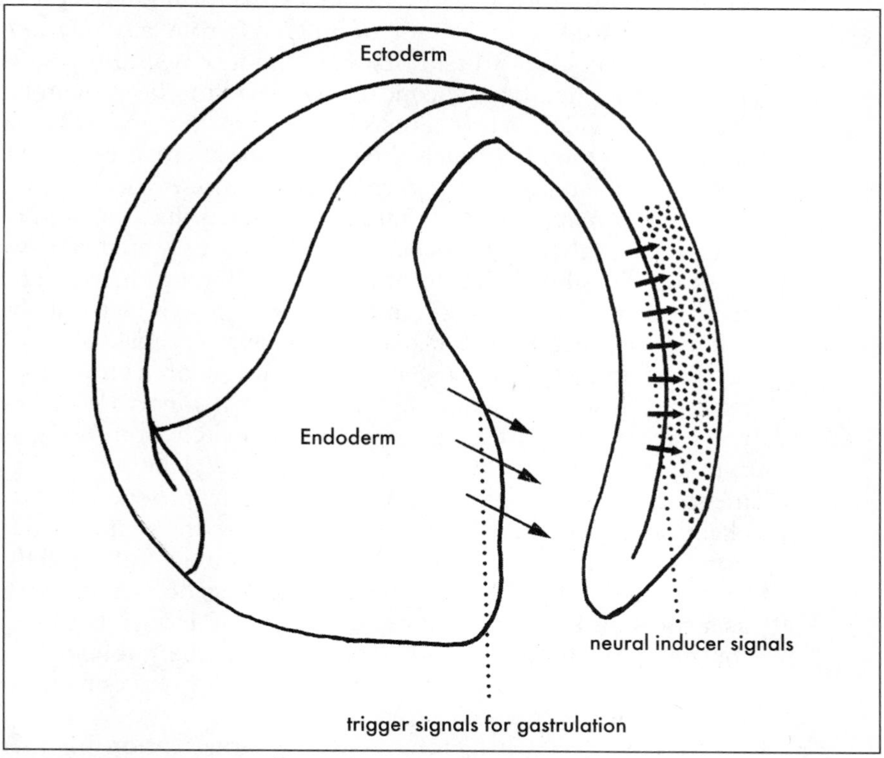

Fig. 9-2 Illustration of the gastrulation stage of embryogenesis, in which chemical signals released from early endoderm trigger the formation of the gastrula, and other chemical signals from dorsal mesoderm trigger the transformation of part of the ectoderm into nervous tissue. The diagram is a cross section of the ball-shaped gastrula.

Some chemical signals, usually proteins, nurture young and adult neurons alike. These neurotrophic factors help keep neurons alive and healthy and help them to grow cytoplasmic processes and axon terminals. Among the best known of such factors are Nerve Growth Factor (NGF), Brain-derived Neurotrophic Factor (BDNF), insulin-like growth factor (IGF), and Fibroblast Growth Factor (FGF). The latter two factors were first discovered in other tissues, but were later found to nurture neurons as well.

Chemical inducers commonly act by regulating gene expression via RNA.

Example

The existence of chemical inducers has been known for about 50 years. The early studies used transplantation techniques, where cells were removed from a developing embryo and put into another embryo in different places, at different times of development. These studies revealed that a variety of chemicals could induce nervous tissue formation, but none of these chemicals seemed to be a specific inducer. Typically, they acted indirectly, by stimulating dorsal mesoderm development, which in turn releases an inducer to form the neural tissue.

Recent studies of a pool of mRNAs from gastrulation-stage frog embryos revealed a protein that could induce neural tissue formation in embryos that had been chemically pretreated to keep them from making it on their own. This protein has been called "noggin." The gene for noggin has been found in rodents. Noggin's specificity is indicated by the fact that it can induce neural formation in pure ectoderm that had been removed from embryos, so that mesodermal influence could be ruled out.

Interestingly, noggin occurs in the brain of adults rats, which leaves open the possibility that it may have a different function after differentiation. Perhaps the noggin in adults is a growth factor or helps to sustain normal metabolic turnover in neuronal organelles.

Some inducers affect genetic expression of neurotransmitters. For example, when the sympathetic nerves that innervate rat sweat glands reach their targets, they are induced to switch from using norepinephrine as their neurotransmitter to using acetylcholine. During development, the norepinephrine acts to stimulate production of an inducer chemical in the glands that in turn promotes the switch to acetylcholine. For example, studies have shown that culture of gland cells with sympathetic neurons (but not other neurons that do not use norepinephrine as a transmitter) induce the switching factor. Drugs that block receptors for norepinephrine prevent acquisition of the cholinergic phenotype. Sweat glands from animals whose sympathetic nerves have been cut lack the switching factor.

Recently, growth factors are attracting attention for their potential clinical value in treating certain diseases. The progressive motor-nerve degenerative disease known as ALS (Lou Gherig's disease) may someday be treatable with trophic factors or synthetic analogs. BDNF, IGF, and CNTF seem to aid in the healing of injured motor neurons in rats. These factors also improve the condition of mice with inherited neurological disorders that are similar to ALS of humans. Other diseases currently being considered for neurotrophic factor therapy include Parkinson's disease, Alzheimer's disease, and peripheral neuropathy.

TERMS

Differentiation A change in cellular characteristics during early development into the permanent, adult form and features. Before differentiation occurs, some cell types are plastic; i.e., they can develop into more than one type of cell.

Ectoderm The embryonic cells that eventually differentiate to form nervous tissue.

Gastrulation The first embryonic stage in which the nervous system begins to develop.

Inducer A chemical substance in embryonic tissue, which, when released, triggers specific kinds of cellular differentiation.

Mesoderm The embryonic cells that eventually differentiate to form muscle, bones, and blood.

mRNA Messenger ribonucleic acid, which directs the synthesis of proteins from their amino acid constituents.

Neurotrophic Factor Chemicals (typical protein) that keep neurons healthy and help them grow.

Sympathetic Nervous System See Visceral Control (Motor Activity and Control).

Related Principles

Epigenetics
Neuron Division
Plasticity
Programmed Development
Visceral Control (Motor Activity and Control)

References

Habecker, B.A. and Landis, S.C. 1994. Noradrenergic regulation of cholinergic differentiation. Science 264:1602-1604.

Jessell, T.M., Siegel, R.E., and Fischbach, G.D. 1979. Induction of acetylcholine receptors on cultured skeletal muscle by a factor extracted from brain and spinal cord. Proc. Natl. Acad. Sci. USA. 76:5397-5401.

Kessler, D.S. and Melton, D.A. 1994. Vertebrate embryonic induction: mesodermal and neuronal patterning. Science 226:596-604.

Keynes, R. and Lumsden, A. 1990. Segmentation and the origin of regional diversity in the vertebrate central nervous system. Neuron. 2:1-9.

Kuwada, J.Y., 1986. Cell recognition by neuronal growth cones in a simple vertebrate embryo. Science 233:740-746.

Lamb, T.M., Knecht, A.K. Smith, W.C., Stachel, S.E., Economides, A.N., Stahl, N., Yancopolous, G.D., and Harland, R.M. 1993. Neural induction by the secreted polypeptide noggin. Science 262:713-718.

Rutishauser, U., Acheson, A., Hall, A. K., Mann, D.M., and Sunshine, J. 1988. The neural cell adhesion molecules (NCAM) as a regulator of cell-cell interactions. Science 240:53-57.

Sharma, S.C. (ed.) 1984. Organizing Principles of Neural Development. Plenum, New York.

Yanker, B.A. and Shooter, E.M. 1982. The biology and mechanism of action of nerve growth factor. Ann. Rev. Biochem. 51:845-868.

Citation Classics

Levi-Montalcini, R. and Cohen, S. 1962. Effects of the extract of the mouse sympathetic system of mammals. Ann. N.Y. Acad. Sci. 85:324-341.

Spemann, H. and Mangold H., 1924. Uber induktion von embryonalanlagen durch implantation artfremder organisatoren. Wilhelm Roux Arch. Entwicklungsmech. Org. 100:599-638.

Sperry, R. W. 1963. Chemoaffinity in the orderly growth of nerve fiber patterns and connections. Proc. Natl. Acad. Sci. 50:703-710.

Neuronal Targets

The growing processes of neurons are guided to their targets by physical forces in their environment and by chemical signals "emitted" by the targets. Depriving a neuron of contact with its normal contact may cause the neuron to degenerate or die and its targets to become supersensitive to the chemical secretions of the missing neuron.

Explanation

Young neurons are driven to make their axons and dendrites grow by a genetic program, and the growth can be stimulated by certain environmental chemicals. One of these chemicals that stimulates growth in the peripheral nervous system is a peptide, known as "nerve growth factor." Another stimulus for neuronal growth are certain acidic-sugar-containing lipids, known as gangliosides.

Axons that are growing toward a target are typically unmyelinated. Growth occurs at the tips of developing axons and dendrites. These "growth cones" are in constant motion, sending out protoplasmic probes for attractant and repellant chemicals into the immediate environment. Physical properties of the surrounding environment help determine the spread of neuronal processes by acting as a "neuronal glue." Also important are certain cell-bound adhesion molecules (CAMS). Two main classes of these molecules exist, one that is calcium dependent and another that is not. If axons are destined to receive a myelin coating, this usually occurs after they have reached their targets and completed synapse formation.

Maturation of dendrites is another phase in the life history of newborn neurons. When a neuron first sends out its axon, the dendrites may be nonexistent or very undeveloped. Thus, dendritic proliferation is much more evident than axonal growth after birth. If a cell does not receive its normal density of synapses, its dendrites will be relatively stunted and underdeveloped.

Synaptic formation is indicated by the appearance of little extensions ("spines") as seen in electron micrographs. The density of dendritic spines is a cardinal indicator of maturation of neuronal circuitry.

If inputs to a neural target are disrupted, the dendrites and axons of target cells atrophy. The mechanism for such degeneration is not understood but presumably relates to metabolic changes by the lack of contact and chemical feedback from the normal target cells. This process, regardless of the underlying mechanism, can serve to help sculpt functional circuitry; this is particularly evident during development of the embryonic brain.

Supersensitivity of denervated targets can occur because of up-regulation of membrane receptors in response to the missing neurotransmitter that is ordinarily delivered by the input that has been eliminated.

Examples

The influence of nerve growth factor can be demonstrated in vitro. Adding NGF to a culture dish of young neurons induces them to generate prolific growth, compared to the sparse growth of a control dish of neurons that are cultured under the same circumstances.

Synaptic formation does not necessarily occur at the same time as the major spurt of axonal growth. For example, in the cortex at a time when neurons have already undergone a major growth spurt to produce an extensive dendritic network, there is a renewed spurt of growth that results in synapse formation.

Growth cones are guided by selective adhesion to a substrate. In cell cultures of embryonic neurons, it is easy to demonstrate that growth cones prefer certain substrate materials over others. If a preferred substrate, such as polyornithine, is laid out in a geometrical pattern, neuronal processes will follow that pattern. In such cultures, growth cones exhibit preferential association with certain neurons. For example, in a culture of mixed neuronal cell types, neurons of the retina associate with neurons (or even fractions of their membranes) in the brainstem that are normally part of the visual pathway.

Related Principles

Epigenetics
Neuron Division
Neural Induction and Trophism
Plasticity

References

Allendoerfer, K.L. and Shatz, C.J. 1994. The subplate, a transient neocortical structure: its role in the development of connections between thalamus and cortex. Ann. Rev. Neuroscience. 17:185-218

Dennis, M.J. 1981. Development of the neuromuscular junction: inductive interactions between cells. Ann. Rev. Neurosci. 4:43-68.

Frank, E. and Mendelson, B. 1990. Specification of synaptic connections between sensory and motor neurons in the developing spinal cord. J. Neurobiol. 21:33-50.

Gottlieb, D.I. and Glaser, L, 1980. Cellular recognition during neural development. Ann. Rev. Neurosci. 3:303-18.

Kawada, J.Y. 1986. Cell recognition by neuronal growth cones in a simple vertebrate embryo. Science. 233:740-746.

Landmesser, L.T. 1980. The generation of neuromuscular specificity. Ann. Rev. Neurosci. 3:279-302.

Sanes, J.R. 1989. Extracellular matrix molecules that influence neural development. Ann. Rev. Neurosci. 12:491-516.

Citation Classics

Sperry, R.W. 1963. Chemoaffinity in the orderly growth of nerve fiber patterns and connections. Proc. Natl. Acad. Sci. U.S.A. 703-710.

Weiss, P. 1936. Selectively controlling the central-peripheral relations in the nervous system. Biol. Rev. 11:494-531.

Weiss, P. and Taylor, A.C. 1944. Further experimental evidence against "Neurotropism" in nerve regeneration. J. Exp. Zool. 95:233-257.

Plasticity

The nervous system can exhibit considerable plasticity. Neurons can grow new axonal and dendritic processes, and new synapses can be formed. Moreover, neural circuits can be recruited into existence and existing circuits can be re-organized. Even certain topographical maps can be "sculpted" by the nature of their input.

Explanation

Nerve Regeneration

If a nerve is cut, the cell bodies that supplied the axons in that nerve may be stimulated to transport new metabolic building materials to the cut site and produce a new growth. The regenerating axon may eventually reach its original target, in which case normal function is restored.

How does a regenerating axon find its target? Typically, this occurs best with myelinated axons because the myelin sheath that contained the original axon remains in place, and its old, cut axon degenerates, leaving an empty tube that can guide the regenerating axon to its original target.

However, in complex, multi-fibered systems, such as the spinal cord, very few regenerating axons are likely to find the correct myelin sheath. Regeneration through an incorrect sheath leads the new axon to an unrecognized target. Synapses do not form and normal function is not restored. Also, in the spinal cord and brain, certain glial cells can release chemicals that actually suppress regeneration. Cut axons in the brain and spinal cord do attempt to grow new processes, but these cease after a few hundred micrometers and typically do not regain normal function.

Denervation during development can lead to severe deficits in adulthood because sensory experience is needed for self-organization of neural circuitry.

Circuit Plasticity

A more complex kind of plasticity is also possible in the nervous system. Nervous system capabilities can be changed by experiences; this is particularly demonstrable in very young animals. The idea goes beyond mere learning; environment can produce fundamental changes in the brain's anatomy and chemistry that lead to changes in an animal's ability to "learn how to learn."

Plasticity presumably results when neuronal circuits change or improve connectivity. Synapses are the preeminent location for such changes to occur. Sensory input promotes synapse formation, as indicated by the degree of dendritic spines. Spines are greatly reduced if the sensory pathways are blocked during embryogenesis or even during early postnatal life. There is a critical period during development when the nervous system is maximally sensitive to environmental stimuli. The exact age at which this occurs, the stimuli to which it applies, and the degree of criticality vary with the species.

Topographical maps are not necessarily fixed. As evidence that they are modified by experience and neural input, it has been observed that maps differ in detail in different individuals. Even within an individual, maps can be reorganized by changed input (see below).

The ability to form new neuronal connections and to modify existing ones seems to decline inevitably with age (it *is* hard to "teach an old dog new tricks"). The first and most conspicuous sign of aging is faulty short-term memory. Older animals have little trouble performing old learned behavior but have more difficulty in consolidating new learning experiences.

The neurophysiological reasons for these effects no doubt include the fact that the total number of neurons in the brain decreases dramatically with age. Neurons die as aging progresses, and they are not replaced.

Development and aging can be thought of as two sides of the same coin. Aging of the brain is manifest in a loss of neurons, reduction of neuronal branches and dendritic spines, accumulation of lipofuscin pigment, and increased glial proliferation. These signs can be partially delayed by three neural stimulant treatments: (1) adrenalectomy, which causes sustained increase of ACTH, (2) daily injections of ACTH fragments 4-9, or (3) daily injection of the stimulant, pentylenetetrazol. None of these provides practical medical prevention of aging.

Examples

Nerve Regeneration

Only in recent years have we learned much about the biochemical processes that direct nerve regeneration. Fish are a good experimental model. Regenerating optic nerves of fish contain an interleukin-2 (IL-2)-like compound that is toxic to the glial cells that otherwise would suppress regeneration. In rats, injection of the enzyme, transglutaminase, which makes IL-2 cytotoxically active, into completely transected optic nerves can promote some functional recovery in about six weeks. Using visual cortex electrical responses to light flashes as the index, researchers demonstrated that cutting the optic nerve completely abolished evoked responses, but after treatment with transglutaminase, some evoked responses re-occurred after 6 weeks. Only minimal change occurred in control rats that were treated only with buffer solution.

Circuit Plasticity

Animals are not born with fully developed nervous systems. The environment helps to sculpt neural circuitry as neurons are growing new processes and making new contacts during early postnatal development. As an example, monkeys with one eye kept shut for the first six months after birth will be blind in that eye. Environmental effects are even evident with complex, multisensory processes. Rats raised in "enriched" environments with lots of "toys" and activities were found to have more developed cerebral cortex than did rats raised in traditional laboratory cages. Young monkeys raised without the normal social interactions with their mothers grow up lacking in many social skills and have serious emotional disturbances.

One reason we know that neuronal circuits can reorganize comes from studies in animals in which the cortical representation of body parts is studied by recording sensory evoked responses in specific cortical areas from known body regions. For example, one can stimulate regions of the skin on the hand and digits of monkeys while recording action potentials in the sensory cortex. What is seen is that the hand and digits are "mapped" on the cortex; i.e., certain regions of the cortex respond to and represent certain portions of the hand or digits. If one of the digits is amputated, the cortical area that normally responds to it obviously becomes unresponsive. But with time, this cortical area assumes new functions and begins to respond to another portion of the hand or digits. In other words, its circuitry has re-organized to create a new sensory receptive field.

As another example, if one cuts a main sensory nerve from the hand, the median nerve, large areas of sensory cortex become "silent." However, these silent areas immediately respond to dorsal digital areas supplied by the radial nerve, which were normally ineffective; that is, these inputs are "unmasked." Over time, these cortical areas that are newly responsive to radial nerve input develop a complete topographical repre-

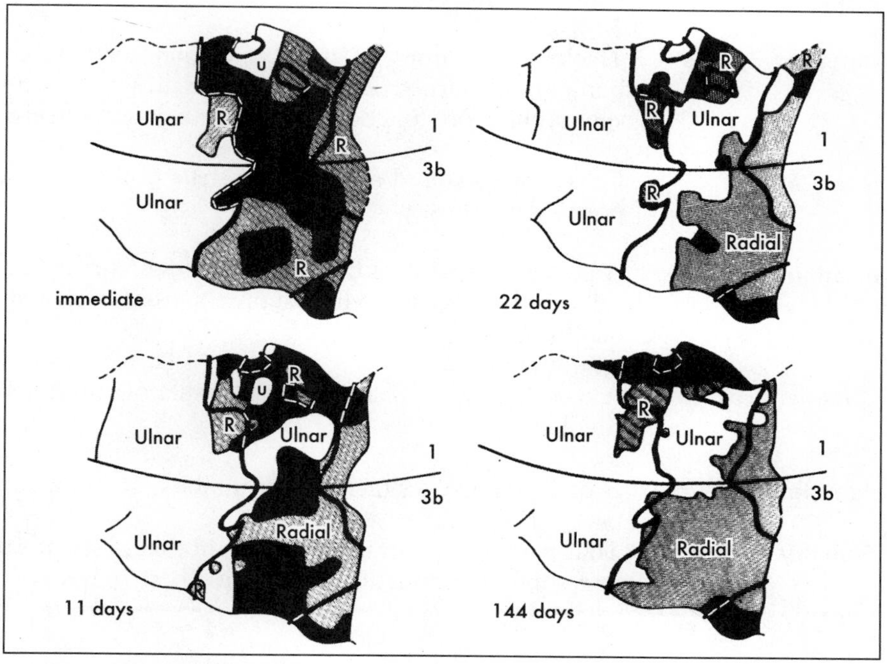

Fig. 9-3 Illustration of experiments that showed that depriving sensory cortex of its normal sensory input causes an unmasking of connections with other sensory fields and a reorganization of the topographical map. Shown are topographic maps of the cerebral cortex's representation for sensation of the hand of adult monkeys, immediately after and at various times after cutting a major sensory nerve (median nerve) of the hand. Immediately after cutting the nerve, cortical areas that received median nerve input became unresponsive to hand stimulation. This zone (in the center, between the heavy black and broken white lines) included regions that responded also to input from other hand nerves, the ulnar (U) and radial (R) nerves. There were also large areas (black) that were unresponsive to any kind of hand input. As time progressed after cutting the median nerve, more of the former median-nerve responsive cells became responsive to stimulation of hand regions that are innervated by ulnar and radial nerves. By 144 days, most of the cortex that was formerly responsive to median nerve input became responsive to ulnar and radial nerve input, at the expense of the large cortical expanse that was formerly unresponsive. (Reprinted from Merzenich et al. 1983a, with the kind permission of Elsevier Science Ltd., The Boulevard, Langford Lane, Kidlington, OX5 1GB, UK.)

sentation of dorsal skin surfaces. Ulnar nerve representations expanded their representation into the former median nerve representation zone of cortex. Within 22 days, almost all of the former median nerve field in the cortex was driven by new inputs (Figure 9-3).

TERMS

ACTH Adrenocorticotrophic hormone. A hormone released from the anterior pituitary that stimulates the release of steroid hormones from the adrenal cortex.

Dendritic Spines Thickened portions of membrane that form small bulges or bumps on dendrites. These spines contain synapses, and the more spines, the more functional synaptic contacts a neuron has.

Glial Cells Cells that surround and provide structural and metabolic support for neurons.

Interleukin-2 A protein, about 15 kDalton in size, that is manufactured in T-cell lymphocytes and which suppress proliferation of glia (oligodendrocytes).

Oligodendrocyte A type of glial cell that forms the insulating myelin around neuronal axons.

Pentylenetetrazol A drug that excites (actually disinhibits) certain neurons.

Sensory Cortex That part of the surface of the brain that is specifically mapped to respond to stimulation of specific body parts.

Related Principles

Early Death
Ensembles of Dynamic Neural Networks (Learning and Memory)
Epigenetics
Memory Consolidation (Learning and Memory)
Neuron Numbers and Types (Cell Biology)
Sensory Modalities and Channels (Senses)
Topographical Mapping (Overview)

References

Baudry, M., Thompson, R.F., and Davis, J.L., (eds.) 1993. Synaptic Plasticity: Molecular, Cellular, and Functional Aspects. MIT Press, Cambridge, Massachusetts.

Boothe, R.G., Dobson, V., and Teller, D.Y. 1985. Postnatal development of vision in human and nonhuman primates. Ann. Rev. Neurosci. 8:495-545.

Bregman, S. et al., (eds.) 1991. Plasticity of Development. MIT Press, Cambridge, Massachusetts.
Brown, M.C., Holland, R. L., and Hopkins, W.G. 1981. Motor nerve sprouting. Ann. Rev. Neurosci. 4:17-42.
Cotman, C.W. (ed.) 1985. Synaptic Plasticity. Guilford Press, New York.
Eitan, S., Solomon, A., Lavie, V. et al. 1994. Recovery of visual response of injured adult rat optic nerves treated with transglutaminase. Science. 264:1764-1767.
Finger, S. et al. 1988. Brain Injury and Recovery. Plenum Press, New York.
Kass, J.H., Merzenich, M. M., and Killackey, H.P. 1983. The Reorganization of somatosensory cortex following peripheral nerve damage in adult and developing mammals. Ann. Rev. Neurosci. 6:325-56.
Lund, J.P., Sun, G. -D., and Lamarie, Y. 1994. Cortical reorganization and deafferentiation in adult Macaques. Science. 256:546-547.
Merzenich, M. M., Nelson, R. J., Stryker, M.P., Cynader, M.S., Schoppmann, A., and Zook. J.M. 1984. Somatosensory cortical map changes following digit amputation in adult monkeys. J. Comp. Neurol. 224:591-605.
Merzenich, M. M. Kaas, J. H., Wall, J. T., Sur, M., Nelson, R. J., and Felleman, D. J. 1983a. Progression of change following median nerve section in the cortical representation of the hand in areas 3b and 1 in adult owl and squirrel monkeys. Neuroscience. 10:639-665.
Merzenich, M. M. Kaas, J. H., Wall, J., Nelson, R. J., Sur, M., and Felleman, D. 1983b. Topographic reorganization of somatosensory cortical areas 3b and 1 in adult monkeys following restricted deafferentation. Neuroscience. 8:33-55.
O'Leary, D.D.M., Schlagger, B. L., and Tuttle, R. 1994. Specification of neocortical areas, and thalamocortical connections. Ann. Rev. Neurosci. 17:419-439.
Tsukahara, N. 1981. Synaptic plasticity in the mammalian central nervous system. Ann. Rev. Neurosci. 4:351-79.

Citation Classics

Harlow, H.F. 1958. The nature of love. Amer. Psychol. 13:673-685.
Jensen, A.R. 1969. How much can we boost IQ and scholastic achievement? Harvard. Educ. 39:1-123.
Maisonpierre, P.C., Belluscio, L., Squinto, S., Ip, N.Y., Furth, M. E., et al. 1990. Neurotrophin-3:a neurotrophic factor related to NGF and BNDF. Science. 247:1446-1451.
Rakic P. 1976. Prenatal genesis of connections subserving ocular dominance in the rhesus monkey. Nature 261:467-71.
Rakic, P. 1977. Prenatal development of the visual system in rhesus monkey. Phil. Trans. Roy. Soc. London B. 278:245-260.
Rosenzweig, M.R., Bennett, E.L., and Diamond, M.C. 1972. Brain changes in response to experience. Sci. Press. 226:22-29.
Singer M. 1952. The influence of the nerve in regeneration of the amphibian extremity. Quart. Rev. Biol. 27:169-200.
VanderLoos, H. and Woolsey, T.H., 1973. Somatosensory cortex: structural alterations following early injury to sense organs. Science. 179:395-398.

Development: Study Questions

1. List and explain each of the principles in this category.
2. For each of the principles, provide an example *that is not mentioned in this text.*
3. What would be the problem if adult neurons died and replaced themselves, as happens in many other tissues?
4. What is different about the environment that neurons live in, compared to other tissue types?
5. Speculate on why so many neurons die during the early developmental process.
6. Why is early neuronal death likened to an evolutionary natural selection process?
7. Why do we say that the programmed development of the nervous system is under the influence of genetic and epigenetic forces?
8. What are some of the known epigenetic influences on developing neurons?
9. Why do most neurons have to migrate to their final destinations?
10. What do we know about the factors that guide migrating neurons?
11. Why do we say that the development of nervous tissue from ectoderm cannot occur without chemical inducers released from mesoderm.
12. What are some of the physical and chemical influences that affect neuronal growth cones?
13. What does the density of dendritic spines signify?
14. What determines whether a cut axon can regenerate back to its original target?
15. What mechanisms might account for the re-organization of a cortical sensory map after some of its input has been interrupted?

INDEX

F

G

H

T

DATE DUE

APR 29 1997
AUG 09 1997
FEB 19 1998
OCT 1 2004

Demco, Inc. 38-293

University of Arkansas
LIBRARY
for Medical Sciences